Radioactive Risk for Humans

First published 2024 in Great Britain and the United States by ISTE Ltd and John Wiley & Sons, Inc.



ISTE Ltd
27-37 St George's Road
London SW19 4EU
UK

www.iste.co.uk

John Wiley & Sons, Inc.
111 River Street
Hoboken, NJ 07030
USA

www.wiley.com



Library of Congress Control Number: 2024944826

British Library Cataloguing-in-Publication Data
A CIP record for this book is available from the British Library
ISBN 978-1-78630-884-9

# Contents

# Preface

## P.1. Our beliefs

When writing this book, we were guided by a certain number of convictions: independence, freedom of spirit, competence, transparency, scientific rigor, absence of conflict of interest and the right to err in all sincerity.

### P.1.1. *Independence*

Our book aims to be scientific and therefore neither pro- nor anti-nuclear, detailing our current scientific knowledge with its strengths and shortcomings. To be independent and responsible, it soon became clear that we needed a limited number of authors. They assume full responsibility for their texts.

### P.1.2. *Freedom of spirit*

There is a big difference between fundamental research organizations such as the CNRS, which until recently was one of our employers, and institutional organizations such as the CEA or IRSN, where the hierarchy is very strong. In the former case, researchers are totally free to publish under their own responsibility, whereas in the latter, the hierarchy controls the writings of its employees, resulting in the risk of censorship or self-censorship.

### P.1.3. *Competence*

Both authors can legitimately claim competence in this field. One of them has completed two theses (specialist doctorate and state doctorate) involving research in marine radioecology. He has produced some 50 scientific publications in this field, most of them in international peer-reviewed journals. He did not abandon the field altogether, however, as he worked as an expert for CNRS Life Sciences division of the GRNC (*Groupe Radioécologie Nord Cotentin*) from 1997 to 2010. The GRNC has published numerous reports. Since then, he has joined the Scientific Committee, then the Expert Group of ANCCLI (*Association Nationale des Comités et Commissions Locales d'Information*). The Scientific Committee's missions include advising and assisting Local Information Commissions, Local Committees and the ANCCLI in their expert appraisals, acting as an advisory body for Local Information Commissions and ANCCLI actions and publications, and acting as a point of contact for expert committees set up by various French and foreign bodies.

The other author was an engineer at the *Commissariat à l'énergie atomique* (CEA) at the Saclay center. He has devoted his career to protection against ionizing radiation (radioactive measurements, radiation protection of X-ray generators and particle gas accelerators, remediation of contaminated sites). He took part in teaching radiation protection at INSTN (1980–1998) and in the IAEA's international radiation protection courses (1994–1996). In 1996, he was appointed "Senior Expert" at the CEA on the advice of an external scientific commission.

He has taken part in a number of national commissions, including the Castaing commission on the reprocessing of irradiated fuels (1981–1984), the Jean Bernard commission on cancers at the Pasteur Institute (1986–1990), the Radioecology group of the North Cotentin region (GRNC) on the dosimetric impact of radioactive releases from plants (1997–2008), the pluralistic expert group (GEP) on uranium mines in the Limousin region (2006–2013), and the Ministry of Labor commissions on occupational diseases and chemical, physical and biological hazards (1983–1998). He is the author of around 100 articles and co-author of several books on nuclear and occupational health issues.

### P.1.4. *Transparency*

Among the multitude of books, scientific publications and gray literature, a choice had to be made. This choice was dictated by scientific quality. Our choices were based on a hierarchy, with French-language publications being the most accessible to the majority. Secondly, work published in peer-reviewed scientific journals, i.e. where peers, other scientists, evaluate and criticize the work before deciding whether its quality justifies publication. As all experts are fallible, this does not certify an absolute value of veracity and quality, but it does contribute significantly to it. Next, all information from official national and international bodies directed our choice. Finally, when we feel that it reinforces information, we include so-called "gray literature", i.e. documents that are much less easily accessible to the public, especially if this literature is old and in the form of printed reports. In this way, anyone can return to most of the sources used in this book.

### P.1.5. *Scientific rigor*

Scientific rigor obviously depends on the choice of information and the way in which it is presented. We read a large part of the literature published on this subject, both pro- and anti-nuclear, as well as literature considered to be scientific. We then formed our own opinion. This is what appears in this book.

### P.1.6. *Conflicts of interest*

The authors have no conflict of interest in the nuclear field, as they have not carried out any research in this area for many years, and have no shareholdings in companies operating in this niche.

### P.1.7. *The right to err in all sincerity*

On such a vast subject, two authors cannot master everything, and they must necessarily trust their peers and the information published. They have not necessarily assimilated all the information correctly, and may therefore have made biased interpretations. All these deviations, if they exist, have been made in all sincerity and the authors apologize in advance to the reader. We undertake to correct them should a second version of this book be published.

## P.2. Public opinion

Public opinion on the nuclear phenomenon has evolved over time. Initially, the public's enthusiasm for radium was very strong, and the most far-fetched applications, even dangerous for consumers, were developed, such as the addition of radium-226 to toothpaste, beauty creams and chocolate, alongside more "useful" applications such as luminescent paints. The next applications were military, with the military bombing of Hiroshima and Nagasaki. This is associated with the use of "defense secrecy" for the majority of nuclear applications. The result is strong public distrust and even opposition. The various accidents that have occurred, particularly the most serious ones such as Three Mile Island in 1979, Chernobyl in 1986 and Fukushima in 2011, have accentuated the opposition that now arises whenever a new basic nuclear infrastructure (BNI) is created or modified.

### P.2.1. *Public perception of radioactive risk*

The public generally fails to distinguish between danger and risk. Furthermore, risk is measured in terms of probability. As a result, the hazard–risk pair is perceived very differently from one individual to another. Some hazards with uncertain risk potential will be perceived as paramount by the public. Conversely, hazards associated with proven risks will be considered derisory by the public. In the first group we find GMOs, and in the second group alcohol, road accidents, tobacco, etc.

In 2021, the rising concern of the French population was the management of nuclear waste in France: 41% of French people believe that nuclear power plants are a source of high risk, and 48% have the same perception of nuclear waste. This is in line with the 31% of French people who believe that the leading cause of accidental risk is that of a major nuclear accident, such as those at Chernobyl and Fukushima, and that this is the main obstacle to the use of nuclear energy. The second most important potential risk is the storage of radioactive waste (21% of opinions) [IRS 21].

### P.2.2. *People trust science, not researchers*

Do you trust scientists to tell the truth about the results and consequences of their work on nuclear energy? The image of scientific experts remains positive for 50% of French people surveyed. The main qualities sought are competence, honesty and independence. Five organizations – the CNRS, ASN, IRSN, HCTISN and CEA – working in the nuclear field had a public confidence rating of over 70% in November 2020 [IRS 21].

### P.2.3. *The creation of independent official bodies*

In the past, the protection of humanity and the environment, as well as nuclear safety, were ensured by various services that were too closely linked to industrial interests, such as the CEA. Others, such as OPRI, had had a critical attitude during the Chernobyl accident, and were therefore completely disqualified in the eyes of the public. To compensate for these serious drawbacks, the French government created a number of new bodies that were dependent on the state but independent of the nuclear lobby. These included the IRSN and ASN. The qualities expected of these organizations are competence, independence, rigor and transparency. The gamble has largely paid off.

Alongside institutional experts, there are independent experts who often work for various associations (ANCCLI, ACRO, Global Chance, etc.). Unfortunately, their perfectly independent work is largely neglected, even denigrated or scorned. This is probably due to the fact that independent experts are very often volunteers, and that unpaid work is undervalued.

## P.3. The expert and the biases of scientific expertise

### P.3.1. *Choice of experts*

The selection of experts for scientific appraisals varies from one field to another. In some cases, experts are co-opted, as in the case of the ICRP, while in others, as in the case of UNSCEAR, the choice is made by political leaders. In all cases, however, scientific criteria are not the only ones involved, but political, industrial and other criteria may also be taken into account.

### P.3.2. *The expert's competence*

The scientific subject under discussion in an expert group is generally very broad, and no individual expert can possibly have a complete scientific knowledge of the field. Expertise is therefore necessarily collective, and each expert must at one time or another have confidence in their colleagues.

### P.3.3. *The expert's integrity*

Scientific experts are first and foremost people with their own limitations and weaknesses. Before being an expert, a scientific researcher's primary mission (in principle) is to carry out fundamental or applied research. To this end, they are paid employees of a public or private organization. In addition, either to finance their research or to enrich themselves personally, they may accept funding from various public or private organizations. It is therefore not uncommon for conflicts of interest to arise between these institutional or occasional funders and the expert's mission. In the latter case, their "good faith" and independence are far from total.

### P.3.4. *Selecting scientific information*

For a long time, the only sources of scientific literature in the nuclear field were the organizations directly involved, which were therefore "biased". It was only after the Chernobyl accident in 1986 that university scientists began to invest in this field of research. The subjects addressed were much more diversified and the concepts renewed. In order to eliminate these independent sources of information, it became common practice in many fields of physical and chemical risk assessment to retain only publications based on "Good Laboratory Practice" (GLP). These standards had been imposed on laboratories dependent on industry to limit scientific drift. However, while academics respect GLP in spirit, they do not apply it sensu stricto. As a result, not all their publications are taken into account. This is particularly true of the summaries produced by the European Food Security Authority (EFSA) [AMI 17].

### P.3.5. *Censorship of scientific publications*

#### P.3.5.1. *Control of nuclear organizations*

Most organizations working in the nuclear field have a pyramid-shaped management structure, and all publications are subject to authorization by the management. As a result, any "disturbing" information can be blocked. Few organizations allow their staff to publish without constraint.

#### P.3.5.2. *"Defense secrets" and "industrial secrets"*

In the nuclear field, many organizations are military or industrial. They can therefore censor information that could embarrass them on the grounds of "defense secrecy" or "industrial secrecy".

In France, the enactment of the "Transparency, Nuclear Safety" (*transparence et sécurité en matière nucléaire* – TSN) law has considerably changed attitudes (at least for the majority) and made it easier to obtain information about nuclear energy and, in particular, safety. However, there are still a number of gray areas, particularly where national defense is concerned, which are covered by "defense secrecy". We can hope for a positive change in the future, as much of the information retained has no military value, and some of it is an open secret published in specialized journals, but of course cannot be verified.

### P.3.6. *Scientific truth*

Scientific truth is by definition provisional, since new scientific advances can call into question our current certainties. In this book, the authors have taken into account the most recent findings.

In many areas, especially those where economic interests are at stake, we can observe divergent interpretations. This is particularly true of human health issues.

For example, the toxicity of tobacco has long been denied by scientists, often linked to the tobacco industry. This is still the case in some countries for asbestos. How can this attitude be explained? Knowledge – and this is no different for toxicology – is gradually gaining ground. So, at the outset, there is a period of time that can unfortunately extend for years, when an isolated piece of information reveals a certain phenomenon, such as the toxicity of tobacco. But does this single result represent an exception or a general phenomenon? New results are often divided between confirming and refuting the first result. This is because, in order to occur, the phenomenon must meet certain conditions (e.g. inter-individual variations in susceptibility to carcinogens, dietary habits).

In science, contrary to popular belief, results are rarely perfectly clear-cut (black and white, yes and no), but are associated with uncertainties. These uncertainties are expressed by a range within which the true answer may lie, associated with a certain probability of being true (95%, 99%). This means that the answer is not absolute; there are 5% or 1% of cases where the answer may be different. Our certainties are therefore only probabilities.

As with any human activity, research involves a number of researchers who lack rigor, and are capable of biasing the interpretation of their observations, or even inventing their results, often in line with the interests of the funders of their work. Limited though it may be, this type of behavior is enough to cause a significant proportion of the population to lose confidence in the profession as a whole. As is the case with the political class, the result for the public is "that we are being lied to", "that serious things are being hidden from us". Internet culture allows all kinds of rumors to be spread quickly and widely.

Biological responses are generally highly variable and often follow a normal (or Gaussian) distribution. The same applies to the responses of organisms to radionuclides, whether in terms of bioaccumulation, elimination or damage caused by ionizing radiation. This is true for all living beings, including humans. As a result, it is extremely difficult to identify the main laws and predict the real impact of this type of aggression.

### P.3.7. *Our conception of the expert's role*

A researcher must make available to society, in an accessible form, scientific knowledge concerning the health and environmental benefits, dangers and risks associated with major societal choices, such as the use of nuclear energy.

Our view of the expert's role is that they are not a substitute for decision-makers or society as a whole, but have a duty to enable as many people as possible to make informed choices.

The expert must be able to keep their distance from the generally preconceived opinions of the protagonists, and must assess the case without preconceived ideas and with complete serenity.

## P.4. Project objectives

The aim of this book is to estimate the radioactive risk to humans. To do this, the authors will follow the steps of the classic approach presented above. On the one hand, radioactive danger was recognized at the same time as the discovery of radioactivity. On the other hand, the estimation of the radioactive risk to humans is still the subject of lively debate. In this volume, the authors will summarize the scientific work, past and present, that has made it possible to estimate the radioactivity of the anthroposphere and the radioactive contamination of humans. They will list the various routes of exposure to ionizing radiation (external, internal, dietary) and estimate the radiation doses suffered by humans under various conditions (natural for the public and professionals, accidental). Also, the harmful effects of ionizing radiation at various biological levels (molecular, cellular, tissue) and the health effects at the individual level will be reported. The focus will be on occupational diseases caused by radiation. The relationship between doses and adverse effects of ionizing radiation will be discussed for high, medium and low doses. Controversies on this subject will be explained. International and French regulatory values will be provided. Finally, an estimate of the radioactive risk to humans will be proposed.

## P.5. Drafting the manuscript

The Preface, Introduction and Chapters 2, 4, 5, 8, 11, 12 and 13 were written by Jean-Claude Amiard, and Chapters 1, 3, 6, 7, 9 and 10 by Jean-Claude Zerbib. Each chapter has been reviewed, corrected, amended, completed and approved by the other co-author.

We wish to thank Professor Philip Rainbow (former Keeper of Zoology, Natural History Museum, London, UK) for reading the English version of the book. We warmly thank him for his time and efforts.

## P.6. References

[AMI 17] AMIARD J.C., *Les risques chimiques environnementaux. Méthodes d'évaluation et impacts sur les organismes*, 2nd edition, Lavoisier, Tec&Doc, Paris, 2017.

[IRS 21] IRSN, Baromètre IRSN 2021 sur la perception des risques et de la sécurité par les Français, IRSN, 26/05/2021.

# Acronyms and Abbreviations

ABCC: Atomic Bomb Casualty Commission

AChE: Acetyl Choline Esterase

ACRO: *Association pour le Contrôle de la Radioactivité dans l'Ouest* (French Association for the Control of Radioactivity in the West)

AFCN: *Agence Fédérale de Contrôle Nucléaire* (Belgian Federal Nuclear Control Agency)

ALARA: As Low As Reasonably Achievable

ALL: Acute Lymphocytic Leukemia

AMAD: Activity Median Aerodynamic Diameter

AML: Acute Myelogenous Leukemia

ANCCLI: *Association National des Comités et Commissions Locales d'Information* (French National Association of Local Information Committees and Commissions)

ANDRA: *Agence nationale pour la gestion des déchets radioactifs* (French National Radioactive Waste Management Agency)

ARNA: Argentine Nuclear Regulatory Authority

ARPANSA: Australian Radiation Protection and Nuclear Safety Agency

ASN: *Autorité de Sûreté Nucléaire* (French Nuclear Safety Authority)

ASND: *Autorité de Sûreté Nucléaire Défense* (French Defense Nuclear Safety Authority)

BBB: Blood–Brain Barrier

BE: Bystander Effect

BEPN: Binding Energy Per Nucleon

BER: Base Excision Repair

BIOMASS: Biosphere Modelling and Assessment

BIR: Break-Induced Replication

BNI: Basic Nuclear Infrastructure

CCHEN: *Comisión Chilena de Energía Nuclear* (Chilean Nuclear Energy Commission)

CCSN: *Commission Canadienne de Sûreté Nucléaire* (French name of the CNSC)

CDF: Cation Diffusion Facilitator

CEA: *Commissariat à l'Energie Atomique et aux Energies Alternatives* (French Atomic Energy and Alternative Energies Commission)

CLL: Chronic Lymphocytic Leukemia

CNAM: *Caisse Nationale d'Assurance Maladie* (French National Health Insurance Fund)

CNEN: *Comissao Nacional de Energia Nuclear* (Brazilian National Nuclear Energy Commission)

CNSC: Canadian Nuclear Safety Commission

COGEMA: *Compagnie Générale des Matières Nucléaires*

CSN: Spanish Council for Nuclear Safety

CVD: Cerebrovascular Disease

DAM: *Direction des Applications Militaires* (French Directorate of Military Applications)

DCN: *Direction des Constructions Navales* (French Directorate of Shipbuilding)

DDR: DNA Damage Response

DDREF: Dose and Dose-Rate Effectiveness Factor

DFD: *Deutsch-Französischer Direktionausschuss* (German-French Management Committee)

DNA: Deoxyribonucleic Acid

DS86: Dosimetry System from 1986

DSB: Double Strand Break

DTPA: Diethylene Triamine Penta-acetic Acid

DU: Depleted Uranium

EDF: *Électricité de France* (French multinational electric utility company owned by the government of France)

EMRAS: Environmental Modelling for RAdiation Safety

ENSI: Swiss Federal Nuclear Safety Inspectorate

ERR: Excess Relative Risk

EU: European Union

FAO: Food and Agriculture Organization

GRNC: *Groupe Radioécologie Nord-Cotentin* (Radioecology Group of the North Cotentin Region)

GRS: *Gesellschaft für Anlagen und Reaktorsicherheit* (Society for Device and Reactor Safety)

HAEA: Hungarian Atomic Energy Authority

HBNRA: High Background Natural Radiation Areas

HCTISN: *Haut Comité pour la Transparence et l'Information sur la Sécurité Nucléaire* (French High Committee for Transparency and Information on Nuclear Safety)

HLNRA: High-Level Natural Radiation Areas

HOS: Human Osteoblast Cells

HR: Hazard Ratio

HTO: Tritiated Water

IAEA: International Atomic Energy Agency

IAEC: Israel Atomic Energy Commission

ICRP: International Commission on Radiological Protection

ICRU: International Commission on Radiation Units and Measurements

IHD: Ischemic Heart Disease

ILO: International Labour Organization

INFN: *Istituto Nazionale di Fisica Nucleare*

ING: Incorporation by Ingestion

INH: Incorporation by Inhalation

INJ: Incorporation by Injection

INRS: *Institut National de Recherche et de Sécurité pour la Prévention des Accidents du Travail et des Maladies Professionnelles* (French National Research and Safety Institute for the Prevention of Occupational Accidents and Diseases)

INSERM: *Institut National de la Santé et de la Recherche Médical* (French National Institute for Health and Medical Research)

INSTN: *Institut National des Sciences et Techniques Nucléaires* (French National Institute for Nuclear Science and Technology)

INWORKS: International Nuclear Workers Study

IPSN: *Institut de Protection et de Sûreté Nucléaire* (French Institute for Nuclear Protection and Safety)

IR: Ionizing Radiation

IRA: Integrated Risk Assessment

IRSN: *Institut de Radioprotection et de Sûreté Nucléaire* (French Institute for Radiation Protection and Nuclear Safety)

ITER: International Thermonuclear Experimental Reactor

KI: Potassium iodide

KINS: Korea Institute of Nuclear Safety

LDIR: Low-Dose Ionizing Radiation

LDRIR: Low-Dose-Rate Ionizing Radiation

LET: Linear Energy Transfer

LNT: Linear Non-Threshold

LRWT: Linear Relationship Without Threshold

LSS: Life Span Study

MAAD: Median Active Aerodynamic Diameter

MRCP: Mesh-type Reference Computational Phantoms

NAS: US National Academy of Sciences

NDK: *Nükleer Düzenleme Kurumu* (Turkish Nuclear Regulatory Authority)

NEA: Nuclear Energy Agency

NNSA: China's National Nuclear Safety Administration

NRA: Bulgarian Nuclear Regulatory Agency

NRA: Japanese Nuclear Regulation Authority

Nramp: Natural resistance-associated macrophage proteins

NRC: National Research Council

NCRP: National Council on Radiation Protection and Measurements

NTE: Non-Targeted Effects

OAP: Thai Office of Atoms for Peace

OBT: Organically Bound Tritium

OD: Occupational Diseases

OECD: Organization for Economic Cooperation and Development

ONR: UK Office for Nuclear Regulation

OPT: Oligo Peptide Transporters

OR: Odds Ratio

PAH: Polycyclic Aromatic Hydrocarbons

PAR: Population Attributable Risk

PCB: Polychlorobiphenyl

RBE: Relative Biological Effectiveness

RERF: Radiation Effect Research Foundation

RIBE: Radiation-Induced Bystander Effect

RIFE: Radioactivity in Food and the Environment

RIGI: Radiation-Induced Genomic Instability

RNA: Ribonucleic Acid

ROS: Reactive Oxygen Species

RPL: Radio-Photo-Luminescent

RR: Risk Ratio

SAF: Specific Absorbed Fractions

SCE: Sister Chromatid Exchanges

SFRP: *Société Française de Radioprotection* (French Radiation Protection Society)

SIR: Standardized Incidence Ratio

SMR: Standardized Mortality Ratio

SSA: Single-Strand Annealing

SSB: Single-Strand Break

STUK: Finnish Radiation and Nuclear Safety Authority

SUJB: Czech State Office for Nuclear Safety

SY: Systemic Sector

TB: Tracheobronchial

TEL: Linear Energy Transfer

UJD: Nuclear Regulatory Authority of the Slovak Republic

UN: United Nations

UNSCEAR: United Nations Scientific Committee on the Effects of Atomic Radiation

USAEC: US Atomic Energy Commission

USTUR: United States Transuranium and Uranium Registries

VARAM: Latvian Ministry of Environmental Protection and Regional Development

VARANS: Vietnam Agency for Radiation and Nuclear Safety

VATEST: Lithuanian Nuclear Power Safety Inspectorate

WHO: World Health Organization

WLM: Working Level Months

ZIP: Zinc–Iron Permease family

# Introduction

## Methodology in Radioactive Risk Assessment

### I.1. Radioactive risk assessment method

The radioactive risk assessment method has already been developed in Volumes 3 [AMI 20] and 6 [AMI 22] of the "Radioactive Risk" series. This method, which is applied to all physical and chemical risk assessments [AMI 17], is developed in four stages (Figure I.1). The first step is hazard identification. The second step is to establish the radiation dose to humans, based on the estimated contamination of the anthroposphere and the multiple routes of exposure. The third step is to assess the harmful effects of ionizing radiation on humans, the relationship between doses received and harmful effects for various exposures (low, medium and high), and the selection of doses with no effect on human health. Steps 2 and 3 are carried out in parallel. The final step is to characterize the radioactive risk.

As with all physical and chemical hazards, particularly those that exist naturally outside human activity, it is not easy to provide doses that have no effect on human health, or on the health of all living beings. This explains why, as scientific knowledge of harmful effects increases, the choice of doses recognized as having no effect is evolving. For the vast majority of physical and chemical hazards, the trend is towards low doses. This is particularly true for ionizing radiation, and also for mercury and cadmium.

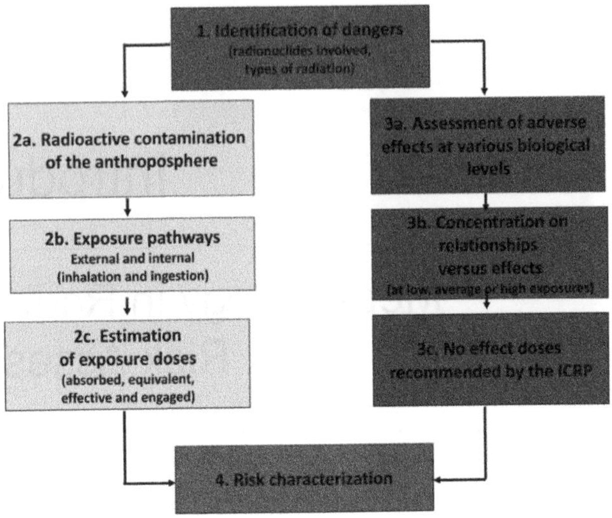


**Figure I.1.** *The general principle of human radioactive risk estimation (according to [AMI 21])*

Risk assessment is part of a wider picture (Figure I.2) that includes consideration of socio-economic and socio-behavioral impacts. These two aspects run in parallel and inform problem formulation as well as risk management and risk communication processes [SUC 18].

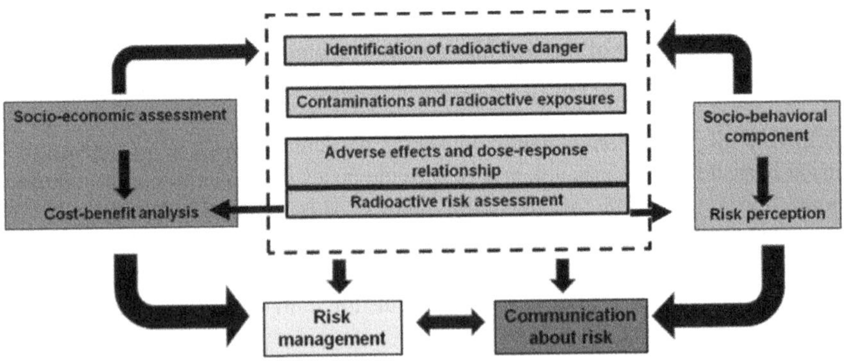


**Figure I.2.** *Conceptual overview of an integrated risk assessment (IRA) (according to [SUC 18])*

## I.2. International institutions responsible for nuclear safety

Mainly three international institutions, the IAEA, ICRP and UNSCEAR, are responsible for making recommendations on nuclear safety. Each state is independent in its use of nuclear energy, and may or may not adopt these recommendations in their entirety, or adapt them to its own needs.

The United Nations Scientific Committee on the Effects of Atomic Radiation (UNSCEAR, United Nations Scientific Committee on the Effects of Atomic Radiation) was established by the United Nations General Assembly in 1955. Its mandate within the UN system is to assess and report on the levels and effects of radiation exposure. For governments and organizations around the world, the Committee's estimates provide the scientific basis for assessing radiation risks and establishing safety measures. The General Assembly has designated 27 countries to delegate scientists as members of the Committee. France's designated member is currently D. Laurier (with A. Lebaron-Jacobs as alternate). The Committee comprises a Chairman, three Vice-Chairmen and a Rapporteur. Since its creation, UNSCEAR has published 25 major studies. These reports are highly regarded as fundamental sources of authoritative information.

The International Commission on Radiological Protection (ICRP) is an independent international organization that advances the science of radiation protection for the public good, in particular by providing recommendations and advice on all aspects of protection against ionizing radiation. It is a charity registered with the Charity Commission of England and Wales (registration number 1166304).

The ICRP is a community of over 250 recognized experts in radiation protection science, policy and practice in more than 30 countries. The ICRP's head office and scientific secretariat are located in Ottawa, Canada. The structure of the ICRP comprises the Main Commission, the Scientific Secretariat, four Standing Committees (Committee 1 on the Effects of Radiation, Committee 2 on Doses from Radiation Exposure, Committee 3 on Radiation Protection in the Medical Field, Committee 4 on the Application of ICRP Recommendations) and a series of Working Groups. Committee 5 on Environmental Protection, set up in 2005, was dissolved in 2017.

The International Atomic Energy Agency (IAEA) is part of the United Nations (UN). Its Statute was approved on October 23, 1956 by the Conference on the Statute of the IAEA, held at UN Headquarters. It came into force on July 29, 1957. The Statute has been amended three times in accordance with the procedure laid down in paragraphs A and C of Article XVIII.

Historically, the IAEA was created in 1957 in response to the great concerns and hopes raised by the discoveries and diverse uses of nuclear technology. Its genesis can be traced back to President Eisenhower's "Atoms for Peace" speech to the United Nations General Assembly on December 8, 1953. President Eisenhower's ratification of the Statute on July 29, 1957 marked the official birth of the IAEA.

The IAEA is closely linked to nuclear technology and its controversial applications, either as a weapon or as a practical and useful tool. The ideas expressed by President Eisenhower in his 1953 speech led to the Statute of the IAEA, unanimously approved by 81 nations in October 1956. The Agency was created as a global "atom for peace" organization within the UN system. From the outset, its mandate has been to work with its Member States and its many partners around the world to promote safe, secure and peaceful nuclear technologies. The objectives of the IAEA's dual mission – to promote and control the atom – are defined in Article II of its Statute. There are currently three distinct themes in the IAEA's work. These are 1) all nuclear technology applications in various fields (agriculture, health, sterilization, industry, water desalination, research, etc.), 2) the safety of nuclear infrastructures (reactors and plants), and 3) proliferation and safeguards compliance (with rogue states such as Iran, which has been playing cat-and-mouse with IAEA inspectors for two decades)[1].

In October 1957, delegates at the first General Conference decided to establish the IAEA Headquarters in Vienna, Austria. Until the opening of the Vienna International Centre in August 1979, the former Grand Hotel next to the Vienna Opera House served as the Agency's temporary headquarters. The IAEA also has two regional offices, one in Toronto (Canada) (since 1979) and the other in Tokyo (Japan) (since 1984), as well as two liaison offices, one in New York (USA) (since 1957) and the other in Geneva (Switzerland) (since 1965). The Agency operates specialized nuclear technology laboratories in Vienna and Seibersdorf (Austria), inaugurated in 1961, and, since 1961, in Monaco.

The IAEA's governing bodies decide on the Agency's programs and budgets. They comprise the General Conference, which brings together all Member States (in 2022, 173 members or states), and the Board of Governors, made up of 35 members. The General Conference meets annually at IAEA headquarters in Vienna, usually in September. The Board meets five times a year, also in Vienna.

Unlike the cases of various physical and/or chemical hazards affecting human health, the World Health Organization (WHO) has no involvement in radioactive

1 In February 2003, during IAEA visits to the workshops of the Kalaye Electric Company in Tehran, inspectors discovered that these premises had been used to produce enriched uranium using centrifuges. Other major violations (secret nuclear facilities) were discovered in the following months.

hazards. This is because, on May 28, 1959, the WHO signed the "WHA12-40" agreement with the IAEA. This agreement sets out the responsibilities, consultation and cooperation arrangements between the two agencies. In particular, it stipulates that both agencies recognize that they may be called upon to take certain restrictive measures to safeguard the confidentiality of information supplied to them. In practice, the WHO has done very little to combat the danger of radioactivity. For example, the WHO abolished its radiation department a few years ago [ARC 14, p. 1784]. The confidentiality clause is a cross-cutting provision of UN agreements, and it is indeed the WHO that is responsible for assessing health risk, including in the event of a nuclear accident. Furthermore, in 1986, the IAEA's control was reinforced by a number of conventions. This situation is not reassuring in terms of transparency concerning the health consequences of radioactivity [ROS 14, p. 6962]. It is also curious that a UN agency such as the IAEA should have a "commercial" character. In fact, the IAEA's prime objective is to promote the safe use of nuclear energy. The latest report by its Director [IEA 21] makes repeated use of the term "commercial". On May 5, 2022, the IAEA launched a new roadmap for the commercial deployment of nuclear-powered hydrogen (see: https://www.sfen.org/rgn/laiea-prepare-une-feuille-de-route-pour-un-deploiement-commercial-de-lhydrogene-grace-au-nucleaire/). The IAEA has created a structure (INIR, Integrated Nuclear Infrastructure Review) whose mission is to help first-licensing states in the development of nuclear infrastructure.

## I.3. National nuclear safety institutions

Each country has its own regulatory body (Table I.1). In France, these are the ASN, ASND and IRSN. According to IAEA recommendations, these bodies must be independent. However, following the Fukushima accident, the Japanese authorities were obliged to create a new institution to make it more independent of the Japanese nuclear industry. Each state defines its own rules, generally following the recommendations of international institutions, but there are differences.

In conclusion, radioactive risk assessment does not follow the same procedures as for all other physical and chemical hazards. In particular, the international and national bodies in charge of this risk are specific to the nuclear sector. However, the fact that the organization of safety in the face of the danger presented by ionizing radiation has been able, thanks to the existence of several international and European organizations, to develop in a relatively homogeneous way throughout the world is a significant and widespread development that many other chemical, physical or biological hazards have not experienced.

| State | Code | Institution |
|---|---|---|
| Germany | GRS | *Gesellschaft für Anlagen- und Reaktorsicherheit* |
| | DFD | *Deutsch-Französischer Direktionausschuss* |
| Argentina | ARN | *Argentine Nuclear Regulatory Authority* |
| Australia | ARPANSA | *Australian Radiation Protection and Nuclear Safety Agency* |
| Belgium | AFCN | *Agence Fédérale de Contrôle Nucléaire* |
| Brazil | CNEN | *Brazilian Comissao Nacional de Energia Nuclear* |
| Bulgaria | NRA | *Bulgarian Nuclear Regulatory Agency* |
| Canada | CNSC | *Canadian Nuclear Safety Commission* |
| Chile | CCHEN | *Comisión Chilena de Energía Nuclear* |
| China | NNSA | *China's National Nuclear Safety Administration* |
| South Korea | KINS | *Korea Institute of Nuclear Safety* |
| Spain | CSN | *Spanish Council for Nuclear Safety* |
| United States | NRC | *US Nuclear Regulatory Commission* |
| Finland | STUK | *Finnish Radiation and Nuclear Safety Authority* |
| France | ASN | *Autorité de Sûreté Nucléaire* |
| | ASND | *Autorité de Sûreté Nucléaire Défense* |
| | IRSN | *Institut de Radioprotection et de Sûreté Nucléaire* |
| Hungary | HAEA | *Hungarian Atomic Energy Authority* |
| Israel | IAEC | *Israel Atomic Energy Commission* |
| Italy | INFN | *Istituto Nazionale di Fisica Nucleare* |
| Japan | NRA | *Nuclear Regulation Authority, Japan* |
| Latvia | VARAM | *Latvian Ministry of Environmental Protection and Regional Development* |
| Lithuania | VATEST | *Lithuanian Nuclear Power Safety Inspectorate* |
| Russia | Rostekhnadzor | *Rostechnadzor* |
| Slovak Republic | UJD | *Nuclear Regulatory Authority of the Slovak Republic* |
| Czech Republic | SUJB | *Czech State Office for Nuclear Safety* |
| United Kingdom | ONR | *UK Office for Nuclear Regulation* |
| Sweden | SSM | *Swedish Radiation Safety Authority* |

| Switzerland | ENSI | *Swiss Federal Nuclear Safety Inspectorate* |
|---|---|---|
| Thailand | OAP | *Thai Office of Atoms for Peace* |
| Turkey | NDK | *Nükleer Düzenleme Kurumu* |
| Vietnam | VARANS | *Vietnam Agency for Radiation and Nuclear Safety* |

**Table I.1.** *National authorities in charge of nuclear safety and radiation protection*

## I.4. References

[AIE 21] AIEA, Rapport d'ensemble sur la technologie nucléaire 2021. Rapport du Directeur général, IAEA/NTR/2021 report, IAEA, Vienna, 2021.

[AMI 17] AMIARD J.C., *Les risques chimiques environnementaux. Méthodes d'évaluation et impacts sur les organismes*, 2nd edition, Lavoisier, Tec&Doc, Paris, 2017.

[AMI 20] AMIARD J.C., *Nuclear Accidents: Prevention and Management of an Accidental Crisis*, vol. 3, ISTE, London, and John Wiley & Sons, New York, 2020.

[AMI 22] AMIARD J.C., *Marine Radioecology*, vol. 6, ISTE, London, and John Wiley & Sons, New York, 2022.

[ARC 14] ARCHIMBAUD A., "Question orale no. 0847S", *Journal officiel du Sénat de la république française,* 31 July 2014.

[IRS 21] IRSN, Baromètre IRSN 2021 sur la perception des risques et de la sécurité par les Français, IRSN, 26 May 2021.

[ROS 14] ROSSIGNOL L., "Réponse du Secrétariat d'État, auprès du ministère des affaires sociales, de la santé et des droits des femmes, chargé de la famille, des personnes âgées et de l'autonomie à la question orale no. 0847S de Mme Aline ARCHIMBAUD (Seine-Saint-Denis - ECOLO)", *Journal Officiel du Sénat de la République française,* 15 October 2014.

[SUC 18] SUCIU N.A., PANIZZI S., CIFFROY P. et al., "Evolution and future of human health and environmental risk assessment", in P. CIFFROY, A. TEDIOSI, E. CAPRI (eds), *Modelling the Fate of Chemicals in the Environment and the Human Body, The Handbook of Environmental Chemistry*, vol. 57, Springer, Cham, pp. 1–21, 2018.

# 1

# Radioactive Danger

## 1.1. Introduction

The concept of the atom[1], the unbreakable unit of matter, was first put forward over 24 centuries ago by Leucippus (460–370 BC) and his disciple Democritus (460–370 BC). That atoms are the ultimate constituents of matter was cast into doubt with the discovery of radioactivity in 1896 by Henri Becquerel (1852–1908), followed by the Curies' work on uranium ore processing residues. In 1900, Marie Curie (1867–1934) wrote: "radioactive matter is not an ordinary chemical state; atoms are not constituted in a stable state, since particles smaller than the atom are radiated. The chemically indivisible atom **is divisible here**, and the **sub-atoms** are in motion".

In 1903, Frédéric Soddy (1877–1956) and William Ramsay (1852–1916) enclosed a 20 mg sample of radium bromide in a closed container, and after a few months discovered that it contained helium. This proved that the radium atom had split by expelling a helium atom. The residue is necessarily another element [FER 06, p. 50]. In her thesis, Marie Curie shows that radioactive decay (known as the law of deactivation) is an exponential law [CUR 04, pp. 118, 162]. She underlined the importance of Soddy and Ramsay's discovery of helium and confirmed that radium and thorium emanations[2] do not seem to be altered by various energetic chemical agents, and for this reason, Rutherford and Soddy assimilated them to gases of the argon family.

1 The word atom is borrowed from ancient Greek, derived from **tomos** meaning to *cut, divide*, with the prefix -a- privative, giving **atomos** the meaning *unbreakable* or *uncut*.

2 These are radon-222 (descendant of radium-226) and radon-220 (descendant of thorium-232).

Several physicists began to suggest that atoms might themselves be compound objects. They also drew up structural diagrams of the atom. Jean Perrin (1870–1942), in February 1900, imagined a structure that is striking in retrospect:

> Each atom would be made up, on the one hand, of one or more masses charged with positive electricity...and on the other hand, a multitude of corpuscles, a kind of small negative planet, all the masses gravitating under the action of electrical forces and the total charge being exactly equivalent to the total positive charge, so that the atom would be electrically neutral...the negative corpuscles always appear identical to each other, whatever the chemical nature of the atom from which they are detached.

The text of his lecture was published, but Jean Perrin did not develop this luminous hypothesis and never returned to the subject.

In contrast, Rutherford (1871–1937) specified a structure on March 7, 1911, in a two-page note. His model of the atom consisted of a central electric charge concentrated in one point and surrounded by a uniform spherical distribution of electricity of opposite sign in equal quantity. In an article dated August 16, 1912, he even used the Latin word *nucleus* [FER 06, p. 92], the nucleus of the atom. However, Niels Bohr (1885–1962) reminded us that there could be no static equilibrium between the nucleus and the electrons, and that they therefore rotate (1912). The blueprint became clearer.

When moving from one element to another in Mendeleev's classification (1834–1907), a fundamental quantity increases by a constant amount. For the English physicist Henry Moseley (1887–1915), this quantity could only be the charge of the positive central nucleus. He thus discovered that what was known as the atomic number corresponded to the number of elementary charges in the atom's nucleus [FER 06, p. 96].

The dangers presented by ionizing radiation were highlighted by Henri Becquerel and Marie Curie at the very time of their discovery. In a note to the Académie des Sciences (French Academy of Sciences) in 1901, Henri Becquerel and Pierre Curie (1859–1906) [BEC 01] recount the story of the first illnesses that affected scientists working with radium, who were not yet aware of the deleterious effects of extensive exposure to ionizing radiation.

It was two German researchers, the physicist Otto Walkoff and the chemist Friedrich Giesel (1852–1927), who published their first observations in October 1900. After placing radium wrapped in a sheet of celluloid on his arm for two hours, Mr. Giesel observed a slight reddening which increased and, after three weeks, caused the skin to “inflame” and then fall off. It was the first voluntary radiodermatitis!

Pierre Curie [CUR 01] then decided to reproduce Giesel’s experiment, placing a thin sheet of gutta-percha on his arm for 10 hours, on which he had deposited “radiferous barium chloride”. The skin became red over an area of 6 $cm^2$, a burn-like redness that increased in intensity and on the 20th day turned into a sore. After 42 days, the epidermis began to reform, but 52 days later, a 1 $cm^2$ wound remained, with a grayish appearance, indicating deep damage to the epidermis and dermis. This was the first severe radiological burn.

Henri Becquerel’s experiment was “involuntary”. For around six hours, he placed a box containing a small sealed glass tube[3] that contained radiferous barium chloride in his vest pocket. Ten days later, he noticed a reddening of the tube’s oblong shape on his skin at the bottom of the pocket. Three weeks after exposure, a wound formed. It closed seven weeks after exposure, leaving a scar. While this burn was being treated, a second oblong red spot appeared, opposite the other corner of his vest pocket. The vial, which had been moved in the pocket, had produced a second erythema.

The authors also recall in this note that Marie Curie, carrying for less than half an hour a small sealed glass tube containing a few centigrams[4] of “the same very active material”, enclosed in a thin metal box, had “after 15 days a red stain which gave a blister similar to that of a superficial burn”. Henri Becquerel and Pierre Curie concluded by noting that the duration of the alterations varies with the intensity of the active rays and the duration of the excitatory action, and that the extremities of the fingers, which have held tubes or capsules containing highly active products, become hard and sometimes very painful. In one case, the inflammation of the fingertips lasted a fortnight before the skin fell off, but the painful sensitivity had not completely disappeared after two months.

As we shall see in Chapter 10, in addition to these short-term effects, caused by high local doses and dose rates[5] of ionizing radiation, radiation-induced diseases can be caused by doses and dose rates much lower than those producing radiation burns, administered to all or part of the whole organism. These conditions often develop late in life, with latency periods measured in years or even decades.

3 A tube 20 mm long and 4 mm in diameter [BER 07, p. 81].

4 One centigram of radiated barium chloride.

5 The dose rate represents the dose delivered per unit of time. It is expressed in grays per second (Gy/s). This is a very important unit. For practical reasons, measurements taken in the workplace are generally expressed in milligrays per hour (mGy/h).

For more than a century, our knowledge of the dangers of radioactivity has made great strides forward. Protecting people and the environment requires knowledge of the harmful effects of the different types of radiation emitted by radionuclides and by electrical ionizing radiation generators.

## 1.2. Radionuclides and radioelements

Chemical elements frequently have several isotopes, the majority of which are stable, while others are radioactive. Some chemical elements have a radioactive isotope, such as carbon with isotope 14 ($^{14}C$) or potassium with $^{40}K$, while the two heaviest natural elements, uranium and thorium, produce a chain of radioactive isotopes. Natural uranium is made up of three radioactive isotopes, uranium 234 ($^{234}U$), uranium 235 ($^{235}U$) and uranium 238 ($^{238}U$), which are, respectively, in the following proportions: 0.0055%, 0.720% and 99.2745%, while thorium has only one nuclide, thorium 232 ($^{232}Th$), and is radioactive. $^{234}U$ is a descendant of $^{238}U$, while $^{235}U$ is the origin of a family of descendants independent of $^{238}U$. As uranium and thorium are naturally occurring chemical elements consisting only of radioactive isotopes, they are referred to as radioelements. The radioactive isotope of an element is called a radionuclide or radioisotope. Hydrogen, for example, has three isotopes, two stable (hydrogen and deuterium) and one radioactive (tritium). In the case of carbon, carbon-14 is a radionuclide, while carbons-12 and 13 are stable isotopes.

Radionuclides can be natural or artificial, i.e. created by a deliberately activated nuclear reaction (fission, neutron activation). They all have an unstable nucleus due to an excess of protons, neutrons or both. The greater their instability, the shorter their physical half-life. Eventually, they regain stability by emitting radiation that releases their excess energy. This is what characterizes radioactivity. However, when a radionuclide decays, it may not regain stability, but instead gives rise to a new radionuclide by filiation. $^{235}U$ and $^{238}U$ as well as $^{232}Th$ are natural radioactive families with many daughter elements. Radium-226 ($^{226}Ra$), discovered by Marie Curie, and uranium-234 ($^{234}U$) are among the 19 descendants of the $^{238}U$ family. The last descendant of the $^{238}U$ family is stable: lead-206. $^{232}Th$ has 11 sons, the last of which is stable: lead-208.

There are approximately 2,000 nuclides, including the current 120 elements[6] of the periodic table, and only 325 nuclides exist naturally (274 stable and 51 radioactive nuclei), the others, all unstable, being created artificially. Only 21 elements have no isotope, since they possess only one natural nuclide ($^{9}Be$, $^{19}F$, $^{23}Na$, $^{27}Al$, $^{31}P$, $^{45}Sc$, $^{55}Mn$, $^{59}Co$, $^{75}As$, $^{89}Y$, $^{93}Nb$, $^{103}Rh$, $^{127}I$, $^{133}Cs$, $^{141}Pr$, $^{159}Tb$,

6 The latest, heaviest, artificially produced element (Z = 118) was named *Oganesson* in August 2016 by the International Union of Pure and Applied Chemistry, in honor of the Armenian-born Russian physicist Yuri Oganessian, a pioneer of transuranium research.

$^{165}$Ho, $^{169}$Tm, $^{197}$Au, $^{209}$Bi, $^{232}$Th) [SEE 81]. Xenon has nine isotopes and tin is the element with the most, with 10 natural isotopes. Technetium, with atomic number 43, and the six chemical elements between bismuth (z = 83) and thorium (z = 90) (polonium, astatine, radon, francium, radium, actinium) have no stable isotope.

Only hydrogen isotopes have been given specific names and symbols. The most abundant (99.985%) is hydrogen ($^1$H), with only one proton. Deuterium[7] ($^2$D), or "heavy" hydrogen, is only 0.015% present in the hydrogen in air. It has one proton and one neutron. Tritium ($^3$T) is the only unstable isotope, with one proton and two neutrons. In oceans, seas and surface waters, there is semi-heavy water, water ($H_2O$) in which a deuterium atom has replaced the hydrogen (HDO), far more abundant than heavy water ($D_2O$). Semi-heavy water is naturally present in water in a proportion of 0.03125%, or one molecule for every 3,200 molecules of water.

## 1.3. Radionuclide-related dangers

In theory, the danger from internal radionuclide contamination is threefold: chemical stress, radioactive stress and combined stress. Chemical stress is caused by stable or radioactive metal ions, which cause cellular toxicity by binding to sensitive biomolecules such as DNA or proteins, or by substituting for biologically useful elements, thus blocking various enzymes. Metals also act indirectly by causing the formation of reactive oxygen species (ROS), notably as a result of oxidation–reduction reactions.

Radiation stress has two modes of exposure, external and internal. In the case of external exposure, only high-energy, penetrating radiation will reach all biological tissues. On the contrary, in the case of low-energy radiation and internal contamination, the low penetrating power of the rays is an amplifying factor, as all their energy is deposited very locally and in a highly concentrated manner, i.e. a high linear energy transfer (LET). In addition, the dose rate can also influence damage[8], with a high dose over a short period of time having a greater effect than a low dose spread over the long term. In the latter case, detoxification and repair mechanisms[9] will be able to take effect. Damage caused by ionizing radiation can be direct (ionization of the target) or indirect (generation of ROS). Damage depends on the nature of the radiation and its energy.

7 The discovery of this hydrogen isotope was announced on December 5, 1931 by three American physicists: H. Urey, F. Brickewedde and G. Murphy of Colombia University (NY).

8 The ICRP has recognized that chronic exposure of workers to ionizing radiation is, for the same dose, half as damaging as a dose delivered at a high dose rate.

9 For example, the induction of certain proteins such as metallothionein, which bind certain elements and thus reduce their chemical toxicity.

Combined stress is due to the fact that most radioactive contaminants combine radiative and chemical effects. Furthermore, in some cases, decay of the radioactive element is accompanied by atomic transformation (transmutation), sometimes resulting in a chain of radioactive or stable descendants with their own toxicity. In most cases of radioactive contamination, there is not just one type of radionuclide, but multiple radioactive elements. These elements may or may not interact with each other. In the latter case, the effects may present three alternatives: simple addition, antagonism or synergy. Low doses of toxic substances can induce potentiation, or on the contrary, trigger adaptation.

In practice, radiative effects are far more toxic than chemical effects, as they appear at much lower concentrations. The only known exception is uranium, for which chemical toxicity outweighs radiological toxicity. The radioactive limits for uranium intake by inhalation and ingestion result from the conversion[10] of the number of milligrams of the limit values adopted for chemical toxicity, into uranium becquerels. The values in Bq/mg of natural uranium (25 Bq/mg) change if the uranium is enriched or depleted: 63 Bq/mg for uranium enriched to 3% in $^{235}U$ and 15 Bq/mg for uranium depleted to 0.2% in $^{235}U$.

## 1.4. Elements of nuclear physics

Nuclear physics is the science concerned with all physical phenomena involving the atomic nucleus. Matter is made up of molecules, themselves made up of atoms. Atoms are made up of a central nucleus surrounded by electrons [BON 02].

In response to the approach of Rutherford, who in March 1911 sketched out the first model of the atom (a central electric charge concentrated in a point surrounded by a uniform spherical distribution of electricity of opposite sign in equal quantity), Bohr clarified on July 6, 1912, in a letter to Rutherford, that there could be no static equilibrium between the nucleus and electrons, and that the latter must therefore necessarily rotate. Moreover, a rotating ring is not stable for more than seven electrons, suggesting a structure of successive rings. He added that he saw a possible explanation for the periodic law of chemical properties of elements (assuming that chemical properties depend on the stability of the outermost ring, the "valence electrons"). A little later, Arnold Sommerfeld (1868–1951) clarified that if electrons obey classical mechanics in their orbits, they must describe *ellipses* with the nucleus as one of the *foci*.

After the question of the atom's structure, we will discuss the question of the nucleus with the discovery of the neutron in 1932. It was two German–Jewish

10 One milligram of natural uranium has an activity equal to 25 becquerels.

physicists, Walter Elsasser (1904–1991) and Kurt Guggenheimer (1896–1940), who took refuge in Paris[11] in 1933, who suggested that protons and neutrons could be *layered* in the nucleus, like electrons in the atom. Elsasser even specified that each particle, neutron or proton moves in an orbit, and that the particles must then be organized in successive layers, which he called envelopes.

The applications of nuclear physics, and in particular the use of the properties of ionizing radiation, are highly varied, in fields as diverse as astrophysics, archaeology (radioactive dating), medicine (radiology, radiotherapy, radioactive tracers), controlled energy production (electricity generation or nuclear propulsion) and uncontrolled energy production (military applications, bombs), food processing (food sterilization), chemical analysis by activation, non-destructive testing (search for anomalies in materials, constructions), conservation and restoration of objects, and cultural goods or archaeological remains[12] in organic materials (wood, leather, fibers) by gamma irradiation.

### 1.4.1. *Nuclear structure*

The atomic nucleus is made up of two types of nucleons[13], protons and neutrons. Protons are particles with a positive elementary electric charge, while neutrons are electrically neutral particles. Four elementary interactions, or fundamental forces, are responsible for all physical phenomena observed in the universe. These are strong nuclear interaction, electromagnetic interaction, weak nuclear interaction and gravitation. The strongest of these forces, strong nuclear interaction, holds nucleons together in the nucleus. The amount of energy that holds the nucleus together is called the binding energy of the nucleus.

When the number of protons in the nucleus is large, the Coulombic (or electromagnetic) interaction takes precedence over the strong interaction, and nuclei become unstable.

---

11 Fleeing Nazi Germany like some 1,500 Jewish scientists, mostly in 1933, Elsasser obtained a position at the Centre de Recherches, forerunner of the CNRS, and Guggenheimer found refuge at the Collège de France, in Paul Langevin's laboratory. Many of these scientists (including James Franck, winner of the Nobel Prize in Physics in 1925) went into exile in the United Sates, where they took part in the Manhattan Project. Between 1937 and 1991, seven of them were awarded the Nobel Prize.

12 The mummy of Ramses II, which was attacked by various parasites, arrived in France (September 1976) for treatment by global gamma irradiation, carried out at the Centre d'études nucléaires de Saclay (May 1977), which destroyed the parasites.

13 Name given to protons and neutrons when we are interested in their similar characteristics: mass, spin, interactions between them. For example, deuterium has two nucleons: a proton and a neutron.

The number of protons in a nucleus defines its Z value (Figure 1.1). Nuclei with the same Z value, but different numbers of neutrons, are called isotopes. For nuclei naturally present in the universe, their mass number, A, varies from 1 to 238, and their atomic number, Z, from 1 to 92. The heaviest natural nucleus is uranium-238, composed of 92 protons and 146 neutrons, or 238 nucleons.

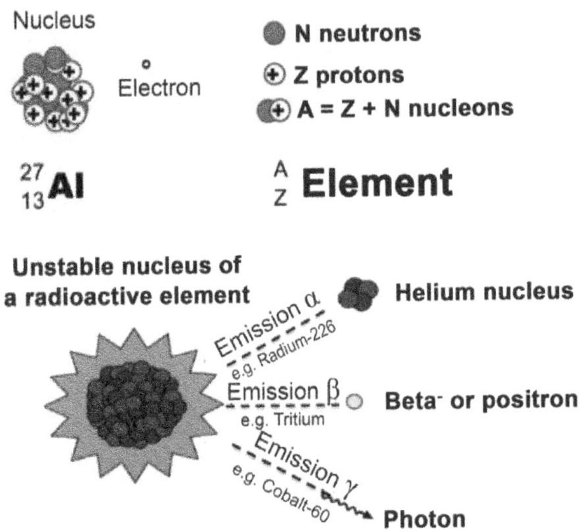


**Figure 1.1.** *Nuclear structure, symbolism and electrical charges of the three types of ionizing radiation. For a color version of this figure, see www.iste.co.uk/amiard/risk.zip*

The atomic mass unit (a.m.u.) is 1 dalton, or 1 g. $(6.10\ )^{23-1}$ or $1.667.10^{-27}$ kg. The mass of a proton is $1.6726.10^{-27}$ kg (or 1.0074 a.m.u.) and its charge is + $1.602.10^{-19}$ C. The mass of a proton is 1,836 times that of an electron. The mass of a neutron is very close to that of a proton ($1.6748.10^{-27}$ kg or 1.0087 a.m.u.) and its charge is zero.

In 1931, the 27-year-old physicist Georges Gamow (1904–1968), a professor at the University of Leningrad, first put forward the idea that the atomic nucleus could be represented as a liquid drop. This liquid drop model is still one of the foundations of our understanding of the nucleus [FER 06, p. 263].

Just as the atom, thought to be unbreakable, was shown to be made up of protons, neutrons and electrons, physicists have also discovered that protons and neutrons, thought to be elementary particles, are themselves each made up of three quarks. As no substructure has yet been discovered at the level of quarks or electrons, quarks are the new elementary particles that make up the neutron and proton.

The matter that surrounds us, stable or almost stable, does not involve all these fermions[14], as some are only involved in extremely energetic processes. In fact, ordinary matter is made up of just three particles: protons, neutrons and electrons. The proton, an assembly of two u quarks and a d quark with a positive elementary electric charge ($1.602.10^{-19}$ Coulomb). The neutron, made up of one u quark and two d quarks, has a zero electric charge (hence its name, a reference to electrical neutrality). The electron is the lightest of the leptons and carries a negative elementary electric charge [NAU 08].

### 1.4.2. *Electrons*

Atoms are made up of a series of electrons orbiting in ellipses around a nucleus (one of the foci of the ellipses). These electrons are about 100,000 times smaller than the nucleus. They are tiny corpuscles with a negative electric charge $e^-$ which is the smallest known electric charge. All observed electric charges are a positive or negative integer multiple of this charge e, which is called elementary.

The number of electrons in an atom ranges from one to almost a hundred (98 for $^{238}U$). This defines the atomic number of the stable atom, which is equal to the number of protons Z. The electric charge of the electron cloud is exactly balanced by the sum of the positive electric charges present in the nucleus, due to the presence of protons, which carry an electric charge equal and opposite to that of the electron. In a stable atom, there are as many electrons as protons in the nucleus: it is therefore electrically neutral. The simplest atom is hydrogen, which has a single electron. The oxygen atom has 8.

Around a nucleus, electrons can only have a limited number of states. An electron's state is characterized by its energy and the way it occupies space. Different states of the same energy are grouped together under the name of layers. An atom's electrons belong to various layers characterized by energy. Atomists call these layers K, L, M, N, etc. The K layer corresponds to the strongest attraction to the nucleus. It is the first to fill up, and only two electrons can occupy this layer. A third electron must necessarily move to the next layer, the L-layer, where it will be less bound to the nucleus than its K-layer counterparts. The binding energies of the electrons on the subsequent layers become progressively lower. When a space becomes available on a more inner layer, an electron on an outer layer fills the gap

14 A fermion is an elementary particle with bosons.

by jumping to occupy the space left free on this inner layer. It thus becomes more closely bound to the nucleus.

This transition of the electron from the L to the K layer is accompanied by the emission of an electromagnetic energy ($E_X$), which inherits the difference in binding energy between the two layers, reduced by the energy $E_L$ required for the electron in the L layer to extract itself from its layer (i.e. $E_X = E_K - 2 E_L$).

These tiny electromagnetic waves are called photons. For a given atom, the photon's energy is unique. Since the photon's wavelength is related to its energy, this wavelength is also unique. As a result, the photons emitted by an atom have energies that characterize it.

The outermost layers are responsible for the atom's chemical properties. This is illustrated by the vertical columns of Mendeleev's periodic table. The first column corresponds to alkalis with an outer layer possessing only one electron, the second column to alkaline earths with two electrons on the first outer sub-layer and the last column corresponding to rare gases where the second outer sub-layer is completed with its six electrons (Chelet, 1968).

When Z is equal to 2, 10, 18, 36, 54 and 86, the layers are complete and stability is maximum; these are the rare or noble gases [BAS 02]. Similarly, atomic nuclei seem to possess a layered structure, and can be built up, like atoms, by filling successive layers. This results in discontinuities with particular values for the number of neutrons N and protons Z called magic numbers (2, 8, 20, 50, 82 and 126, to which we must add 38 and 40) [BAS 02].

### 1.4.3. *Stability of the atomic nucleus and nuclear reactions*

According to Einstein's theory of special relativity ($E = m.c^2$ with E in joules, m in kilograms and c the speed of light[15], $3.10^8$ $m.s^{-1}$), there is an equivalence between energy and mass, meaning that mass can be transformed into energy and vice versa. In the case of the helium nucleus, for example, its mass is 4.002 a.m.u., whereas the sum of the masses of two protons and two neutrons in isolation is 4.032 a.m.u., i.e. a difference of $0.03.10^{-3}$ kg for one mole of helium. The energy released in the formation of a helium nucleus is therefore $2.7.10^{12}$ $J.mol^{-1}$ $(0.03.10^{-3}).(3.10)^{82}$.

The energy released when a nucleus is formed depends on the number of associated nucleons. It is expressed in BEPN (binding energy per nucleon). It is minimal for lithium, then gradually increases to a maximum for $^{56}Fe$ (close to

15 According to the National Bureau of Standards, the most precise value for the speed of light in a *vacuum* is 299,792,456.2 ± 1.1 $m.s^{-1}$.

9 MeV), then slowly decreases for heavy nuclides. For helium, it is $1.12.10^{-12}$ J, or approximately 7 MeV. When measuring the kinetic energy of an electron, the unit used will be the electron volt (eV) rather than the joule (general unit), with 1 eV = $1.6.10^{-19}$ J or conversely 1 J = $6.25.10^{18}$ eV.

Core stability is achieved through three processes. For light nuclei, by nuclear fusion of two light nuclei; for heavy nuclei, either by radioactive decay, which takes place in stages, or by nuclear fission, which corresponds to the breakup of a heavy nucleus into two lighter nuclei. This mechanism is explained by the average BEPN expressed in million electron volts, MeV (Figure 1.2).

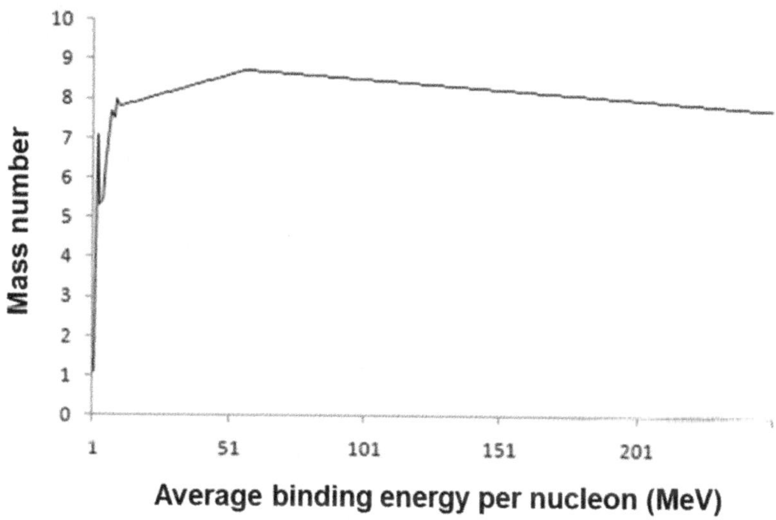


**Figure 1.2.** *Comparison of nuclide stability as a function of mass number and average binding energy per nucleon (MeV)*

A reaction is said to be nuclear when there is a change in the quantum state of the nucleus. The reaction involves not only the nucleons that make up the nucleus, but also other particles such as $e^-$ electrons and $e^+$ positrons.

Nuclear reactions are varied. The main ones are fission, fusion and radioactivity.

Fission corresponds to a heavy nucleus breaking into two fragments. This is the type of reaction used in type A atomic bombs and nuclear power plants. Fission does not generally give rise to two identical isotopes. It is accompanied by the emission of neutrons[16], which maintain the chain reaction. The fission of uranium-235 by thermal neutrons, for example, produces strontium-90 and cesium-137 with high yields and half-lives of approximately 30 years.

Fusion occurs when several light nuclei fuse together. This is how stars produce energy. It is also the type of reaction used in so-called hydrogen bombs (H-bomb). The use of fusion for civilian energy production has not yet been mastered in the international project ITER (International Thermonuclear Experimental Reactor).

### 1.4.4. *Various types of ionizing radiation*

Radioactivity refers to a nucleus that emits one or more particles spontaneously. The type of radioactivity is defined by the particle emitted during the disintegration or transmutation of an atomic nucleus. There are four types of radioactivity and one mode of radiation:

i) $\alpha$ radioactivity, where a helium nucleus, formed of four nucleons, is emitted. This type of radioactivity concerns heavy nuclei (A>200) with an excess of nucleons. They lose four nucleons to stabilize. The Z is changed, resulting in the formation of a lighter chemical element. The alpha particle emitted along a straight trajectory is positively charged, and will be directly ionizing, with an increase in ionization density at the end of the trajectory (Bragg peak).

ii) $\beta^-$ radioactivity, where an electron and an electron antineutrino are emitted. This type of radioactivity occurs in neutron-rich nuclei. A neutron is transformed into a proton, followed by the emission of an electron and an antineutrino. The electron emitted does not pre-exist in the nucleus, but is born from the transformation of a neutron in the nucleus. This electron or $\beta^-$ particle is relativistic (emitted with a speed close to that of light) and its path through matter (several mm) is non-rectilinear. As the Z changes, this transmutation produces another chemical element.

iii) $\beta^+$ radioactivity, where a positron and an electron neutrino are emitted while a proton is transformed into a neutron. This type of radioactivity occurs in nuclei with too many protons. When the positron reaches the end of its path, it couples with an $e^-$ electron, causing the two electrons to annihilate, followed by the formation of a pair of photons of 0.511 MeV (the electron's rest mass). As their momentum was zero, the two gamma photons were emitted in opposition. This mode of decay was discovered by physicist Lise Meitner. This property is used for medical imaging: positron emission tomography (PET scan).

16 In the case of fission of uranium-235 by thermal neutrons, an average number of 2.47 neutrons is produced. The average energy of these neutrons is 2 MeV [BAS 11, p. 66].

iv) Electron capture, which is a mode of nuclear decay that concerns nuclides with an excess of *protons*, and for which $\beta^+$ decay is impossible. During this decay, the neutron-deficient nucleus absorbs an electron from one of the atom's electronic layers. This operation leads to transmutation, because by incorporating the outer peripheral electron, the proton is transformed into a neutron and an electron neutrino is emitted. The atom's nucleus is modified, as it loses a proton and gains a neutron. The electron gap leads to the emission of X-ray photons in competition with Auger electrons.

$\gamma$ emission is a de-excitation, not a decay. The daughter nuclei resulting from $\alpha$, $\beta^-$ or $\beta^+$ decay are very often in an excited energy state. They regain their *stable* ground state by emitting energy in the form of $\gamma$ radiation. This is not material particle emission, but very high-frequency electromagnetic radiation ($f > 10^{18}$ Hz) which is far more penetrating of matter than alpha or beta particles.

The following example shows the decay scheme of cobalt-60 by $\beta^-$ decay which leads to an excited nickel-60 level of 2.505 MeV. Two gammas, of 1.173 MeV and 1.332 MeV, are emitted to reach the stable nickel-60 level (Figure 1.3).

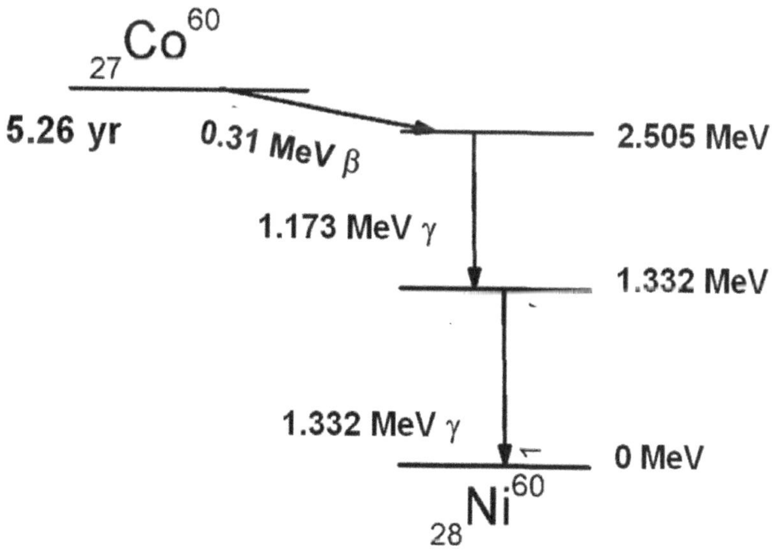


**Figure 1.3.** *Diagram of cobalt-60 decay*

A beam of radiation emitted by a radioactive source and subjected to an electric field created between two metal plates charged + and -, respectively, will cause a deviation in the beam's trajectory depending on the type of radiation. $\alpha$ rays, attracted by the - charged plate, are made up of positively charged particles. $\beta$ rays, attracted by the + charged plate, are made up of negatively charged particles. $\gamma$ rays, undeflected, are not made up of charged particles (Figure 1.1.). Depending on the type of radiation, the power to penetrate matter is very different (Figure 1.4.).

If we classify particles according to the length of their path through matter, alpha rays, with their rectilinear path[17], are the least penetrating, as the particle is heavy and ionizing, followed by $\beta^-$ and $\beta^+$ electrons, which have a complex curvilinear path with energy cessions to other electrons present in matter, which then have their own trajectories. Gamma rays have the longest paths. The paths they take depend on the energy of the particles and the nature (density and z) of the materials they pass through. This conditions their paths in air or living matter (Tables 2.1 and 2.2).

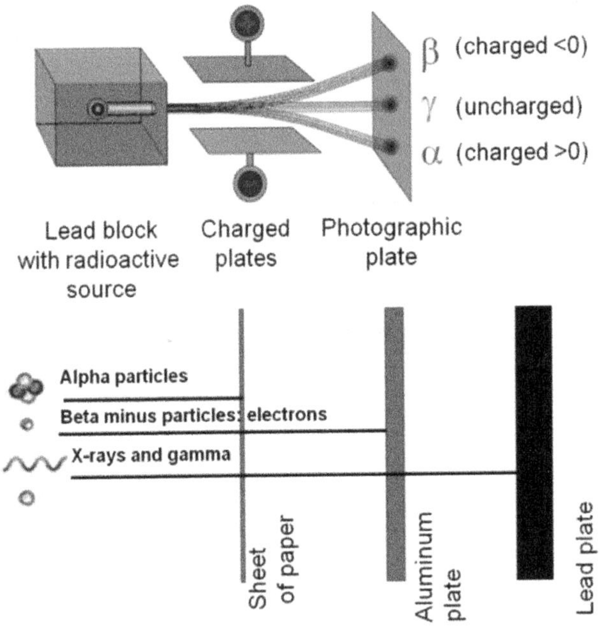


**Figure 1.4.** *The three types of ionizing radiation produced by nuclear reactions and the penetrating power of the various types of ionizing radiation. For a color version of this figure, see www.iste.co.uk/amiard/risk.zip*

17 Becquerel shows that the path of alphas is rectilinear, by making on a photographic plate, subjected to a beam of alphas, a very sharp image of a copper wire 1.5 mm in diameter [CUR 04, p. 85].

| Type of radiation | Energy (MeV) | HLV (cm) |
|---|---|---|
| Gamma | 6 | 16 |
| Gamma ($^{60}Co$) | 1.33 | 11 |
| Gamma ($^{137}Cs$) | 0.663 | 8.05 |

**Table 1.1.** *Half-layer values (HLV) in biological tissue*

Other types of ionizing radiation include particles emitted during beta+ decay and electron capture (positive electrons or positrons), and X-rays. X-rays and $\gamma$ are the most energetic electromagnetic waves, which explains their ionizing power towards matter (Figure 1.5).

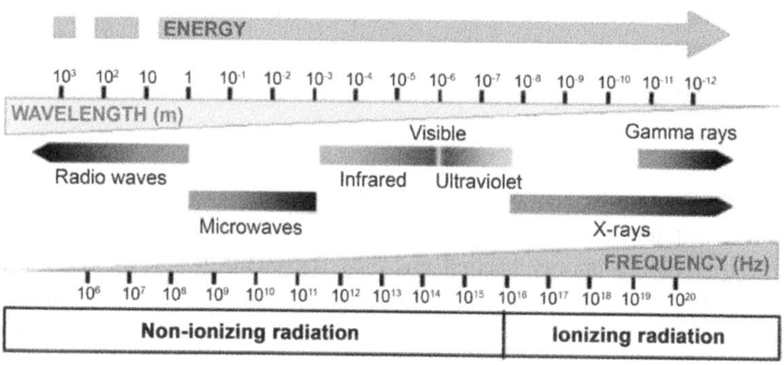


**Figure 1.5.** *Spectrum of electromagnetic waves. For a color version of this figure, see www.iste.co.uk/amiard/risk.zip*

### 1.4.4.1. *Alpha ionizing radiation*

The alpha particle is heavy, as it consists of four nucleons: two protons and two neutrons. It carries two positive charges, making it highly ionizing. The alphas emitted by emitting radionuclides are highly energetic, but within a relatively small energy range[18]; it extends from 4 to 9 MeV. Alpha decay occurs in heavier nuclei and concerns radionuclides with masses greater than 200[19]. Being heavy and positively charged, their mean free path is very short (on average 80 µm, i.e. about

18 The lowest alpha energy emitted by natural radionuclides is that of lead 210 (3.720 MeV, T = 22.23 years) and the highest is that of bismuth 214 (9.080 MeV, T = 19.8 min).
19 Tellurium-106 is often cited as the lightest isotope that undergoes alpha decay, but it is an artificial radionuclide with a half-life of 0.06 milliseconds.

twice the diameter of a cell), so they rapidly lose their kinetic energy at a short distance from the emission source.

In the event of internal contamination, alpha particles deliver high doses. Their capacity to cause harm is estimated at 20 times that of betas, X-rays and gammas.

In the event of external contamination, the alphas' path is, in most cases, too short to reach the basal layer of the epidermis. However, the risk of internal contamination persists in the case of irritated skin. Skin decontamination should therefore be carried out with mild soaps to avoid transcutaneous passages.

Accidental situations: the modes of contamination encountered in the workplace involve contaminated wounds or burns on the hand. A fraction remains locally fixed, but the remainder is rapidly absorbed into the circulating blood. In such cases, the victim must be injected with a chelating agent[20] adapted to the radionuclide in question, to speed up the natural elimination process. In the case of plutonium, the chelator used is DTPA[21].

#### 1.4.4.2. *Beta ionizing radiation*

Beta radiation, emitted by a radioactive atom, consists of electrons with energies varying between zero and the maximum beta energy $E_{\beta max}$, which are emitted in $4\pi$. The variability of the energy carried away by the electron results in a continuous spectrum of the distribution from zero to $E_{\beta max}$. The average energy of the electron spectrum $E_{\beta av}$ is approximately equal to one-third of the maximum beta energy $E_{\beta max}$. For a given electron in the energy spectrum, the complement to the maximum beta energy is carried away by the antineutrino. The antineutrino is an elementary corpuscle with no electric charge, virtually zero mass and very little interaction with matter.

Accidental situations: in some cases, contamination of the body can cause radiological burns. This was the case during the accident that occurred on the night of May 26, 1986 at Chernobyl reactor No. 4 (Ukraine). The firefighters, who came in light uniforms to extinguish the fire in the reactor's graphite mass, were heavily contaminated by combustion aerosols. Radiological burns, caused by the presence of a large number of fission products in the smoke and dust, severely aggravated their health situation, due to the severe external irradiation caused by the fuel in the

20 A chelator is a chemical substance that can take hold of a metal and form a stable, inert complex, which can be eliminated from the body much more quickly than the toxic metal.
21 DTPA = diethylene triamine penta-acetic acid.

reactor core, which had been stopped but virtually laid bare by the energies delivered by the enormous power spike[22] and the ensuing explosion.

#### 1.4.4.3. *Gamma ionizing radiation*

Gamma radiation is composed of high-energy photons. This radiation will penetrate more matter and the body than alpha and beta radiation. There are three types of interaction between gamma rays and matter, depending on their energy: the photoelectric effect, the Compton effect and pair formation (two 0.511 MeV photons are formed). In the case of the photoelectric effect and pair formation, the incident photon disappears.

The photoelectric effect, which predominates at low X and gamma energies, takes place between the incident photon and a K or L layer electron. This requires the photon energy to be equal to or greater than the binding energy. The electron is ejected with an energy equal to the energy of the incident photon minus the binding energy. A rearrangement of the electron pattern ensues, with the emission of X-rays and/or Auger electrons.

The Compton effect also takes place between the incident photon and an electron of the K or L layer. The incident photon tears an electron from its layer, imparting energy to it, while the photon is scattered with an energy reduced by the binding energy ($E_K$ or $E_L$) and the kinetic energy transferred to the electron. The energy of the scattered photon depends on the energy of the incident photon and the scattering angle. For incident energies of a few tens of keV, the energy transferred to the electron is low. Table 1.2 gives the kinetic energy carried away by the electron and that of the photon scattered at an angle of 90° to the incident photon.

| Radionuclide | Eγ of the incident photon (keV) | Compton electron energy (keV) | Eγ of the scattered photon (keV) |
|---|---|---|---|
| $^{125}$I | 27.5 | 1.4 (5.1%) | 26.1 |
| $^{99m}$Tc | 141.0 | 30.5 (21.6%) | 110.5 |
| $^{131}$I | 364.0 | 151.4 (41.6%) | 212.6 |
| $^{60}$Co | 1.332 | 961.0 (72.1%) | 369 |

**Table 1.2.** *Energy of the Compton electron and the photon scattered at 90° (according to [GAM 97, p. 61]). Note: in brackets, fraction of energy transferred to the electron*

22 Calculations have shown that the reactor's instantaneous power would have reached 100 times its nominal value in four seconds [LIB 96, page 313].

Table 1.2 shows that at low incident photon energies, the kinetic energy imparted is low, resulting in a scattered photon with low attenuated energy. The higher the incident photon energy, the higher the energy transfer to the electron: 5.1% for photons of 27.5 keV and 72.1% for 1.332 MeV. For incident photons of a few tens of keV, the scattered photon's energy is only slightly reduced, enabling multiple scattering. A cloud of scattered photons then surrounds the target. At energies in excess of 1 MeV, Compton electrons are predominantly scattered forward, and the angle of the scattered photon cone is reduced. This is why radiotherapy treatments are carried out using high-energy photons (approximately 5 MeV).

| Radionuclide (period) | Gamma energy in keV (emission rate) | Dose rate mSv/h for 1 GBq at 1 meter | Lead thickness in cm for a 95% attenuation |
|---|---|---|---|
| Chromium-51 (27,703 days) | 320.08 (9.87%) | $6.32.10^{-3}$ | 0.78 |
| Cesium-137 (30.05 years) | 661.66 (84.99%) | 0.102 | 2.45 |
| Cobalt-60 (5.2711 years) | 1,173.2 and 1,332.5 (99.85% and 99.98%) | 0.37 | 4.41 |

**Table 1.3.** *Dose rate and photon attenuation of a gamma point source*

Table 1.3 shows the dose rates delivered by gamma photons of increasing energy, emitted by point sources of radionuclides [AIR 12]. Energy transfer is achieved by the photoelectric and Compton effects for chromium-51 and cesium-137. For cobalt-60, the pair-forming effect is partially added to the other two effects. Dose rates, along with the lead thicknesses needed to attenuate 95% of the γ emitted, increase with photon energy.

Pair formation is only possible for incident photons of energy $E\gamma \geq 1.02$ MeV. In the electric field of the nucleus, the photon disappears, giving a negative electron and a positive electron (positron). The conservation of electric charge is respected (as the photon has no charge), as is that of energy. The difference in energy ($E\gamma$ -1.02 MeV) is shared between the electron and the positron in the form of kinetic energy. At the end of its path, the positron couples with an electron, and their annihilation produces two 0.511 MeV photons, emitted in phase opposition (momentum being zero at the end of the path).

Accidental situations: these occur in a number of scenarios, including sealed radioactive sources which accidentally become disconnected from a portable irradiator. Found by chance and picked up by a member of the public, who puts it in their pocket, where it remains for several hours. This type of exposure can result in one or more amputations [AMI 19, ZER 94]. More serious, however, is the case where the person leaves the source at home, since exposure times are no longer measured in hours, but in days and sometimes weeks. The discovery of the source usually follows the nausea and vomiting of family members. In these situations, fatal injuries are also reported (28 deaths listed by Amiard for 1968–2002) [AMI 19, pp. 195–202].

In addition to these exposures of members of the public, there are also cases, generally very serious, of exposure of professionals who enter an irradiation room while the sources or the gas accelerator beam are in irradiation position [AMI 19, pp. 206–210, ZER 93, ZER 94].

#### 1.4.4.4. *Neutrons*

The neutron is not electrically charged, so it does not produce ionizations as it passes through matter. It only interacts with a nucleus, but as the latter occupies a very small volume in the atom, this limits interactions. A neutron flux can therefore be extremely penetrating. Free neutrons do not therefore directly form ionizing radiation, but by provoking nuclear interactions, they can generate ionizing particles (gammas). When a neutron interacts with a nucleus, it usually bounces back, and in a few cases is incorporated into the nucleus. This neutron reaction increases the mass of the target nucleus by one, producing a transmutation that creates a radionuclide. In the case of a heavy nucleus, this neutron integration produces an unstable nucleus that fissions into two unequal parts, emitting neutrons. In the case of uranium-235, the capture of a neutron by the nucleus produces uranium-236, which fissions rapidly, producing two fission products and two or three fast neutrons (2.47 on average). These neutrons keep the nuclear reaction under control in a reactor, but in certain accidental conditions, they can produce a powerful neutron flash that delivers a lethal dose up to several meters away. To stop the neutrons, protective shields rich in hydrogenated materials such as water ($H_2O$), graphite (C) or kerosene ($C_nH_{2n+2}$) are used. Neutron activation can be used to analyze chemical composition or detect impurities in a material.

Accidental situations: cases of accidental neutron irradiation only occur in the professional world. They occur when two metallic masses of plutonium or enriched uranium are mistakenly brought together, thus reaching and then exceeding the critical mass[23] for this material, or when a container is poorly controlled and filled with one of these fissile materials. If the source is in liquid phase, the dose delivered is lethal within a radius of approximately 8 m from the source. In the solid phase,

23 Mass of fissile material nuclei required for a chain reaction to produce neutrons equal to the number of neutrons absorbed.

this radius is increased to 12 m. In addition to the neutron dose, a gamma dose is produced by activation reactions in the victim's body and environment.

Since 1945, there have been 60 accidents worldwide, of which only six occurred after 1978. The first victim was a 24-year-old physicist, Harry Daghlian, associated with the Manhattan program, who two weeks after the bombing of Hiroshima on August 6th, inadvertently brought together two masses of plutonium metal. The critical mass was reached, and the critical excursion, with its neutron flash, led to his death 20 days later, in excruciating pain [BER 22, p. 185]. The last three accidents took place in Japan, at the Tokai-Mura site. On September 30, 1999 at Tokai-Mura, a criticality accident occurred in a uranium processing plant where a neutron safety procedure had not been observed. An excessive quantity of enriched uranyl nitrate was poured into a tank instead of a safe geometry dissolver. This accident caused severe global irradiation syndrome[24] in three workers, resulting in the death of one of the victims despite 83 days of intensive care [IRS 00]. It should therefore be noted that the mortality risks associated with criticality accidents are high. Depending on the author, they range from 17 to 20, i.e. almost one death for every three accidents [AMI 19, pp. 14 and 15].

In the absence of individual dosimetry, biological dosimetry is applied, using gamma spectrometry to measure the $^{24}$Na produced by activation (n,γ) of $^{23}$Na, the stable sodium present in blood. In order to determine the angular position of the victim during the fast neutron flash, the victim's skin (nails, hair) is sampled for sulfur, which produces phosphorus-32 ($^{32}$P) by reaction (n,p[25]) with the isotope $^{32}$S, the majority being sulfur (i = 95.3%). The activities induced in the skin make it possible to determine the angle the person was at, in relation to the incident neutron beam. This data is useful for calculating the dose, or for reconstructing the accident using a Phantom[26] equipped with dosimeters. The rate of chromosomal abnormalities can also be used to assess the dose received.

#### 1.4.4.5. *Cosmic radiation*

In 1897, physicist Charles Wilson (1869–1959) discovered with his apparatus that about 20 ions per second are formed in 1 cm$^3$ of air, regardless of the location. He assumed that this ionization could be caused by sources outside the atmosphere. Rutherford (1871–1937) confirmed these results in 1902.

24 The Tokai-Mura Center estimated that the three operators had received approximately 9 grays (Gy), approximately 5 Gy and just over 1.2 Gy, respectively [IRS 00].

25 n = neutron, p = proton.

26 An anthropomorphic mannequin, made up of a human skeleton encased in a plastic material equivalent to human tissue. It takes the form of a stack of 1-inch-thick (25.4 mm) slices pierced by small blind holes into which FLi dosimeters are inserted. All the point doses measured are used to assess the dose to organs and calculate the effective dose received by the exposed victim.

Since the discovery made by Austrian physicist Viktor Franz Hess (1883–1964) on August 7, 1912, when he measured the rate of air ionization (number of ions created per second and per $cm^3$ of air) as a function of altitude, a number of studies have been carried out to discover the exact nature of these cosmic rays, which propagate at speeds close to that of light.

In 1938, French astronomer Pierre Auger (1898–1993) discovered that the penetration of particles into the Earth's atmosphere, which collide with the oxygen and nitrogen in the air[27], triggers large "sheaves of particles", which are secondary particles in contrast to primary particles of cosmic origin. The vast majority are atomic nuclei, positively charged heavy ions. Virtually all chemical elements are represented, from hydrogen to iron and beyond, but at lower intensities. The intensity of this radiation depends on latitude (intensity increases closer to the poles) and altitude. As altitude increases, the layer of air protecting us from this radiation decreases.

#### 1.4.4.6. *Auger electrons*

Their penetration is identical to that of electrons of the same energy. Auger electrons are mono-energetic electrons, emitted after a rearrangement of peripheral electrons, caused by the disappearance of a K-layer electron. This non-radiative transition phenomenon, which competes with the emission of X-ray lines, was discovered and published in 1923 by Lise Meitner, but it was Pierre Auger, who also became interested in the phenomenon a few years later, who gave it its name.

### 1.4.5. *Radiation properties*

The two main properties of radiation are its power to penetrate matter and ionizing power (Table 1.3). Radiation thus has a greater or lesser tendency to penetrate obstacles placed in its path. This penetration power depends on the nature of the radiation, with $\alpha$ emissions penetrating very little and $\gamma$ rays penetrating a great deal. Ionizing power translates into a greater or lesser degree of ionization of the material traversed along the path of the radiation. Under the impact of collisions, certain atoms lose electrons and are transformed into cations, and these stripped electrons are captured by other atoms.

---

27 It is these secondary particles that produce carbon-14, the radioactive isotope of the long-lived element carbon (5,700 years), by interacting with nitrogen in the air (mainly with the nitrogen-14 isotope, which makes up the majority of the element nitrogen, chemical symbol N).

| Capacity | α rays | β rays | γ rays |
|---|---|---|---|
| Penetration power | Very weak. All radionuclide alphas are stopped by 10 centimeters of air or a thin sheet of paper. | Quite large. All β rays from radionuclides can be stopped by a 12 mm plate of Plexiglass*. | Very large. To stop 95% of γ rays from all radionuclides, an 8-centimeter lead shield is required. |
| Ionizing power | Very high | Significantly lower than α | Weak compared to β rays |

*To stop β-radiation, it is best to choose a screen with a low atomic mass, because as electrons slow down in matter, they produce braking gamma radiation (*Bremsstrahlung*). The intensity of this γ-radiation increases with the atomic mass of the screen.

**Table 1.4.** *Penetration and ionization power of various types of radiation*

## 1.5. References

[AIR 12] AIREY P., HINTON T., TWINING J., "The scientific basis", in TWINING J.R., *Tropical Radioecology*, vol. 18, pp. 1–57, 2012.

[AMI 19] AMIARD J.C., *Industrial and Medical Nuclear Accidents: Environmental, Ecological, Health and Socio-economic Consequences*, ISTE, London, and John Wiley & Sons, New York, 2019.

[BAS 02] BASDEVANT J.L., RICH J., SPIRO M., *Energie nucléaire,* Les éditions de l'école Polytechnique, Palaiseau, 2002.

[BAS 11] BASDEVANT J.-L., *Maîtriser le nucléaire, Que sait-on et que peut-on faire après Fukushima ?* Éditions Eyrolles, Paris, 2011.

[BEC 01] BECQUEREL H., CURIE P., "Action physiologique des rayons du radium", *CR des Séances de l'Académie des sciences*, vol. 132, pp. 1289–91, 3 June 1901.

[BER 07] BERGET A., *Le radium et les nouvelles radiations*, Librairie universelle, Paris, 1907.

[BER 22] BERNAS H., *L'île au bonheur, Homme, atomes et cécité volontaire*, Editions Le Pommier, Paris, 2022.

[BON 02] BONCHE P., *Le nucléaire expliquée par des physiciens*, EDP Sciences, Les Ulis, 2002.

[CUR 01] CURIE P., BECQUEREL H., "Action physiologique des rayons du radium", *Comptes rendus de l'Académie des Sciences*, vol. 132, pp. 1289–1291, 3 June 1901.

[CUR 04] CURIE M., Recherche sur les substances radioactives, Thesis, 2nd reviewed and corrected edition, Imprimerie Gauthier-Villard, Paris, 1904.

[FER 06] FERNANDEZ B., *De l'atome au noyau. Une approche historique de la physique atomique et de la physique nucléaire*, Editions Ellipse, Paris, 2006.

[GAM 97] GAMBINI D.J., GRANIER R., *Manuel pratique de radioprotection*, Lavoisier, Tec&Doc, Paris, 1997.

[IRS 00] IRSN, Tokaï-Mura, le point sur l'accident de septembre 1999, 2000. Available at: https://www.irsn.fr/actualites/tokai-mura-point-sur-laccident-septembre-1999.

[LIB 96] LIBMANN J., *Éléments de sûreté nucléaire*, IPSN, Les Editions de physique, Les Ulis, 1996.

[MET 01] METIVIER H., "L'uranium, propriétés et toxicité", *Journée de la SFRP, l'Uranium sous forme appauvri*, IPSN, Paris, 2001.

[NAU 08] NAUDET G., REUSS P., *Energie, électricité et nucléaire*, EDP Sciences, Les Ulis, 2008.

[SEE 81] SEELMANN-EGGEBERG W., et al., *Nuklidkarte, Chart of the Nuclides*, Institut für Radiochemie, Kernforschungszentrum, Karlsruhe, 1981.

[ZER 93] ZERBIB J.-C., "Forbach, une certaine logique industrielle ?", *Revue de préventique*, pp. 33–39, August–September 1993.

[ZER 94] ZERBIB J.-C., "Accidents radiologiques : Les retours d'expériences", *Annales de l'Association belge de radioprotection*, vol. 19, no. 4, pp. 621–643, 1994.

# 2

# Radioactive Contamination of the Anthroposphere

## 2.1. Introduction

As with chemical risk assessment [AMI 17], radioactive risk assessment is carried out in successive stages (Figure 2.1). The first steps are to estimate natural radioactivity and anthropogenic releases into the environment, and the resulting radioactive contamination of the anthroposphere. The next step is to identify the ways in which humans are exposed to radionuclides. The final stages involve estimating the doses of irradiation suffered by humans, and assessing the risk of adverse effects incurred by comparison with the known effects of irradiation.

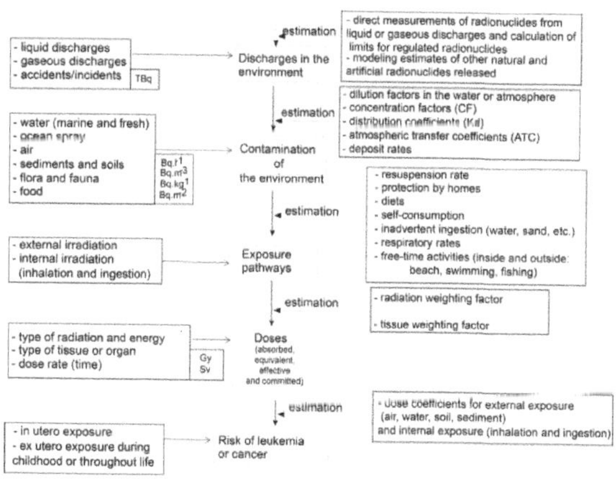


**Figure 2.1.** *The various stages of radiological risk assessment, the parameters to be quantified and the main factors for estimating releases, environmental contamination, exposure routes, doses and leukemia risks (modified from [GRN 99])*

Radioactivity metrology has reached very low detection limits, as well as very small measurement uncertainties. A detailed review of the various methods and techniques for measuring radioactivity is presented by Saha [SAH 13].

## 2.2. Sources of radioactive contamination of the anthroposphere

The environment is made up of four major compartments: atmosphere, lithosphere, hydrosphere and biosphere. The anthroposphere is the part of the environment created or modified by humans for use in human activities and habitats. Human-cultivated ecosystems, or agroecosystems, are highly modified ecosystems, with biodiversity reduced to a minimum. Their role is to exploit part of the organic matter they produce, generally for food purposes. Human settlements are generally grouped together in cities of varying density and size.

The presence of radionuclides in the environment, and therefore in the anthroposphere, has two origins: natural and anthropogenic (Table 2.1).

The natural origin of the Earth's radioactive contamination is mainly due to the presence of radionuclides in the Earth's crust since the origin of various radionuclides descending from four radioactive families, of which the leading radionuclide (uranium-235, uranium-238, thorium-232) has a very long physical half-life (respectively, $704.10^6$, $4,468.10^9$ and $1,402.10^{10}$ years). In addition to the naturally occurring radionuclides in these three families, there are five other naturally occurring radioactive elements with very long half-lives: potassium-40 ($1.25.10^9$ years), rubidium-87 ($4.934.10^{10}$ years), samarium-152 ($2.5.10^{11}$ years), lutetium-176 ($3.79.10^{10}$ years) and rhenium-187 ($4.310^{10}$ years).

There are many anthropogenic origins of radionuclides, including military, industrial and medical origins. Historically, the first deliberate discharges into the environment came from the manufacture of atomic bombs, with the Manhattan Project in the United States. The main discharges were from the Hanford site in the United States, into the Colombia River, flowing into the Pacific Ocean. Then, there were the discharges associated with the atomic bomb tests and the two combat firings at Hiroshima and Nagasaki. These tests introduced large quantities of uranium, plutonium, cesium-137 and strontium-90 into the atmosphere, especially in the northern hemisphere, which have fallen, and continue to fall, to the Earth's surface. The main Soviet discharges were from the Mayak site into the Tetcha River, which flows into the Kara Sea via several other rivers, including the Ob River.

| Category | Origin | Radionuclides |
|---|---|---|
| Natural | Cosmic | $^{7}Be$, $^{10}Be$, $^{14}C$, $^{22}Na$, $^{24}Na$, $^{32}Si$, $^{36}Cl$ |
| | Earth's crust | $^{235}U$, $^{238}U$, $^{232}Th$, $^{40}K$, $^{226}Ra$, $^{222}Rn$, $^{210}Po$ |
| Artificial | Fission products | $^{90}Sr$, $^{131}I$, $^{137}Cs$ |
| | Activation products | $^{51}Cr$, $^{54}Mn$, $^{59}Fe$, $^{58}Co$, $^{60}Co$, $^{65}Zn$, $^{110m}Ag$ |
| | Transuranics | $^{237}Np$, $^{239}Pu$, $^{240}Pu$, $^{243}Am$, $^{231}Pa$ |
| Natural and artificial | Mixed | $^{3}H$, $^{235}U$, $^{14}C$ |

**Table 2.1.** *Origin of the main radionuclides*

The use of nuclear energy also releases radionuclides into the atmosphere and hydrosphere, mainly into rivers. Normal operating releases from basic nuclear infrastructures (BNIs) are generally limited. This is particularly true of nuclear power reactors, especially the latest generations. On the contrary, other stages in the nuclear fuel cycle release larger quantities of radionuclides into the environment. This is the case for spent fuel reprocessing plants, in particular the French La Hague plant and, until recently, the British Windscale-Sellafield plant[1]. These two plants discharge their radionuclides mainly into the marine environment.

After atomic bomb testing, accidental releases into the environment account for the largest quantities. Major accidents involve nuclear power reactors (Chernobyl and Fukushima for the two most recent) and spent fuel reprocessing plants (Mayak, Windscale).

On the contrary, medical discharges have little impact on the environment, since virtually all the radionuclides used have very short physical half-lives, and are stored by medical establishments prior to discharge. However, patients returning home release some of these radionuclides into the environment.

Readers interested in the values of radionuclide discharges into the environment, whether direct or indirect, voluntary or involuntary, should refer to the various volumes in the "Radioactive Risk" series.

Through atmospheric and oceanic circulation and biogeochemical cycles, the behavior and fate of various radionuclides in the environment depend on numerous factors. These include the chemical nature of the radionuclide, the evolution of its physico-chemical form (speciation), its possible biological role, etc. In the end,

1 The British plant is now closed for good.

many radionuclides accumulate strongly in soils and marine sediments, but a negligible quantity bioaccumulates in the biosphere, particularly in humans.

Atmospheric circulation is structured by three cells in each of the two hemispheres: the Hadley cell (0–30°), the Ferrel cell (30–60°) and the polar cell (60–90°). Radionuclides injected into a cell disperse within it, then spread to neighboring cells in the same hemisphere, and finally to cells in the other hemisphere. Gradually, the radionuclides fall back onto land and into oceans.

Oceanic circulations are varied, and alongside a general circulation that takes approximately 1,200 years, there are local and regional currents that disperse all radionuclides coming into the oceans from direct discharges, rivers or atmospheric fallout.

Whether in terrestrial, marine or agrosystems, radionuclides bioaccumulate to a greater or lesser extent in the biosphere and, through the food chain, contaminate all foodstuffs likely to contaminate humans (Figure 2.2).

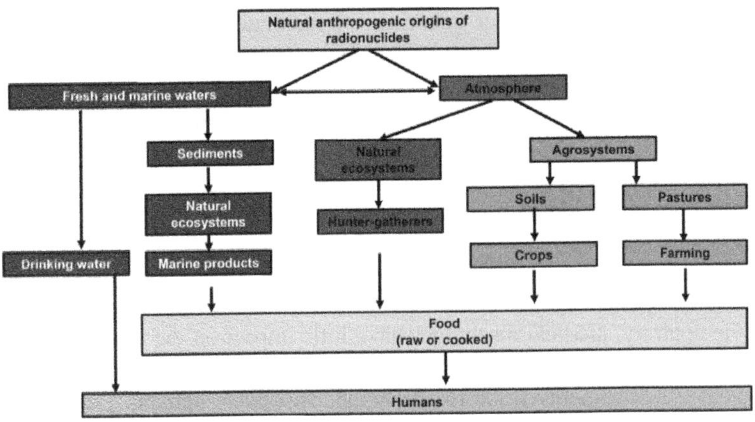


**Figure 2.2.** *The main transfers of radionuclides of natural and anthropogenic origin to humans*

## 2.3. Natural radioactive contamination of the anthroposphere

The environment is naturally contaminated by a variety of radionuclides, which reach the anthroposphere through various pathways where they come into contact with humans.

Natural radionuclide contamination of the environment varies according to the physical environment, with soils generally being the most radioactive (Table 2.2).

| Environment | Contaminant | Activity ($Bq.L^{-1}$ or $*Bq.kg$ )$^{-1}$ |
|---|---|---|
| Rainwater | Total activity | 0.3–1 |
| River water | $^{226}$Ra and descendants | 0.07 |
| | $^{40}$K | 0.07 |
| | Tritium | 11 |
| Seawater ($^{40}$K mainly) | Total activity | 14 |
| Mineral water | $^{226}$Ra, $^{222}$Rn | 1–2 |
| Milk | Total activity | 60 |
| Sedimentary soil | Total activity | 400* |
| Granitic soil | Total activity | 8,000* |
| Human body | Total activity | 12,000* (6,000* due to $^{40}$K) |

**Table 2.2.** *Radioactivity of various natural environments (modified from [GAM 07])*

### 2.3.1. *Natural contamination of physical environments*

The atmosphere is naturally contaminated mainly by radioactive radon and secondarily by tritium ($^{3}$H), beryllium-7 ($^{7}$Be) and carbon-14 ($^{14}$C).

The airborne radon concentration in $Bq.m^{-3}$ varies from 1 to 10 in the atmosphere (but from 1 to 1,000 in valley bottoms), from 10 to 1,000 in houses, from 1,000 to 10,000 in closed cavities in the Earth's crust, from 10 to 1,000 in conventional mines or underground workings, but from 1 to 10 million for closed cavities in uranium ore zones. Levels vary widely in uranium mines, depending on the ventilation of the galleries [GAM 07]. In France, radon concentrations vary greatly according to soil type, with the highest concentrations in the granitic rocks of the Hercynian massifs (Massif Central, Corsica, Monts d'Arrée, Vosges). Arithmetic averages for each department range from 19 to 297 $Bq.m^{-3}$ [BIL 04]. The IRSN measured 12,641 houses in France, and 60 (0.5%) exceeded 1,000 $Bq.m^{-3}$. The maximum measurement was 4,964 $Bq.m^{-3}$ [IRS 12]. A measurement campaign organized between 1982 and 2003 by the Ministry of Health and IRSN on over 10,000 buildings showed that the average for France as a whole was 90 $Bq/m^3$. There were, however, major disparities from one department to another, within a department, and from one building to another. The average was 24 $Bq/m^3$ in Paris and 264 $Bq/m^3$ in Lozère [IRS 12].

Activities of $^{7}$Be in aerosols, of the order of 3,200 $\mu Bq\ m^{-3}$, are similar in mainland France and Tahiti [BOU 04]. Lead-210 ($^{210}$Pb) (a descendant of radon-222) is present in air in mainland France at between <0.3 and 1 $mBq.m^{-3}$ [IRS 11].

Water is contaminated by radionuclides from the uranium and thorium radioactive families. Natural contamination of drinking water in France averages (in $mBq.L^{-1}$) from 4.4 to 930 for $^{238}U$, from 0 to 4.2 for $^{232}Th$ and from 7 to 700 for $^{226}Ra$ [UNS 08]. The radioactivity of groundwater is linked to the nature of the rock (Table 2.3). Certain thermal waters in particular are radioactively rich.

| Aquifer lithology | $^{222}$Rn in water ($Bq.m^{-3}$) |
|---|---|
| Granites | 500,000 |
| Metamorphic rocks | 150,000 |
| Basalts | 15,000 |
| Sand | 10,000 |
| Limestone | 1,000 |

**Table 2.3.** *Radon-222 concentrations in groundwater as a function of aquifer rock type (modified from [MET 88])*

Natural soil contamination is highly variable, depending on the nature of the rock. The main natural radionuclides are potassium-40, rubidium-87 and the various descendants of the radioactive families (uranium-235, uranium-238 and thorium-232). Average values, in $Bq.kg^{-1}$, vary from country to country, from 140 to 850 for $^{40}K$, from 10 to 114 for $^{238}U$, from 8.5 to 70 for $^{226}Ra$ and from 5 to 82 for $^{232}Th$ [UNS 08].

Natural soil contamination is highest in the primary massifs (Hercynian or Variscan massifs) formed during the Primary Era (Table 2.4).

| Site | $^{40}$K | $^{238}$U | $^{226}$Ra | $^{232}$Th |
|---|---|---|---|---|
| Brazil (R.G. Norte) | 941 | | 50 | 69 |
| Czech Republic | 986–1599 | 68–220 | 76–225 | 74–159 |
| Indonesia (West Java) | 48–252 | | 2.4–422 | 0.5–66 |
| Iran (Ramsar) | 300–945 | | 80–50,000 | 15–47 |
| Iran (Mahallat) | 364–873 | | 500–7,300 | 15–41 |
| Azerbaijan | 800–1,000 | 100–7,000 | 500–2,500 | 100–1,000 |
| Russia | | | | 100–150 |
| Spain | 810–1,240 | | 60–250 | 42–71 |
| **World average** | **420** | **33** | **32** | **45** |

**Table 2.4.** *Concentrations of naturally occurring radionuclides in various soils ($Bq.kg^{-1}$) [UNS 08]*

### 2.3.2. *Natural food contamination*

Foodstuffs (plants and domestic animals) are naturally contaminated by various radionuclides, notably potassium-40 and carbon-14. Beryllium-7 concentrations in milk range from 34.6 to 68.5 $Bq.L^{-1}$ [IRS 11]. In marine organisms, particularly shellfish, radioactive contamination is mainly due to polonium-210 [FOW 11].

### 2.3.3. *Natural contamination of seafood*

This is a topic that is frequently addressed, with an impressive number of publications.

Radioactivity levels of $^{210}Po$ and $^{137}Cs$ in seawater and biota (fish and shellfish) were estimated for the various FAO fishing zones on the basis of measurements taken for the year 1990, chosen as the reference year. Values in seawater ranged from 0.3 to 125 (Baltic Sea) $Bq.m^{-3}$ for $^{137}Cs$. Mean concentrations of $^{137}Cs$ in fish ranged from 0.04 to 11.90 (Irish Sea) $Bq.kg^{-1}$ dw[2] [IAE 95, AAR 97].

In human food, natural radioactivity is significant, and $^{210}Po$, an alpha emitter, is the main contributor [ALA 11].

In Japan, consumption of seafood and its natural radioactive contamination lead to radiation doses estimated at 0.48–0.69 and 0.022–0.042 $Bq.d^{-1}$ per person, for polonium-210 and lead-210, respectively [YAM 94].

Concentrations of $^{210}Po$ were measured in a range of fish and shellfish species that are important in the context of the Irish seafood diet. Data are presented for plaice, cod, whiting, skate, mackerel, herring, oyster, lobster, shrimp and mussels. Values range from 0.2 to 0.8 $Bq.kg^{-1}$ dw for $^{210}Pb$ and from 0.16 to 39.3 (mussels) $Bq.kg^{-1}$ dw for $^{210}Po$ [POL 98].

In Indonesia, seafood is an important part of the diet with various fish: tuna, mackerel, red snapper, shrimp and green mussels. These species were collected in 2017, and $^{210}Po$ activity in the edible portion of these biota ranged from 9.05 ± 3.05 to 137.37 ± 37 $Bq.kg^{-1}$ dw. The highest activity was found in mussels. The daily intake of $^{210}Po$ from these biota ranged from 0.24 to 14.27 $Bq.yr^{-1}$ [MAK 20].

Contamination by naturally occurring radionuclides ($^{238}U$, $^{232}Th$ and $^{40}K$) in the marine fish croaker (Pseudotolitus typus) widely consumed in two coastal areas of Nigeria is sufficiently high to present a major radiological risk to the public [ADE 20].

2 fw: fresh weight (or wet weight); dw: dry weight.

In China on the Qingdao coast, the levels of naturally occurring radionuclides in the edible parts of various seafood (11 different marine species) were determined. The activities of $^{226}$Ra, $^{232}$Th and $^{40}$K ranged from 0.08 ± 0.03 to 1.65 ± 0.60 Bq kg$^{-1}$ dw, 0.09 ± 0.02 to 1.44 ± 0.10 Bq kg$^{-1}$ dw, 26.89 ± 1.25 to 219.25 ± 5.61 Bq kg$^{-1}$ dw, respectively [YAN 15].

### 2.3.4. *Natural contamination of buildings*

Natural radioactivity in the materials used to construct buildings varies greatly depending on the material and its origin (Table 2.5). As a result, the presence of radon gas in homes will be due in part (21%) to the nature of these materials and to the soil on which the dwelling rests (41%) (Table 2.6).

| | $^{226}$Ra | $^{232}$Th | $^{40}$K |
|---|---|---|---|
| Sand | 3.7–70.3 | 3–180 | 1–1,008 |
| Cement | 5.7–218 | 11–66 | 112–862 |
| Concrete | 7–106 | 1–556 | 23–918 |
| Tiles | 46–63 | 13–99 | 251–7,541 |
| Breeze blocks | 10–48 | | 27–767 |
| Bricks | 8–91 | 12–120 | 196–1,058 |
| Ceramics | 51–73 | 45–66 | 1,000–1,322 |
| Granite | 53–100 | 45–120 | 8–483 |
| Marble | 4–309 | 1–34 | 8–483 |

**Table 2.5.** *Concentrations of naturally occurring radionuclides in building materials (Bq.kg$^{-1}$) (modified from [UNS 08])*

Jelle [JEL 12] developed a model, taking into account the main parameters, for calculating the radon concentration in indoor air.

| **Radon source** | **Mechanism** | **Entry rate (Bq.m$^{-3}$.h$^{-1}$)** | **Percentage** |
|---|---|---|---|
| Building materials | Broadcast | 10 | 21 |
| Underlying soil | Broadcast | 7.5 | 15 |
| | Advection | 20 | 41 |
| Outside air | Infiltration | 10 | 20 |
| Water supply | Degassing | 1 | 2 |
| Natural gas | Combustion | 0.3 | 1 |

**Table 2.6.** *Contribution of different radon sources in a building. Examples given by UNSCEAR [UNS 93] for a typical house in a temperate climate country*

## 2.4. Anthropogenic radioactive contamination

Contamination of the anthroposphere by artificial radionuclides results from two main sources: chronic releases from BNIs and discharges due to accidents such as Chernobyl or Fukushima. The nature and quantity of radionuclides present in people's immediate environment will therefore be extremely variable. Monitoring programs exist in all countries using nuclear energy. For France, the organization is presented in Chapter 5. Annual reports are published by the IRSN [IRS 11]. Additional studies are also carried out at certain sites.

### 2.4.1. *Anthropogenic radioactive contamination of the physical environment*

Artificial radioactive contamination of the atmosphere is mainly due to atomic bomb testing. Four relatively long-lived radionuclides are still present in the atmosphere, and atmospheric fallout is highly dependent on latitude and hemisphere (Table 2.7). Most nuclear explosions have taken place in the northern hemisphere, which remains the most contaminated.

In the atmosphere, $^{137}$Cs concentrations in aerosols in 2002 were of the order of 0.4 µBq.m$^{-3}$ in mainland France and 10 times lower in Tahiti [BOU 04]. More recently, $^{137}$Cs concentrations in air in mainland France were approximately 0.25 µBq.m$^{-3}$ [IRS 11] and recently 0.2 µBq.m$^{-3}$ [IRS 21a].

Radioactive contamination of the lithosphere by artificial radionuclides is also due to atomic bomb testing. This is the case of Bikini Atoll (Table 2.8). Similarly, the first test site in Nevada, USA (NTS, Nevada Test Site) remains contaminated. In 1996, the Bq activity at the site was 7.4.10$^{10}$ in the atmosphere at ground level, and 1.3.10$^{12}$ in surface soils, mainly due to radionuclides of Am, Cs, Co, Pu and Sr [UNS 08].

| Latitude | Integrated deposit of $^{90}$Sr (PBq) | Deposit in the belt (percentage) |
|---|---|---|
| | Northern Hemisphere | |
| 80–90 | 1 | 0.2 |
| 70–80 | 7.9 | 1.7 |
| 60–70 | 32.9 | 7.1 |
| 50–60 | 73.9 | 16.1 |
| 40–50 | 101.6 | 22.1 |
| 30–40 | 85.3 | 18.5 |
| 20–30 | 71.2 | 15.5 |

| 10–20 | 50.9 | 11.1 |
|---|---|---|
| 0–10 | 35.7 | 7.8 |
| **Southern Hemisphere** | | |
| 80–90 | 0.3 | 0.2 |
| 70–80 | 2.5 | 1.7 |
| 60–70 | 8.7 | 4.6 |
| 50–60 | 12.1 | 8.4 |
| 40–50 | 28.1 | 19.5 |
| 30–40 | 27.6 | 19.1 |
| 20–30 | 28.1 | 19.5 |
| 10–20 | 17.8 | 12.3 |
| 0–10 | 21 | 14.5 |

**Table 2.7.** *Latitudinal distribution of atmospheric fallout from atomic testing for $^{90}Sr$ (according to [UNS 08, Table 35])*

| Soil depth (cm) | $^{137}$Cs | $^{90}$Sr | $^{239,240}$Pu | $^{241}$Am |
|---|---|---|---|---|
| | **Inside the atoll** | | | |
| 0–5 | 3.0 | 2.1 | 0.42 | 0.30 |
| 0–40 | 0.91 | 1.5 | 0.21 | 0.14 |
| | **Village area** | | | |
| 0–5 | 2.0 | 2.0 | 0.40 | 0.30 |
| 0–40 | 1.1 | 1.5 | 0.20 | 0.17 |

**Table 2.8.** *Average concentrations of artificial radionuclides in Bikini Atoll soils (in $Bq.g^{-1}$) (from [UNS 08])*

The hydrosphere covers two-thirds of the Earth's surface and therefore receives most of the radioactive fallout from atomic bomb testing. In addition, the ocean has received discharges from the Hanford nuclear power plant in the USA, and receives discharges from the British irradiated fuel processing plants at Sellafield in the Irish Sea and the French plant at La Hague in the English Channel. In addition, most of the accidental discharges from the Fukushima reactors reached the Pacific Ocean. A detailed review of radioactive concentrations in marine waters can be found in Volume 6 of the "Radioactive Risk" series [AMI 22].

In France, radioactive contamination of water and suspended matter in the major French rivers is monitored annually (Table 2.9). Concentrations in rainwater in

France vary from approximately 5 to 30 mBq.m$^{-2}$.month$^{-1}$ but are around three times higher in southern Corsica due to inputs of African desert dust [IRS 11]. Concentrations of $^{137}$Cs measured in seawater in 2009 were generally below 1 mBq.L$^{-1}$. Tritium activities measured in Arcachon (0.16 Bq.L$^{-1}$) were representative of the very low background observed in the Atlantic Ocean. Tritium activities measured in the North Sea and Channel show the influence of nuclear activities at AREVA's La Hague plant, as well as at coastal nuclear power plants, with mean values ranging from 6.4 to 11 Bq.L$^{-1}$ [IRS 11]. In Goury, in 2019, values were recently higher (31.2 Bq.L$^{-1}$) [IRS 21c]. Around nuclear reactors, tritium values are approximately 5 Bq.L$^{-1}$ and all concentrations are falling steadily [IRS 21b].

| | | $^{3}$H | $^{137}$Cs | $^{60}$Co | $^{54}$Mn | $^{131}$I |
|---|---|---|---|---|---|---|
| Porcheville | Seine | 6.87 ± 0.95 | 4.79 ± 0.62 | <2.5 | <2.6 | 75 ± 12 |
| Le Vésinet | Seine | 8.2 ± 1.5 | 4.88 ± 0.82 | <2.3 | <2.5 | 30.6 ± 6.4 |
| Les Ponts-de-Cé | Loire | 14.1 ± 1.2 | 5.95 ± 0.6 | <2.9 | <3.4 | - |
| Kembs | Rhine | 4.91 ± 0.72 | 4.20 ± 0.41 | 0.84 ± 0.18 | <0.72 | - |
| Vallabrègues | Rhône | 7.27 ± 0.49 | 9.43 ± 0.49 | 1.89 ± 0.2 | 1.54 ± 0.26 | <16 |

**Table 2.9.** *Activities measured in water ($^{3}$H in Bq.L$^{-1}$) and settled suspended solids ($^{137}$Cs, $^{60}$Co, $^{54}$Mn and $^{131}$I in Bq.kg$^{-1}$ dw) collected in 2009 by hydrocollectors located in the far field of nuclear facilities (modified from [IRS 11])*

### 2.4.2. Anthropogenic radioactive contamination of seafood products

Several marine areas are particularly radioactively contaminated. Such is the case of the Irish Sea. In this sea, technetium-99 concentrations were measured in fish, lobster, shrimp, mussels and oysters landed in the main fishing ports on the east and northeast coasts of Ireland between 1996 and 1998. In general, concentration factors were higher than those reported in the literature from laboratory studies, but agreed well with values based on field studies [SMI 01].

Welsh people living near the radioactive nuclear fuel reprocessing plant used to eat laverbread made from Porphyra algae. These algae bioaccumulate ruthenium-106 to a high degree. Concentrations were significantly higher near the Sellafield plant (less than 4 km) than further away (40 km). Maximum concentrations of $^{106}$Ru in Porphyra algae varied between 1963 and 1988 from 2,500 to 5,000 Bq.kg$^{-1}$ [WOO 97]. Over the last 10 years, depending on the radionuclides considered, direct discharges of radionuclides into the Irish Sea have been steadily decreasing [RIF 21].

Numerous food processing and culinary preparation methods are applied worldwide. Surveys carried out after Japan's Fukushima accident on food processing generally agree with previously reported values. In its report, the IAEA [IAE 20] details food processing methods specific to Japan.

Concentrations of persistent artificial radionuclides were measured in atolls in the region of the Marshall Islands in 1978. The concentration of $^{90}$Sr in the muscle tissue of various fish species was very low. However, the concentration of $^{239,240}$Pu in the muscle tissue of all reef species was higher than that of pelagic lagoon fish. In contrast, $^{137}$Cs concentrations were lowest in the muscle tissue of bottom-dwelling reef species such as mullet and goatfish, and highest in pelagic lagoon fish [ROB 81].

### 2.4.3. *Anthropogenic radioactive contamination of other foods*

Radioactive contamination of foodstuffs by artificial radionuclides has varied considerably over time. In France, maximum concentrations were observed in 1963, due to heavy atmospheric fallout from atomic testing. Concentrations in salads reached 800 $Bq.kg^{-1}$ fresh weight, and in milk over 500 $Bq.L^{-1}$. Another peak corresponded to the Chernobyl accident of May 1–5, 1986. The concentration in lettuce was almost 3,000 $Bq.kg^{-1}$ fw and in milk over 500 $Bq.L^{-1}$. Currently, the values are 150 $Bq.kg^{-1}$ fw and approximately 20 $Bq.L^{-1}$ for lettuce and milk, respectively [IRS 21b].

The concentration of $^{137}$Cs in milk varies greatly from region to region, from 0.01 to 0.83 $Bq.L^{-1}$, which can be compared with the persistence of this radionuclide in the soils of certain grasslands (variation from 4.3 to 175 $Bq.kg^{-1}$), with maximum values observed in the soils of mountainous areas [IRS 21b].

In milk, $^{90}$Sr concentrations vary on average from 0.030 to 0.208 $Bq.L^{-1}$ and remain below 0.5 $Bq.L^{-1}$ for $^{137}$Cs [IRS 11]. In wheat from the 2009 harvest, concentrations of $^{54}$Mn, $^{60}$Co, $^{90}$Sr, $^{134}$Cs and $^{13}$Cs averaged $<0.6$, $< 1.2$, $<0.5$, 0.10, $<0.5$ and $<0.5$ $Bq.kg^{-1}$ fw [IRS 11]. These values tended to decrease steadily.

### 2.4.4. *Radioactive contamination of polluted sites*

Around the world, many sites have relatively high levels of radioactive soil and environmental pollution. These include sites where atomic bombs were tested, sites around spent fuel reprocessing plants, Soviet radioactive waste repositories and sites where serious nuclear accidents have occurred.

#### 2.4.4.1. *Atomic bomb test sites*

Contamination at the French Pacific atomic bomb test site [IRS 09] is monitored annually, with measurements of radionuclides present in atmospheric aerosols ($^{137}$Cs, $^{7}$Be, $^{22}$Na, $^{40}$K, $^{210}$Pb) and in marine and fresh waters ($^{137}$Cs and $^{3}$H). Artificial radionuclides ($^{137}$Cs, $^{60}$Co, $^{90}$Sr, $^{238}$Pu and $^{239,240}$Pu) are quantified in various biological samples (fish, milk, etc.). Since 1980, clams have been the marine product with the highest concentrations of $^{60}$Co, plutonium and often also $^{90}$Sr. On the contrary, fish, and particularly grouper, concentrate the most $^{137}$Cs. In the 1980–1990s, the maximum activities of $^{137}$Cs, $^{60}$Co and $^{90}$Sr measured in lagoon products from all islands were of the order of a few Bq.kg$^{-1}$ fw. Since the end of the 1990–2000 decade for the same sampling sites, the maximum values measured were 10 times lower (often less than 0.1 Bq.kg$^{-1}$ fw) for $^{60}$Co and $^{90}$Sr and on average half as much for $^{137}$Cs (between 0.2 and 1 Bq.kg$^{-1}$ fw) [IRS 09]. Since 2014, a new direction in the monitoring program has been implemented. It concerns the study of artificial radioactivity in soils, currently at the origin of root transfer to plants. Analyses are mainly focused on $^{137}$Cs, a radionuclide still detectable in over a third of the foodstuffs analyzed, plutonium isotopes 238, 239 and 240 and $^{241}$Am [IRS 21b].

#### 2.4.4.2. *Sites around spent fuel reprocessing plants*

Around the world, five spent fuel reprocessing plants have operated extensively and discharged significant quantities of radioactivity in the past. Three plants were mainly military in nature: Hanford in the USA, which discharged into the Columbia River and the Pacific Ocean; Mayak in the former USSR, which discharged into a tributary of the Ob River and the Kara Sea; and Marcoule in France, which discharged into the Rhône River and the Mediterranean Sea. The other two plants were mainly for civilian use: Windscale (now Sellafield) in the United Kingdom discharging into the Irish Sea, and La Hague in France discharging into the English Channel.

For the military plants, discharges and environmental contamination were detailed in Volume 1 of the "Radioactive Risk" series [AMI 19a]. For the two civilian marine plants, the impact was summarized in Volume 6 of the same series [AMI 22].

For example, for the La Hague irradiated fuel reprocessing plant, annual environmental quality reports are published by the operator (COGEMA, then AREVA, then ORANO). For milk, average concentrations (Bq.L$^{-1}$) at Beaumont-Hague in 2009 for $^{40}$K, $^{3}$H, $^{131}$I, $^{129}$I, $^{134}$Cs, $^{137}$Cs and $^{90}$Sr were 44.5, $<5.1$, $<0.32$, 0.021, $<0.18$, $<0.16$ and 0.055, respectively. The highest iodine-129 value measured since 2010 was $0.144 \pm 0.023$ Bq.L$^{-1}$ [IRS 21c]. For wheat, the averages (Bq.kg$^{-1}$) at Digulleville for 2009 were always below 0.47, 5.05, 3.84, 0.45, 1.43, 0.0011, 0.0010 and 0.0010 for $^{60}$Co, $^{106}$Ru, $^{124}$Sb, $^{137}$Cs, other gamma emitters, $^{241}$Am, $^{238}$Pu and $^{239,240}$Pu, respectively. In shellfish, mean concentrations in 2009 (Bq.kg$^{-1}$ fresh weight) were 0.33, <0.0043, <0.13, <0.0063, <0.0094 and 0.0201, respectively, for $^{129}$I, $^{137}$Cs, $^{60}$Co, $^{238}$Pu, $^{239,240}$Pu and $^{241}$Am [IRS 11]. More recently, values have remained more or less the same [IRS 21b]. Radioactive contamination of milk and shellfish in the La Hague environment is shown in Table 2.10 for the GRNC, i.e. for 1990 to 2000.

| | **$^{129}$I** | **$^{3}$H** | **$^{14}$C** | **$^{90}$Sr** | **$^{241}$Pu** |
|---|---|---|---|---|---|
| Milk (Bq.L$^{-1}$) | 1–70 | <dl in general Max 73 | - | - | - |
| Mollusks (Bq.kg$^{1}$ fw) | 0.25–2.3 | 0.3–3.1 | 300–600 | 0.05–3.6 | 0.11–0.37 |

**Table 2.10.** *Mass and volume activities of radionuclides in samples from Nord-Cotentin and the Channel coast (modified from [GER 04]). dl: detection limit*

### 2.4.4.3. *Soviet radioactive waste disposal sites*

A lot of radioactive waste, mostly military, was dumped in the various Nordic seas by the Soviets, with potential risks of radioactive pollution in the future. The main sites are around Murmansk and Nova Zemlya [AMI 22].

### 2.4.4.4. *Accidental releases of radionuclides into the environment*

These releases are due to military [AMI 19a] or industrial [AMI 19b] nuclear accidents. The radionuclides released during an accident following the explosion of an atomic bomb, the sinking of a nuclear submarine, the core meltdown of a nuclear reactor or an accident at another nuclear facility spread into the atmosphere, soil and water, and are then dispersed throughout the environment, including plants, animals and mankind.

The main accidents involved the Mayak reprocessing plant in Russia, the Ukrainian Chernobyl reactor and the Japanese Fukushima reactors. The Ukrainian nuclear reactor at Chernobyl contaminated vast areas of Ukraine, Belarus and Russia, while the Fukushima reactors polluted Japanese soil and the Pacific Ocean [AMI 22].

At Mayak, plutonium particles in dust remain the main source of air contamination and human intake [ROM 20].

In Chernobyl, 35 years after the accident, soil contamination with $^{137}$Cs is expected to fall below 37 kBq/m$^2$ (1 Ci/km$^2$) in farmland in the Kaluga, Orel and

Tula regions by 2090, and in the Bryansk region by 2018 at the earliest. Radiological standards are exceeded in fodder (4–15% of samples), milk (4–8%) and beef (5–8%) only in five districts in the southwest of the Bryansk region [PAN 21]. Temporal trends for the period 2011–2016 in $^{137}Cs$ content in cow's milk from private farms and households in 14 settlements located in the territories of the Rivne region, Ukraine, highlight that the Ukrainian permissible levels for adults (100 $Bq.L^{-1}$) and children (40 $Bq.L^{-1}$) were exceeded in six and eight settlements, respectively. The areas of Ukraine affected by Chernobyl could persist for many years to come, and at least until 2040 [LAB 18]. Since the beginning of April 2020, forest fires have been raging in the heavily contaminated areas around the Chernobyl nuclear power plant in Ukraine. In total, 0.7–1.2 TBq of $^{137}Cs$ have been released into the atmosphere, resulting in an increase in airborne $^{137}Cs$ of several hundred $\mu Bq.m^{-3}$ in northern Ukraine. Similarly, releases of $^{90}Sr$ (345–612 GBq) and Pu (up to 75 GBq, mainly $^{241}Pu$) have been reported in the Chernobyl exclusion zone [MAS 21].

Several years after the Fukushima Daiichi nuclear power plant accident, cesium-137 is still present in various foods. A study of the concentration of $^{137}Cs$, $^{134}Cs$ and $^{40}K$ in various foods representative of the total diet (TDS, Total Diet Study) was carried out to assess the average dietary exposure levels of a Japanese adult population, before and after the accident. From 2006 to 2010, the volume activity in $^{137}Cs$ of the "fish and shellfish" group was 0.099 $Bq.kg^{-1}$, the highest value obtained. In 2011, "milk and dairy products" from the city of Sendai had a Cs activity concentration of 12 $Bq.kg^{-1}$, which represented the highest value of all food groups studied [TER 20].

A study of the behavior of radioactive cesium during the processing and/or cooking of domestic food products has been undertaken. Losses of radioactive cesium contamination vary considerably from one foodstuff to another [HAC 21].

## 2.5. Conclusions

In addition to the natural presence of radionuclides in the anthroposphere, human activities introduce significant quantities of other radionuclides. This is clearly evident in major accidents (Windscale, Three Mile Island, Chernobyl and Fukushima), as well as in the use of nuclear energy, particularly in the reprocessing of spent nuclear fuel. Radioactive contamination of the environment, and particularly of food, will automatically lead to contamination of the human body (Chapter 5).

## 2.6. References

[AAR 97] AARKORG A., BAXTER M.S., BETTENCOURT A.O. et al., “A comparison of doses from $^{137}Cs$ and $^{210}Po$ in marine food, a major international study”, *Journal of Environmental Radioactivity*, vol. 34, no. 1, pp. 69–90, 1997.

[ADE 20] ADELEYE M.O., MUSA B., OYEBANJO O. et al., “Activity concentration of natural radionuclides and assessment of the associated radiological hazards in the marine croaker (*Pseudotolitus typus*) fish from two coastal areas of Nigeria”, *Science World Journal*, vol. 15, no. 2, pp. 90–95, 2020.

[ALA 11] ALAM L., MOHAMED C.A.R., “A mini review on bioaccumulation of $^{210}Po$ by marine organisms”, *International Food Research Journal*, vol. 18, pp. 1–10, 2011.

[AMI 13] AMIARD J.C., *Le risque radioactif. Devenir des radionucléides dans l'environnement et impacts sur la santé*, Lavoisier, Tec&Doc, Paris, 2013.

[AMI 17] AMIARD J.C., *Les risques chimiques environnementaux. Méthodes d'évaluation et impacts sur les organismes*, 2nd edition, Lavoisier, Tec&Doc, Paris, 2017.

[AMI 18] AMIARD J.C., “Radioactivité et aliments”, in CAMEL V., RIVIÈRE G., LE BIZEC B., (eds), *Risques chimiques liés aux aliments : principes et applications*, Lavoisier, Tec&Doc, Paris, pp. 387–410, 2018.

[AMI 19a] AMIARD J.C., *Military Nuclear Accidents: Environmental, Ecological, Health and Socio-economic Consequences*, vol. 1, ISTE, London, and John Wiley & Sons, New York, 2019.

[AMI 19b] AMIARD J.C., *Industrial and Medical Nuclear Accidents: Environmental, Ecological, Health and Socio-economic Consequences*, vol. 2, ISTE, London, and John Wiley & Sons, New York, 2019.

[AMI 22] AMIARD J.C., *Marine Radioecology*, vol. 6, ISTE, London, and John Wiley & Sons, New York, 2022.

[BIL 04] BILLON S., MORIN A., CAËR S. et al., “Évaluation de l'exposition de la population française a la radioactivité naturelle”, *Radioprotection*, vol. 39, pp. 213–232, 2004.

[BOU 04] BOUISSET P., BARKER E., MASSON O. et al., “Concentration de $^{137}Cs$ et de $^{7}Be$ dans les aérosols en France métropolitaine et à Tahiti de 1959 à 2002”, *Radioprotection*, vol. 39, pp. 367–381, 2004.

[FOW 11] FOWLER S.W., “$^{210}Po$ in the marine environment with emphasis on its behaviour within the biosphere”, *Journal of Environmental Radioactivity*, vol. 102, pp. 448–461, 2011.

[GAM 07] GAMBINI D.J., GRANIER R., *Manuel pratique de radioprotection*, Lavoisier, Tec&Doc, Paris, 2007.

[GER 04] GERMAIN P., HERBELET J., MERLE-SZEREMETA A. et al., “Mises à jour de données scientifiques suite aux recommandations du groupe radioécologie Nord-Cotentin (GRNC) formulées en 1999”, *Radioprotection*, vol. 39, pp. 77–98, 2004.

[GRN 99] GRNC, Estimation des niveaux d'exposition aux rayonnements ionisants et des risqué de leucémies associés de populations du Nord-Cotentin, Summary report, 1999.

[HAC 21] HACHINOHE M., HAMAMATSU S., KAWAMOTO S., "A review of the radioactive cesium behavior in Japanese agricultural, livestock, fishery products and their foods in the decade following the Fukushima nuclear accident", *Food Science and Technology Research*, vol. 27, no. 1, pp. 1–23, 2021.

[IAE 95] IAEA, Sources of radioactivity in the marine environment and is relative contributions to overall dose assessment from marine radioactivity (MARDOS), IAEA-TECDOC-838, IAEA, Vienna, 1995.

[IAE 20] IAEA, Environmental transfer of radionuclides in Japan following the accident at the Fukushima Daiichi Nuclear Power Plant, IAEA-TECDOC-1927, IAEA, Vienna, 2020.

[IRS 09] IRSN, Bilan de la surveillance de la radioactivité en Polynésie française en 2008. Résultats du réseau de surveillance de l'IRSN, DEI/SECURE report no. 2009-42, 2009.

[IRS 11] IRSN, Bilan de l'état radiologique de l'environnement français en 2009. Synthèse des résultats des réseaux de surveillance de l'IRSN, IRSN, Fontenay-aux-Roses, 2011.

[IRS 12] IRSN, Le radon, Collection thématique, December 2012.

[IRS 21a] IRSN, Bilan de la surveillance de la radioactivité en Polynésie française en 2019-2020. Synthèse des résultats du réseau de surveillance de l'IRSN, IRSN/2021-00849 report, IRSN, Fontenay-aux-Roses, 2021.

[IRS 21b] IRSN, Bilan de l'état radiologique de l'environnement français de 2018 à 2020, IRSN 2021-00765 report, IRSN, Fontenay-aux-Roses, 2021.

[IRS 21c] IRSN, Surveillance radiologique de l'environnement en 2019. Mise en œuvre du plan de surveillance régulière de l'IRSN, IRSN 2022-00175 report, IRSN, Fontenay-aux-Roses, 2021.

[JEL 12] JELLE B.P., "Development of a model for radon concentration in indoor air", *Science of the Total Environment*, vol. 416, pp. 343–350, 2012.

[LAB 18] LABUNSKA I., KASHPAROV V., LEVCHUK S. et al., "Current radiological situation in areas of Ukraine contaminated by the Chernobyl accident: Part 1. Human dietary exposure to Caesium-137 and possible mitigation measures", *Environment International*, vol. 117, pp. 250–259, 2018.

[MAK 20] MAKMUR M., PRIHATININGSIH W.R., YAHYA M.N., "Baseline concentration of Polonium-210 ($^{210}$Po) in several biota from Jakarta Bay", *IOP Conference Series: Earth and Environmental Science*, vol. 429, p. 012061, 2020.

[MAS 21] MASSON O., ROMANENKO O., SAUNIER O. et al., "Europe-wide atmospheric radionuclide dispersion by unprecedented wildfires in the Chernobyl exclusion zone, April 2020", *Environmental Science & Technology*, vol. 55, pp. 13834–13848, 2021.

[MET 88] MÉTIVIER H., ROBÉ M.C., *Le radon de l'environnement à l'Homme*, EDP Sciences, Les Ulis, 1988.

[PAN 21] PANOV A.V., RATNIKOV A.N., SVIRIDENKO D.G. et al., "Remediation of Agricultural Lands Exposed to Large-Scale Radioactive Contamination (to the 35th Anniversary of the Chernobyl NPP Accident)", *Russian Agricultural Sciences*, vol. 47, no. 4, pp. 399–403, 2021.

[POL 98] POLLARD D., RYAN T.P., DOWDALL A., "The dose to Irish seafood consumers from $^{210}$Po", *Radiation Protection Dosimetry*, vol. 75, nos. 1–4, pp. 139–142, 1998.

[RIF 21] RIFE, Radioactivity in Food and the Environment, RIFE report no. 26, Environment Agency, UK, 2021.

[ROB 81] ROBISON W.L., NOSHKIN V.E., PHILLIPS W.A. et al., The Northern Marshall Islands radiological survey: Radionuclide concentrations in fish and clams and estimated doses via the marine pathway, UCkL-52853 Pt. 3, 1981.

[ROM 20] ROMANOV S.A., EPHIMOV A.V., ALADOVA E.E. et al., "Plutonium production and particles incorporation into the human body", *Journal of Environmental Radioactivity*, vol. 211, 106073, 2020.

[SAH 13] SAHA G.B., *Physics and Radiobiology of Nuclear Medicine*, 4th edition, Springer, New York, 2013.

[SMI 01] SMITH V., FEGAN M., POLLARD D. et al., "Technetium 99 in the Irish marine environment", *Journal of Environmental Radioactivity*, vol. 56, pp. 269–284, 2001.

[TER 20] TERADA H., IIJIMA I., MIYAKE S. et al., "Total diet study to assess radioactive Cs and $^{40}$K levels in the Japanese population before and after the Fukushima Daiichi Nuclear Power Plant Accident", *International Journal of Environmental Research and Public Health*, vol. 17, 8131, 2020.

[UNS 93] UNSCEAR, Sources and Effects of Ionizing Radiation, United Nations, New York, 1993.

[UNS 08] UNSCEAR, Sources and Effects of Ionizing Radiation, United Nations, New York, 2008.

[WOO 97] WOODHEAD D.S., The behaviour of radionuclides discharged from Sellafield in UK coastal waters, IAEA, Strategies and Methodologies for Applied Marine Radioactivity Studies, IAEA Training Courses Series no. 7, Chapter 9, pp. 195–235, IAEA, Vienna, 1997.

[YAM 94] YAMAMOTO M., ABE T., KUWABARA J. et al., "Polonium-210 and Lead-210 in Marine Organisms: Intake Levels for Japanese", *Journal of Radioanalytical and Nuclear Chemistry*, vol. 178, no. 1, pp. 81–90, 1994.

[YAN 15] YANG B., HA Y., JIN J., "Assessment of radiological risk for marine biota and human consumers of seafood in the coast of Qingdao, China", *Chemosphere*, vol. 135, pp. 363–369, 2015.

# 3

# Human Exposure to Radionuclides

## 3.1. Introduction

After quantifying the radioactive contamination of the environment, the second stage in assessing radioactive risk involves identifying the modes and pathways of human exposure to ionizing radiation. Firstly, the distinction between irradiation and contamination will be explained, followed by that between external and internal contamination. Finally, for these three exposure modes, the various sources, both natural and artificial, will be listed.

## 3.2. Types of exposure to ionizing radiation

Ionizing radiation affects humans in three ways: irradiation, external contamination and internal contamination.

### 3.2.1. *The distinction between irradiation and contamination*

When the radionuclide is far from the individual, the effect of radiation is solely through radiation. When the radionuclide comes into contact with humans, there is external contamination. On the contrary, when the radionuclide enters the body, contamination becomes internal (Figure 3.1). While external contamination is relatively fixed, internal contamination is mobile within the body, and its retention depends on its chemical form.

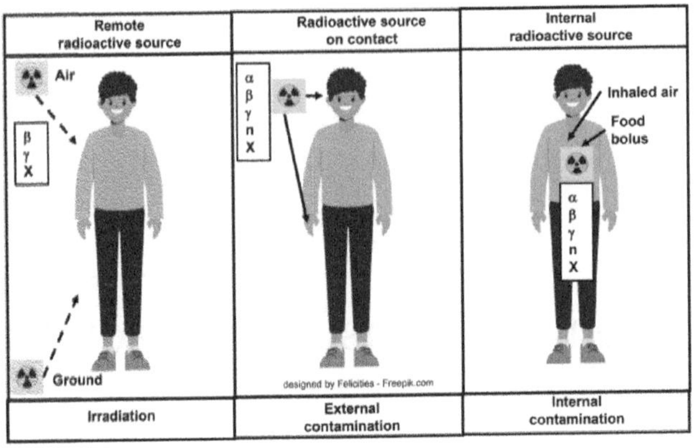


**Figure 3.1.** *Distinction between irradiation and external and internal contamination. The various ionizing radiations involved*

#### 3.2.1.1. *External irradiation*

When the radionuclide is in the individual's environment, its impact is limited to its ionizing radiation. In this case, alpha emitters are not affected by external irradiation, since their path through air or water is very small[1]. However, other emitters (β, γ, neutrons, X-rays, etc.) are concerned.

#### 3.2.1.2. *External radioactive contamination*

Radioactive contamination can be external, often referred to as "on contact", and concerns deposits on body surfaces (skin, hair). It involves several types of emitters (α, β, γ, X). In addition, the sensitive part of the skin is the basal layer of the epidermis, at the level of the cells that regenerate the epidermis. This layer lies on average 0.7 mm from the outer surface, and is therefore beyond the reach of alpha emitters. Only X-rays and beta emitters play a role in irradiating this sensitive part of the skin (risk of radiological burns).

#### 3.2.1.3. *Internal radioactive contamination*

When the radionuclide enters the body, contamination becomes internal. This applies to all emitters. As soon as the radiation emitter is inside the body, it comes into contact with living cells without any intervening shielding. In this case, the higher the linear energy transfer (LET) (alpha), the greater the damage to the nearest cells.

1 The orders of magnitude for alphas are in air approximately 100 mm per MeV and in water approximately 0.06–0.07 mm per MeV.

In radiation protection, two exceptions concern inhaled air and the food bolus. From a strictly physiological point of view, the volume of the lungs and the lumen of the digestive tract are outside the body. However, in the case of radioactive contamination of inhaled air and ingested food, these will be considered as internal contamination.

### 3.2.2. *Multiple exposure pathways*

Due to the varied sources of radioactive contamination (Chapter 1) in the various environmental compartments, and particularly in the anthroposphere, humans are exposed to ionizing radiation via a multitude of potential exposure pathways (Figure 3.2).

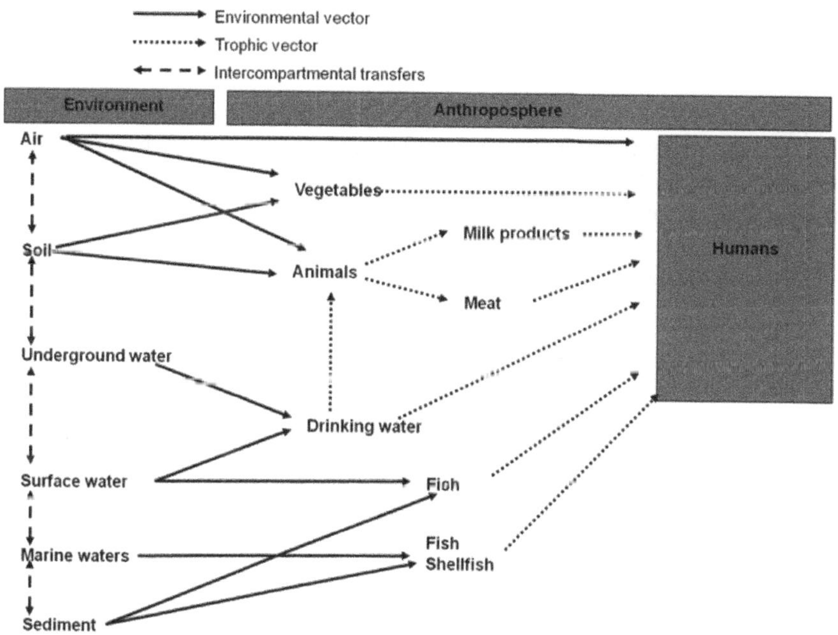


**Figure 3.2.** *Schematic representation of human exposure to radionuclides from the environment (modified from [AMI 17])*

## 3.3. External exposure pathways

The contamination of various physical environments, mainly the atmosphere and lithosphere, by radionuclides of natural or man-made origin, results in human irradiation.

Ambient radioactivity comes from a variety of sources. Part of environmental exposure is due to cosmic and telluric radiation, and to natural radioelements present in the environment, notably in the Earth's crust. Radon is one of the radionuclides of natural origin responsible for a significant proportion of human irradiation[2]. Exposure is also caused by human activity: firstly, fallout from atomic tests and nuclear accidents, such as those at Chernobyl and Fukushima; secondly, residues from past activities; and thirdly, discharges from operating nuclear facilities.

The level of exposure depends on location and lifestyle. The main factors are the resuspension rate of radioactive particles, protection by dwellings, diets, individual consumption rates, inadvertent ingestion of sand or soil, respiratory rates and the time spent on various activities (indoor or outdoor activities; professional or leisure activities).

### 3.3.1. *External exposure from natural sources*

Ionizing radiation irradiating humans originates from the cosmos or the Earth's crust. For the majority of environments, exposures due to natural background are more significant for the population than most exposures from man-made sources, as has been detailed for the situation in Switzerland [MUR 99].

#### 3.3.1.1. *Cosmic exposure*

Cosmic radiation is a stream of very high-energy particles, mainly protons, in the GeV range. This flux is of solar or galactic origin. These high-energy protons collide with the nuclei of atmospheric atoms, creating high-energy fragments (protons, neutrons, muons, neutrinos, mesons, etc.). The atmosphere and the Earth's magnetic field act as a shield, reducing the magnitude of the impact. Near the poles, this protection is less effective, and high-energy cosmic rays are the source of the aurora borealis and aurora australis. This bombardment of atmospheric atoms also generates radioactive elements, known as cosmogenic elements, because they are created in the series of collisions with oxygen and nitrogen in the air. The main radionuclides are carbon-14, tritium, beryllium-7 and beryllium-10.

#### 3.3.1.2. *Land-based exposure*

In the lithosphere, four elements contemporary with the birth and death of stars have survived, as their lifetimes are measured in billions of years. These are two uranium isotopes, uranium-238 and uranium-235, thorium-232 and potassium-40[3],

2 It should be noted that radon exposure will be mitigated by all outdoor activities (professional or leisure).

3 The physical half-lives (or periods) expressed in billions of years are for $^{40}$K > 1.25; $^{238}$U > 4.468; $^{235}$U >0.704 and $^{232}$Th >14.02. It is only 0.2455 million years for $^{234}$U.

which makes up around one ten-thousandth of natural potassium. The energy released during supernova explosions enables the production of nuclei heavier than iron, such as the three radioactive uranium and thorium nuclei. They were therefore present in clouds of interstellar dust whose aggregation gave rise to the formation of stars and planets, such as the Sun and Earth 4.5 billion years ago.

The lithosphere contains some 50 natural radionuclides, most of which belong to the three natural families of thorium, uranium-235 and 238. Thorium is the most abundant radionuclide (10 mg.kg$^{-1}$ on average), followed by uranium (2–3 mg.kg$^{-1}$). Potassium-40 ($^{40}$K), a natural isotope of potassium (0.01167%), is a significant contributor to human external exposure.

The level of radioactivity in the environment has varied greatly over time. It was clearly higher for the first living organisms, which were therefore more exposed to radioactivity than those of the 21st century (Table 3.1).

| Period | Uranium-238 family | | Thorium-232 family | | Potassium-40 (Bq.kg-1) | Total (Bq.kg-1) |
|---|---|---|---|---|---|---|
| | Uranium-238 (Bq.kg-1) | Descendants (Bq.kg-1) | Thorium-232 (Bq.kg-1) | Descendants (Bq.kg-1) | | |
| 4.5 billion years ago | 80 | 1,040 | 50 | 450 | 4,230 | 5,850 |
| Today | 40 | 520 | 40 | 360 | 370 | 1,350 |

**Table 3.1.** *Average radioactivity of the Earth's crust over time (modified from Pradel et al., 2002–2003, in [GAM 07])*

### 3.3.2. External exposure of artificial origin

Since their discovery at the end of the 19th century, radionuclides and radioactivity have been used in a variety of applications, resulting in environmental contamination and increased irradiation.

#### 3.3.2.1. *Medical exposure*

Medical exposures, due to radiology, nuclear medicine and radiotherapy, are the main causes of human exposure of artificial origin [UNS 00]. At the end of the 20th century, in countries such as France, where there is more than one doctor per 1,000 inhabitants (for 26% of the world's population), the number of diagnostic medical, dental and nuclear medicine procedures was 920, 310 and 19, respectively, per year. In the case of therapeutic procedures, teletherapy, brachytherapy and nuclear medicine result in 1.5, 0.2 and 0.2 procedures per year, respectively [UNS 00]. This phenomenon has become even more marked in recent years.

#### 3.3.2.2. *Exposure arising from the normal operation of industries*

Radioactive noble gases (krypton, xenon) only cause external exposure, as rare or noble gases do not bioaccumulate. They are mainly released into the atmosphere during the spent fuel reprocessing stage, notably at La Hague. These releases mainly concern krypton-85, and take place from the very start of reprocessing operations, i.e. fuel shearing and dissolution of the fuel material in nitric acid. On the contrary, numerous radionuclides are released by the various nuclear facilities into the atmosphere and the hydrosphere (seas, rivers, lakes). They contribute to environmental contamination, irradiation and radioactive contamination of humans.

Exposure from industrial sources comes mainly from the nuclear power industry, and is present at all stages of the nuclear fuel cycle. However, radioactive contamination is more prevalent in the first (mining) and last (spent fuel reprocessing) stages. Certain non-nuclear industrial applications are also sources of exposure, such as electric X-ray generators, particle gas accelerators and the various uses of radionuclides as radioactive tracers. Finally, radioactive waste management could lead to exposure of members of the public. On the contrary, nuclear professionals are and will be exposed at all stages of the fuel cycle.

#### 3.3.2.3. *Occupational exposure*

Workers in the nuclear industry and laboratories, as well as radiologists and hospital staff, are exposed to radiation in the course of their work, and receive doses higher than those permitted for the general public, excluding natural radioactivity. However, occupational exposure concerns a limited number of people who are subject to dosimetric monitoring. What's more, in the past, these people were subject to medical surveillance, which is no longer as systematic today.

#### 3.3.2.4. *External exposure due to fallout from atomic testing*

Following the atmospheric testing of atomic bombs, large quantities of radionuclides were introduced into the atmosphere[4] and, to a lesser extent, into the oceans. These radionuclides quickly fell back onto the continents and above all the oceans, making a significant contribution to the radioactive contamination of the environment. In terms of activity, the main radionuclides remaining in the

4 In particular, the pulverized nuclear material (enriched uranium or plutonium) did not take part in the nuclear exhibition. The yields of these bombs, especially the first ones, were below 10%.

environment today are tritium (released in abundance by hydrogen bombs), carbon-14, cesium-137, strontium-90 and plutoniums-239 and 240.

#### 3.3.2.5. *Exposure from domestic sources*

Various devices and products used in everyday life contained radionuclides or produced ionizing radiation, causing minimal radiation exposure, but which, depending on habits, could become non-negligible. These were mainly objects containing luminescent paints (radium-226, promethium-147, tritium), cathode-ray screens, various radioactive materials contained in certain terracottas, certain cell phone batteries, certain Walkman headphones and tobacco. They also included some old radioactive lightning rods [GAM 07].

#### 3.3.2.6. *Exposure due to nuclear accidents*

Nuclear accidents, particularly those considered major, release large quantities of radionuclides into the atmosphere. During atmospheric fallout, these radionuclides enter other environmental compartments [AMI 19a, AMI 19b]. Fallout from the Chernobyl accident has been quantified by UNSCEAR [UNS 88] and by Poïarkov and Robeau [POÏ 01]. Fallout from the Fukushima accident was quantified by Povinec et al. [POV 13].

## 3.4. Internal exposure pathways

The main sources of internal exposure are natural and, to a lesser extent, man-made, through inhalation of contaminated air and ingestion of radioactively contaminated food.

### 3.4.1. *Radionuclide penetration routes (contamination)*

An individual's internal contamination penetrates mainly through the respiratory and alimentary tracts, and secondarily through the skin and its products (hair).

#### 3.4.1.1. *The respiratory pathway*

Inhaled air, if radioactively contaminated, will irradiate the individual. Inhalation of radioactive aerosols depends mainly on their size. The smaller they are, the deeper they penetrate the pulmonary system (bronchi, bronchioles, alveoli).

#### 3.4.1.2. *The food pathway*

Foodstuffs are likely to be radioactive. The impact of the food bolus will depend mainly on its residence time in the digestive tract. Digestive transit (stomach, intestine) depends on the nature of the food and the age of the subject. Liquids (water, etc.) generally transit faster than solids. Elderly individuals have a slower transit (see Chapter 5).

#### 3.4.1.3. *Skin*

Radionuclide penetration is generally limited when the epidermis and dermis are healthy. On the contrary, penetration can be significant in the case of skin lesions.

### 3.4.2. *Internal exposures of natural origin*

Humans inhale and ingest radionuclides naturally present in the Earth's crust or produced by cosmic radiation. These radioactive elements then irradiate the individual from within.

In the case of internal contamination by inhalation, the main radionuclide responsible is radon in the air, which is deposited in the lungs. Certain short-lived radon progeny and alpha emitters will irradiate the basal and secretory cells of the airways. In addition, long-lived alpha emitters such as lead-210 and polonium-210 can be inhaled.

Two naturally occurring radionuclides, potassium-40 and polonium-210, are the main contributors to radioactive contamination of food ingested by humans. Potassium is an important element in our constitution and vital to the proper functioning of our cells, but our food naturally contains it. Polonium-210 internally contaminates large consumers of shellfish. To these radionuclides must be added uranium, whose concentrations vary considerably worldwide in drinking water, reaching up to 100,000 $Bq.L^{-1}$ [UNS 08].

### 3.4.3. *Internal exposure of artificial origin*

Internal exposure from artificial sources is due either to inhalation of air, or ingestion of water or food containing radionuclides, as well as ingestion of soil and sediment particles, which can be significant, particularly in young children.

## 3.5. Conclusions

The ways and means of exposure are highly varied. Each member of the population receives an annual exposure that depends on their lifestyle and activities, and on the presence of radioactivity in the environment. This is shown in Figure 3.3 for the population of the Beaumont-Hague administrative district, France, which is subject to radioactive contamination, in particular from gaseous and liquid discharges from the La Hague irradiated fuel reprocessing plant, as well as from the EDF Flamanville nuclear power plant and from discharges from the French Navy in the military port of Cherbourg.

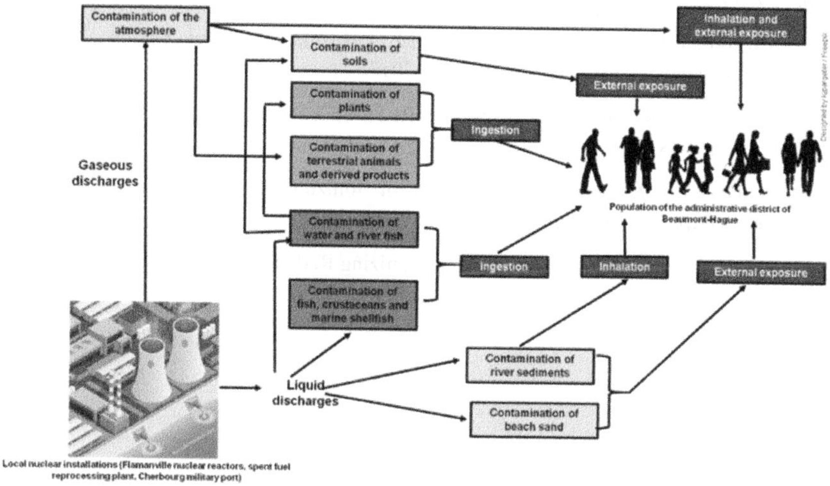


**Figure 3.3.** *The various exposure pathways for the population of the Beaumont-Hague administrative district (modified from [GRN 99])*

## 3.6. References

[AMI 13] AMIARD J.C., *Le risque radioactif. Devenir des radionucléides dans l'environnement et impacts sur la santé*, Lavoisier, Tec&Doc, Paris, 2013.

[AMI 17] AMIARD J.C., *Les risques chimiques environnementaux. Méthodes d'évaluation et impacts sur les organismes*, 2nd edition, Lavoisier, Tec&Doc, Paris, 2017.

[AMI 19a] AMIARD J.C., *Military Nuclear Accidents: Environmental, Ecological, Health and Socio-economic Consequences*, vol. 1, ISTE, London, and John Wiley & Sons, New York, 2019.

[AMI 19b] AMIARD J.C., *Industrial and Medical Nuclear Accidents: Environmental, Ecological, Health and Socio-economic Consequences*, vol. 2, ISTE, London, and John Wiley & Sons, New York, 2019.

[GAM 07] GAMBINI D.J., GRANIER R., *Manuel pratique de radioprotection*, Lavoisier, Tec&Doc, Paris, 2007.

[GRN 99] GRNC, Estimation des niveaux d'exposition aux rayonnements ionisants et des risqué de leucémies associés de populations du Nord-Cotentin, Summary report, 1999.

[MUR 99] MURITH C., GURTNER A., “Les expositions d’origine naturelle en Suisse”, *Radioprotection*, vol. 34, pp. 37–46, 1999.

[POÏ 01] POÏARKOV V., ROBEAU D., “L’accident de Tchernobyl”, in ROBEAU D. (ed.) *Catastrophes et accidents nucléaires dans l’ex-Union soviétique*, EDP Sciences, Les Ulis, pp. 59–99, 2001.

[POV 13] POVINEC P.P., HIROSE K., AOYAMA M., *Fukushima Accident: Radioactivity Impact on the Environment*, Elsevier, Amsterdam, 2013.

[UNS 88] UNSCEAR, Sources, Effects and Risks of Ionizing Radiation, United Nations, New York, 1988.

[UNS 00] UNSCEAR, Sources and Effects of Ionizing Radiation, Annex D Medical Radiation Exposures, United Nations, New York, 2000.

[UNS 08] UNSCEAR, Sources and Effects of Ionizing Radiation, United Nations, New York, 2008.

# 4

# Radioactive Contamination of Food and Trophic Transfer

## 4.1. Introduction

Human foodstuffs can be naturally or artificially contaminated with radionuclides (Chapter 2). This radioactive contamination can be transferred to humans when they eat. How much of the contamination is actually transferred and assimilated by individuals? How important is food processing in terms of contamination levels?

About 11% of human exposure to radionuclides is via food, both liquid and solid [AMI 18].

## 4.2. Transfer of radionuclides from food to humans

A large proportion of human contamination by radionuclides, both natural and artificial, comes from the diet. Some of the radionuclides in food are of natural origin and belong to the uranium and thorium families (Table 4.1). Another part comes from artificial radionuclides released into the environment.

| Radionuclide | Annual ingested activity (Bq) | | |
|---|---|---|---|
| | Newborn | Child | Adult |
| $^{210}Pb$ | 11 | 21 | 30 |
| $^{210}Po$ | 21 | 39 | 58 |
| $^{234}U$ | 1.9 | 3.8 | 5.7 |
| $^{235}U$ | 0.1 | 0.2 | 0.2 |
| $^{238}U$ | 1.9 | 3.8 | 5.7 |
| $^{226}Ra$ | 7.8 | 15 | 22 |
| $^{228}Ra$ | 5.5 | 10 | 15 |
| $^{228}Th$ | 1.0 | 2.0 | 3.0 |
| $^{230}Th$ | 1.0 | 2.0 | 3.0 |
| $^{232}Th$ | 0.6 | 1.1 | 1.7 |

**Table 4.1**. *Activities of uranium and thorium radionuclides in the diet (modified from [UNS 00])*

### 4.2.1. *Human digestion*

Digestion is a complex process that can be summarized in four stages: A, food intake, B, stomach digestion, C, intestinal digestion and D, fecal excretion (Figure 4.1). The absorption of radionuclides contained in food by the consumer is greatly influenced by several parameters: the nature of the food, the quantity of food, the physico-chemical forms in which the radionuclides are stored, etc.

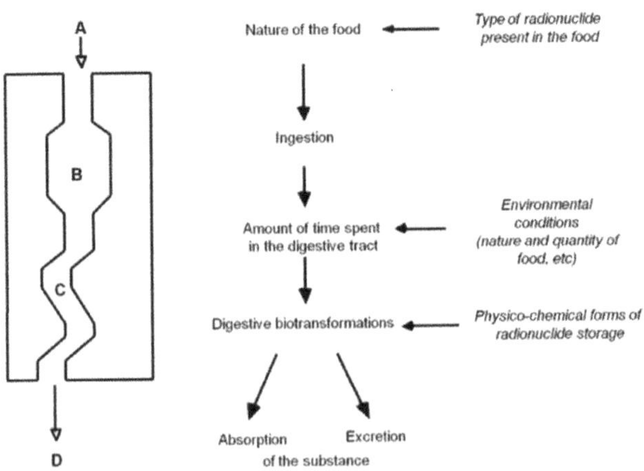


**Figure 4.1.** *The main digestive factors influencing radionuclide absorption*

The length of time the food is kept will have a major impact on internal exposure to ionizing radiation.

### 4.2.2. *Distribution of radionuclides in foodstuffs*

Radionuclides bind to various parts of a food. Many radionuclides are metals. Most authors identify five metal-binding components [WAL 03a, RAI 11]. Here (Figure 4.2), we have added the category of other molecules (which play a minor role) and, on the other hand, grouped the two categories of heat-sensitive and heat-resistant proteins into a single category. Heat-resistant proteins include metallothioneins, which are involved in the regulation of essential metals such as zinc and copper, as well as in protecting cells from transition metal toxicity [AMI 06].

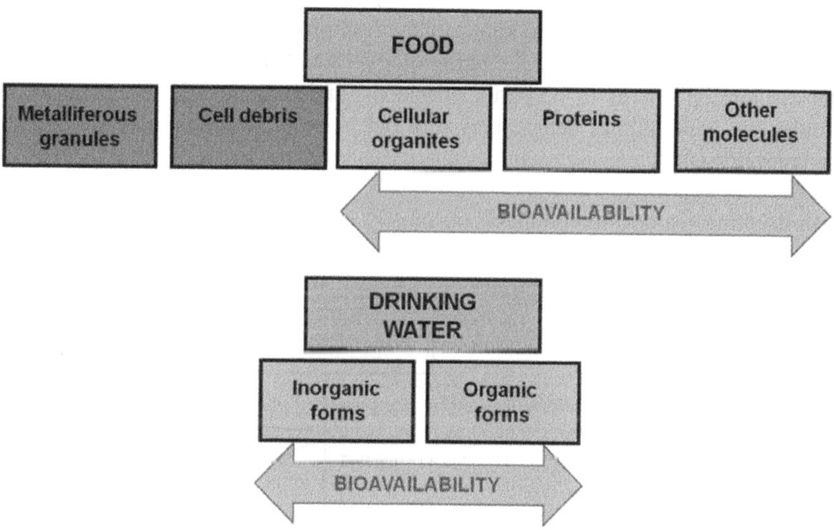


**Figure 4.2.** *Potential bioavailability of metals and radionuclides in meat products and drinking water as a function of tissue distribution*

### 4.2.3. *Bioaccessibility*

The bioaccessibility of a metal is the maximum amount of the metal that is soluble in a synthetic gastric fluid, and therefore potentially available for absorption through the intestinal lumen [HAM 98, WAI 00].

### 4.2.4. *Bioavailability*

Oral bioavailability is the fraction of an administered dose that reaches the central compartment (blood) from the gastrointestinal tract. It is therefore "absolute bioavailability", equivalent to the oral absorption fraction [RUB 99]. Metal that can pass the intestinal barrier is potentially bioavailable. This bioavailability depends on the capacity of the element's physico-chemical forms, in which it is present at the site of absorption, to pass the intestinal barrier [AZH 00].

The difference between bioaccessibility and bioavailability of a radionuclide contained in food or drinking water is shown in Figure 4.3.

According to Reinfelder and Fisher [REI 91], the theoretical bioavailability of metals in copepods concerns only two components, soluble forms, i.e. the two categories of proteins. According to Wallace and Luoma [WAL 03b], bioavailability includes three components: cellular organelles, proteins and organic molecules (Figure 4.2). These authors use the terminology TAM (Tropically Available Metal). This assumption proves to be partly erroneous, and radionuclides bound to metal granules, or even cellular debris, may turn out to be bioavailable according to the consumer [RAI 11].

In drinking water, radionuclides are mostly in soluble form and therefore fully bioavailable (Figure 4.2).

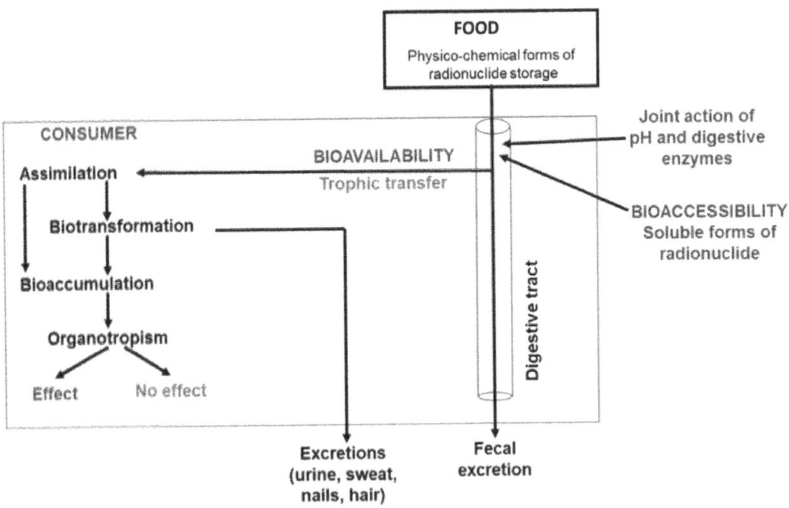


**Figure 4.3.** *Bioaccessibility and bioavailability of metals and radionuclides during the digestive transit of food and drinking of water*

### 4.2.5. *Absorption factors during ingestion*

The ICRP proposes absorption factors ($f_a$) of elements following the ingestion of various particles or chemical forms of the element. The absorbed fraction is the fraction of ingestion or inhalation that is transferred to a specific receptor tissue in an organism, including humans. An "apparent" fraction is calculated from information on the absorption and excretion of the radionuclide by the whole organism, while a "true" absorbed fraction is measured in a metabolic study which involves the injection of a tracer and makes it possible to determine endogenous fecal excretion (i.e. direct transfer from the blood to the intestine) [MAY 96]. Maximum values are shown in Table 4.2.

| **Element** | **H** | **C** | **Co** | **Zn** | **Sr** | **Tc** | **I** | **Cs** |
|---|---|---|---|---|---|---|---|---|
| $f_a$ | 1.0 | 1.0 | 0.1 | 0.5 | 0.25 | 0.9 | 1.0 | 1.0 |
| **Element** | **Po** | **Ra** | **Th** | **U** | **Np** | **Pu** | **Am** | |
| $f_a$ | 0.1 | 0.2 | $5.10^{-4}$ | 0.02 | $5.10^{-4}$ | $5.10^{-4}$ | $5.10^{-4}$ | |

**Table 4.2.** *Maximum values for the absorbed fraction ($f_A$) following ingestion of an element (from [ICR 16; ICR 17; ICR 19])*

Absorbed fractions are highest (100%) for tritium, carbon, iodine and cesium isotopes. Conversely, they are very low for uranium and transuranics.

## 4.3. Transfers from the environment to animal products consumed by humans

Transfers from the environment to animal products consumed by humans are generally expressed as transfer coefficients. Transfer from food to humans can be influenced by food processing, whether industrial or culinary.

The transfer coefficients between grass and milk (*Fm*), between animal feed and meat and eggs (*Ff*) and the food transfer coefficients (*Cft*) of animal products used as food by humans have been defined by the International Commission on Radiation Units and Measurements [ICR 01].

### 4.3.1. *Food transfer coefficients*

Food transfer coefficients ($C_{ft}$) from the environment to animal products used for human consumption have been defined by the ICRU [ICR 01]. This coefficient is the mass or volume activity ($A_m$ or $A_v$) in a receptor tissue or animal product

(Bq.kg$^{-1}$ fresh mass or Bq.L$^{-1}$) divided by the daily intake ($R$ in Bq.d$^{-1}$) of the radionuclide ingested by mouth. They are therefore expressed in d.kg$^{-1}$ or d.L$^{-1}$, where d is the time in days. Previous symbols were $TF$, $TC$, $TE$, $F$, $F_m$ and $F_f$. They vary according to the nature of the food and the radionuclide (Table 4.3).

This notion of transfer coefficient has been defined in order to link the concentration of a radionuclide in a foodstuff (Bq.kg$^{-1}$) and the consumer's daily intake. It is a tricky one to grasp. Its advantages are to provide a less variable transfer coefficient for the migration of radionuclides in the food chain, over a longer period of time corresponding to the establishment of an equilibrium between radionuclide concentrations in the foodstuff and in the consumer's tissues.

Such transfer factors can be measured directly in experimental studies, but they can also be assessed by using biokinetic models for the uptake and retention of radionuclides in animals. Based on a review of the literature, biokinetic models have been developed for the uptake and retention of iodine, technetium, selenium and uranium. These biokinetic models enable $C_{ft}$ values to be evaluated for different types of animals and for different life stages. For each radionuclide considered, reference values and ranges of $C_{ft}$ values are evaluated [THO 03]. These are summarized in Table 4.3.

| Material | $^{60}$Co | $^{90}$Sr | $^{125}$Sb | $^{129}$I | $^{137}$Cs | $^{237}$Np | $^{239,240}$Pu | $^{241}$Am |
|---|---|---|---|---|---|---|---|---|
| Milk | 1.10$^{-2}$ | 2.8.10$^{-3}$ to 5.6.10$^{-2}$ | 2.5.10$^{-5}$ | 1.10$^{-2}$ to 4.9.10$^{-1}$ | 6.10$^{-3}$ to 1.10$^{-1}$ | 0.00001 | 6.10$^{-5}$ to 4.10$^{-4}$ | 1.5.10$^{-6}$ o 1.4.10$^{-5}$ |
| Meat | 7.10$^{-2}$ | 8.10$^{-3}$ to 3.3.10$^{-1}$ | 4.0.10$^{-5}$ | 3.3.10$^{-3}$ to 4.0.10$^{-2}$ | 3.10$^{-2}$ to 3.5.10$^{-1}$ | 0.001 | 6.10$^{-5}$ to 7.10$^{-4}$ | 4.10$^{-5}$ to 4.1.10$^{-4}$ |
| Eggs | | 2.10$^{-1}$ | | 3.0 | 5.7.10$^{-1}$ | | 7.10$^{-3}$ | 4.0.10$^{-3}$ |

**Table 4.3.** *Food transfer coefficients from animal products (j.kg$^{-1}$ or j.L$^{-1}$)*

Howard et al. [HOW 09a] present a compilation of food transfer coefficient values from the Russian literature to complement the work of the IAEA (1994). Three main types of data are listed: a) milk from cows, sheep and goats, b) meat from calves, sheep, goats, pigs and poultry, and c) eggs, with numerous radionuclides taken into account. Compared with the values recommended by the IAEA [IAE 94], three values for milk are at least one order of magnitude higher than those of TRS 364 [Cr, Pu (cow) and Pu (sheep)] and one value is lower [Ni (cow)]. For meat, four values [Am, Cd, Sb (beef), I (pork)] are at least one order of magnitude higher than TRS 364 and eight values are at least one order of magnitude lower [Ru, Pu (beef), Ru, Sr, Zn (sheep), Ru, Sr (pork) and Mn (poultry)].

Howard et al. [HOW 09b] summarize Russian data on food transfer coefficients ($F_m$ to milk, j.L$^{-1}$ or $F_f$ and eggs j.kg$^{-1}$) to domestic animals from their feed and the percentages absorbed from the digestive tract and compare them with IAEA data [IAE 94].

Depending on the element, passage through the digestive tract barrier will be either extremely easy or negligible. This is highlighted in Table 4.4 for ruminants. Tritium and carbon can be added to the first row.

| Order of magnitude of absorbable fraction | Radionuclide |
|---|---|
| $10^{-1}$–1 | I, Cl, Na, Cs, P, Se, Ca, Te, Zn, Sr, Fe |
| $10^{-2}$–$10^{-1}$ | Ag, Ba, Co, Pb, U |
| $10^{-3}$–$10^{-2}$ | Mn, Ru, Cd, Y |
| $10^{-4}$–$10^{-3}$ | Zr, Ce, Pm, Am, Rb |
| $10^{-5}$–$10^{-4}$ | Pu |

**Table 4.4.** *Example values for absorbable fractions of various radionuclides in adult ruminants (modified from [HOW 09c])*

Numerous experiments have been carried out in the USSR on the absorption of radionuclides by the digestive tract of poultry. Fesenko et al. [FES 09] summarize the results. They also identify the main factors influencing the digestive absorption ($f_1$) of radionuclides, such as age, species and chemical form. The authors compare the values obtained by the Soviets with those recommended by the ICRP.

Fesenko et al. [FES 07] list the transfer coefficients ($F_f$, ratio of radioactivity in muscle or eggs, by fresh weight, to radioactivity in the equilibrium daily intake) obtained for poultry by Soviet researchers, and compare them with the data recommended by the IAEA.

Recent reviews by Sheppard et al. [SHE 10] have proposed transfer factor values from plants to domestic animal products (milk, eggs and meat) for 10–15 relevant elements, including $^{36}$Cl, $^{129}$I, $^{222}$Rn, $^{226}$Ra, $^{237}$Np and $^{238}$U. Several transfer factors for neptunium have been revised downwards by a factor of 10, as has the concentration ratio for uranium between fish and water. Specific parameters are discussed in detail, including outgassing of iodine from the soil, soil sorption of Cl, plant-to-soil concentration ratios for U, Ra and Np, and calcium dependence in radium transfers [SHE 06]. Product/substrate concentration ratios (*CR*) were quite consistent across species, while traditional dietary transfer coefficients (*TF*, in $d.kg^{-1}$ or $d.L^{-1}$) necessarily vary with organism mass (food consumption). This suggests that for long-term assessments, it may be advisable to change models to use *CR* concentration ratios rather than *TF* food transfer coefficients [SHE 10].

### 4.3.2. *The influence of product processing*

Humans select the edible parts of raw products for consumption. We also frequently transform raw products, such as cheese from milk or bread from wheat. This has consequences for the level of radioactive contamination in the food we eat.

The bioavailability of pollutants can be limited by the fact that only part of the organism is usually ingested. For example, fish bones and viscera are generally removed before ingestion. There are exceptions, however, when fish is eaten whole for frying, for example, or in the case of shad, where the bones are culinarily dissolved before consumption, or in the case of the liver of certain particularly prized fish (cod, monkfish, red mullet, etc.). Malek et al. [MAL 04] have shown that 30.3% of relative radiocesium activity ($^{137}Cs$) is eliminated in Japanese catfish, by removing inedible parts such as bones, fins, viscera, liver and kidney, and cleaning edible cut parts. Green crab is eaten without the exoskeleton (although in soup the exoskeleton is present during culinary preparation), which potentially reduces the bioavailability of metals, as the exoskeleton contains most of the iron (71%) and manganese (95%) present in crabs, as well as 45% of the cobalt [BJE 02].

Transfer coefficients will also vary according to storage conditions and culinary preparation in particular [GRE 01]. With regard to the effect of culinary processing on contaminants present in food, several studies have been carried out. The preparation and cooking processes of Japanese catfish eliminate 74.7% of the activity of the initial $^{137}Cs$ that was present in live fish [MAL 04]. During cooking, after preparing the fish with curry, the authors found that the radioactivity eliminated from the fish was distributed throughout the curry ingredients. Indeed, among the curry ingredients, the sauce contained on average 34.8% of the activity that was present in the cooked fish. Culinary processes therefore remove a significant fraction of the radioactivity from the fish, but the curry was absorbed by the consumer. On the other hand, Burger et al. [BUR 04] studied the effect of frying on $^{137}Cs$ levels in fish. Concentrations of $^{137}Cs$ were significantly higher in cooked fish than in raw fish. They conclude that deep-frying results in a weight loss of 25–39%, while $^{137}Cs$ levels increase by 32–62%. Consequently, the differences are essentially due to weight loss during the cooking process. This is certainly due to the concentration phenomenon.

Finally, Cabañero et al. [CAB 04] found that selenium levels in cooked tuna are slightly higher than those in raw tuna. Furthermore, the application of heat to this fish accelerates water and vitamin loss, protein degradation, and may also affect chemical contaminants. Protein degradation can then improve protein digestibility, which in turn facilitates the release of bound selenium, thus increasing selenium bioaccessibility.

Besson et al. [BES 09] compared the transfer of two radionuclides, cesium-137 and strontium-90, from cow's milk to whey and cheese in three different types of French cheese production with rennet coagulation. Most of the $^{137}Cs$ was present in the water phase and remained in the whey. Therefore, in the event of cesium contamination of the milk, butter can be made, but the whey should not be consumed. For $^{137}Cs$, the whey processing factor[1] (Pf, i.e. the ratio of activity in cheese or whey to activity in milk) ranged from 0.86 to 1.30 (n = 12). The food processing retention factor (Rf, taking into account the efficiency of the Pe processing, calculated between differences in calcium concentration in milk, cheese and whey), ranged from 0.85 to 1.19 (n = 9). There is therefore no statistical difference between the Pf and Rf of whey for $^{137}Cs$ and cheese products. The Pf of cheese calculated for $^{90}Sr$ ranged from 3.95 to 12.16, with significant differences depending on the type of cheese. In addition, a linear correlation was observed between $^{90}Sr$ and Pf in cheese and cheese calcium concentration (r2 = 0.57). Therefore, Pf is increased in hard cheeses, which are enriched with calcium. This is confirmed by almost constant Rf values, ranging from 0.66 to 0.83.

## 4.4. Transfers from the environment to plant products consumed by humans

The transfer factors from atmosphere to fruit will vary according to radionuclide speciation in the physical medium [OUL 01]. Mitchell [MIT 01] (BIOMASS Fruits Working Group) compared four models of radionuclide transfer to fruit (Fruit tree model, Homeostatic model, SPADE and FARMLAND).

Information on the transfer of radionuclides to fruit was almost absent in the former TRS 364 "Manual of parameter values for the prediction of radionuclide transfer in temperate environments". The revision of the Manual, carried out in accordance with the IAEA Programme on Environmental Modelling for Radiation

1 The various ratios used in Besson et al. [BES 09] are: processing factor: Pf = Activity (cheese or whey)/Activity (milk); the processing retention factor: Rf = Activity (cheese or whey)/Pe x Activity (milk); the processing efficiencies: Pe (cheese) = [Ca (whey) - Ca (cheese)]/[Ca (whey) - Ca (milk)] and Pe (whey) = [Ca (cheese) - Ca (whey)]/[Ca (cheese) - Ca (milk)].

Safety (EMRAS), takes into account the information produced in the years following the Chernobyl accident and the knowledge produced by the IAEA Programme on Biosphere Modelling and Assessment (BIOMASS) of the years 1997–2000. Carini [CAR 09] describes the most important processes concerning the behavior of radionuclides in fruits announced in the IAEA revision TRS 364 [IAE 94] and provides recommendations for research and modeling.

Stewart et al. [STE 01] quantified the fate of $^{3}H$, $^{14}C$ and $S^{35}$ in fruit. After deposition on fruit, the elimination of these three radionuclides differs greatly according to the radionuclide, with tritium lost rapidly in the form of HTO, but retained longer when converted to OBT; losses of $^{14}C$ are less, and those of sulfur are minimal. When we consider the fruit harvest alone, the amount of information is still reduced, but predictions on the possible behavior of these radionuclides can be made with our current knowledge.

Critical pathways for radionuclides and critical foodstuffs in Asian countries differ from those in Western countries because agricultural products and diets are different. Consequently, safety assessments for Asian countries must consider rice as a critical food. As rice is mostly produced under flooded conditions, the uptake of radionuclides by rice is affected by soil conditions. Uchida et al. [UCH 09] summarize the soil-to-plant transfer factors for rice of various radionuclides and stable elements. According to field data on $^{137}Cs$ fallout and naturally occurring stable Cs, the highest transfer factors are observed for $^{137}Cs$ fallout, compared with those for Cs, which have been stable for several decades. However, the geometric means remain approximately identical for stable and radioactive cesiums (transfer factor for $^{137}Cs$ being $3.6.10^{-3}$ and for stable Cs being $2.5.10^{-3}$). Although there are some limitations, particularly for radionuclides recently introduced into the environment and therefore not yet at equilibrium, to the use of transfer factor for stable elements in these few circumstances, these values can be used to assess the long-term transfer of long-lived radionuclides into the environment. The data compiled showed that transfer factor values were higher in brown rice than in white rice because the distribution patterns for elements were different in the bran and white parts of rice grains [UCH 09]. In the event of environmental contamination, various radiation protection measures must be taken, such as not consuming whole-grain rice, mechanically processing rough rice to produce white rice, and washing white rice that is covered with fine rice dust.

Recently, studies have been initiated for plants produced in tropical countries. For paddy rice produced in Malaysia, the estimated transfer factors for $^{226}Ra$ and $^{232}Th$ were found to be well above the value reported by the IAEA for rice. Among the radionuclides detected, potassium-40 had the highest transfer factor at all the study sites, but remained close to the range reported by the IAEA [ASA 15].

In Nigeria, the mean transfer factor values were 0.51, 0.53 and 1.36 (for cowpea) and 0.42, 0.43 and 1.33 (for sesame), respectively, for $^{226}$Ra, $^{232}$Th and $^{40}$K [ELS 20]. For various crops in the Kogi region (Nigeria), transfer factor values obtained for $^{40}$K, $^{232}$Th and $^{238}$U ranged from 0.45 to 0.79 [OCH 20]. The transfer factors from soil to food for $^{238}$U, $^{232}$Th and $^{40}$K were calculated from the volumetric activities of soil and food, while the averages were 0.73, 0.61 and 0.39, respectively, in the study area (Fashina village, Ile-Ife, Osun State, Nigeria) [OLU 18]. For cassava (*Manihot esculenta*), transfer values for $^{226}$Ra, $^{232}$Th and $^{40}$K ranged from 0 to 1.81, 0 to 3.41 and 0.68 to 4.5, respectively. For the authors of the study [ONO 19], the use of the land on these farms and their crops may pose a threat to the bone marrow and general health of the inhabitants of this region of Nigeria.

In Syria, the highest transfer factors for wheat from $^{238}$U, $^{232}$Th and $^{40}$K are 0.32, 0.70 and 3.44, respectively [MOS 21].

The geometric mean transfer factors of natural radionuclides studied in wheat grains were in the order of 0.371 ± 0.07 for $^{40}$K, 0.003 ± 0.002 for $^{226}$Ra and 0.001 ± 0.001 for $^{232}$Th in an Indian region prior to the building of a nuclear reactor [YAD 17].

In China, the $^{210}$Po transfer factors for various mushroom species (*Boletus edulis*, *Tylopllus holloul*, *Boletus brunneissimus*, *Boletus* sp. and *Tylopilus felleus*) were 5.40 ± 1.11, 6.32 ± 1.5, 8.71 ± 0.31, 10.0 ± 0.8 and 9.18 ± 0.65, respectively [PAN 19].

## 4.5. Transfer from drinking water to humans

Every day, humans ingest radionuclides present in food and drinking water, particularly potassium-40. Sanitary control of the radiological quality of water is based on the monitoring of four indicators: overall alpha activity, with a guide value of 0.1 Bq.L$^{-1}$; overall residual beta activity, with a guide value of 1 Bq.L$^{-1}$; tritium activity, with a quality reference of 100 Bq.L$^{-1}$ (the volume activity value set by the WHO) and the total indicative dose (TID), which is calculated on the basis of analyses of the various natural and possibly artificial radionuclides investigated when the overall alpha activity and/or the overall residual beta activity exceed(s) its guide value or the tritium activity exceed(s) its quality reference.

Based on data from the SISE-Eaux database, an analysis of all health monitoring results (51,866 water samples) taken at resource, production and distribution levels between 2005 and 2007 reveals the following observations: no exceedance of the tritium quality reference (considered an indicator of man-made contamination) was observed. More than 95% of the samples taken showed an overall alpha activity less than or equal to the guide value (4.55% of samples showed an overall alpha activity greater than 0.1 $Bq.L^{-1}$). Over 99.9% of samples taken showed residual overall beta activity less than or equal to the guide value (0.09% of samples showed overall beta activity greater than 1 $Bq.L^{-1}$) [ASN 09].

## 4.6. Environmental monitoring of products consumed by humans

Following the Chernobyl accident, there has been environmental monitoring in many countries. For example, the average transfer coefficients to milk[2] for $^{137}Cs$ and $^{90}Sr$ have been estimated in the Austrian Alps at $0.0071 \pm 0.0009$ $d.L^{-1}$ and $0.0011 \pm 0.0004$ $d.L^{-1}$, respectively [LET 07].

Caribou meat (North America) and reindeer meat (Eurasia) are highly prized by Arctic populations. Could the radioactive cesium content of these animals present a health hazard? These deer eat an abundance of lichen, the only food available during the long winter months. Cesium is highly concentrated and persistent in fungi, mosses and lichens, far more so than in higher plants. As a result, the tissues of caribou, particularly their muscles, have high concentrations of cesium-137, since cesium is an alkaline metal, such as potassium, which is preferentially concentrated in muscle mass. Macdonald et al. (2007) quantified cesium-137 levels in 1,200 muscle tissue samples taken from caribou in Northern Canada, Alaska and Greenland between 1958 and 2000. In particular, they revealed a strong seasonal variation linked to the winter consumption of lichens, replaced in spring by young shoots and leaves. Concentrations were five times higher in February than in September of the same year. The highest cesium-137 concentrations were found in the 1960s, with 2,000–3,000 $Bq.kg^{-1}$ in the muscles of caribou, a period during which many nuclear tests were carried out around the world. These concentrations fell gradually until April 1986, when the Chernobyl accident triggered a transient rise of approximately 20%. They then declined to between 200 and 300 $Bq.kg^{-1}$ in the 1990s, below the limits set by international health authorities (the WHO's recommended limit for cesium-137 contamination of foodstuffs is 1,000 $Bq.kg^{-1}$). However, these figures should be compared with the maximum of 150,000 $Bq.kg^{-1}$ taken from reindeer muscle in Scandinavia after the Chernobyl accident [STR 92].

Since this accident, international regulations have evolved considerably. Following the Fukushima accident, European maximum limits for the sum of cesium-134 and cesium-137 in $Bq.kg^{-1}$ have been set at 50 for infants, young children and milk, 10 for beverages and 100 for other foodstuffs [UE 21].

---

2 This unit is recommended by the ICRU [ICR 01] to express the food transfer coefficient to replace the old transfer coefficient. It corresponds to the activity in the receptor ($Bq.L^{-1}$) divided by the transfer rate ($Bq.d^{-1}$).

The Direction Générale de l'Alimentation [DGA 00] monitors radioactive contamination by $^{137}Cs$ and $^{134}Cs$ annually in French and imported plant products, and in samples taken around nuclear power plants. In 2000, domestic plant products showed no contamination, except for wild mushrooms from various regions. On the other hand, 51% of imported mushrooms, more than half of them from Eastern Europe, were contaminated, notably with $^{137}Cs$, although this was still below the regulatory limit of 600 Bq.kg$^{-1}$. Similarly, the Direction Générale de la Concurrence et de la Consommation et de la Répression des Fraudes [DGC 00] quantifies cesium-134 and -137 activities in various foodstuffs imported into France. More recently, the IRSN has been measuring radioactivity in foodstuffs to calculate the average effective dose for the French population (Table 4.5).

| Category | Average | Min. | Max. |
|---|---|---|---|
| Dairy products | 122 | 102 | 175 |
| Meats | 78 | 47 | 98 |
| Cereal derivatives | 86 | 66 | 123 |
| Leafy vegetables | 34 | 22 | 40 |
| Other vegetables and fruit | 211 | 128 | 337 |
| Fish* | 17 | 6 | 19 |
| Fish** | 35 | 15 | 88 |
| Crustaceans* | 5 | 2 | 6 |
| Crustaceans** | 12 | 12 | 46 |

**Table 4.5.** *Average, minimum and maximum values for the annual quantities of cesium radioisotopes consumed from different food categories (kg.year$^{-1}$), derived from the results of 11 food surveys conducted in France from 1991 to 2015 (from [IRS 21]) * Sites far from the sea. ** Seaside sites.*

In baby food marketed in Italy, volumetric activities ranged from 0.005 to 0.238, from 0.0082 to 1.65 and from 0.0003 to 0.015 from <13.6 to 233.3 Bq.kg$^{-1}$ for $^{210}Po$, $^{238}U$, $^{232}Th$ and $^{40}K$, respectively, while they are below the detection limit for $^{137}Cs$ and $^{226}Ra$ [DES 19].

The subarctic food chain of lichen reindeer–human in Finland has been extensively studied for cesium and strontium radionuclides, but little for transuranics. Yet, radionuclide enrichment in this food chain can lead to exceptionally high body loads in indigenous Sami and Inuit populations consuming large quantities of reindeer meat and edible organs. The existence of neptunium, plutonium, americium and curium isotopes in the environment of northern Finland has been noted, the sources being atmospheric fallout from atomic testing and the

Chernobyl accident [PAA 20]. The inhalation intake of $^{239,240}$Pu by male Lapps in the period 1954–1978 was approximately 13 pCi (0.48 Bq) and by southern Finns approximately 18 pCi (0.67 Bq). Estimated values for dietary intake were 535 pCi (19.7 Bq) and 36 pCi (1.3 Bq), respectively. The absorption factor of $^{239,240}$Pu in the human gastrointestinal tract was estimated at $8–9.10^{-4}$, which is somewhat higher than that currently suggested by the ICRP for soluble plutonium compounds ($1.10^{-4}$) [MUS 84].

## 4.7. References

[AMI 06] AMIARD J.-C., AMIARD-TRIQUET C., BARKA S. et al., "Metallothioneins in aquatic invertebrates: their role in metal detoxification and their use as biomarkers", *Aquatic Toxicology*, vol. 76, no. 2, pp. 160–202, 2006.

[AMI 18] AMIARD J.C., "Radionucléides et aliments", in CAMEL V., RIVIÈRE G., LE BIZEC B. (eds), *Risques chimiques liés aux aliments : principes et applications*, Lavoisier, Tec&Doc, Paris, pp. 387–410, 2018.

[ASA 15] ASADUZZAMAN K., KHANDAKER M.U., AMIN Y.M. et al., "Uptake and distribution of natural radioactivity in rice from soil in north and west part of peninsular malaysia for the estimation of ingestion dose to man", *Annals of Nuclear Energy*, vol. 76, pp. 85–93, 2015.

[ASN 09] ASN, La qualité radiologique de l'eau mise en distribution en France 2005-2007, 2009.

[AZH 00] AZHENA M.A.G.O., VASCONCELOS M.T.S.D., "Assessment of the Pb and Cu *in vitro* availability in wines by means of speciation procedures", *Food and Chemical Toxicology*, vol. 38, pp. 899–912, 2000.

[BES 09] BESSON B., POURCELOT L., LUCOT E. et al., "Variations in the transfer of radiocesium ($^{137}$Cs) and radiostrontium ($^{90}$Sr) from milk to cheese", *Journal of Dairy Science*, vol. 92, pp. 5363–5370, 2009.

[BJE 02] BJERREGAARD P., DEPLEDGE M.H., "Trace metal concentrations and contents in the tissues of the shore crab *Carcinus maenas*: effects of size and tissue hydration", *Marine Biology*, vol. 141, pp. 741–752, 2002.

[BUR 04] BURGER J., GAINES K.F., BORING C.S. et al., "Effects of cooking on radiocesium in fish from the Savannah River: exposure differences for the public", *Archives of Environmental Contamination and Toxicology*, vol. 46, pp. 231–235, 2004.

[CAB 04] CABAÑERO A.I., MADRID Y., CÁMARA C., "Selenium and mercury bioaccessibility in fish samples: an *in vitro* digestion method", *Analytica Chimica Acta*, vol. 526, pp. 51–61, 2004.

[CAR 09] CARINI F. "Radionuclide transfer to fruit in the IAEA TRS 364 Revision", *Journal of Environmental Radioactivity*, vol. 100, pp. 752–756, 2009.

[DES 19] DESIDERI D., BATTISTI P., GNIARDIN, I. et al., "Assessment of radioactivity in Italian baby food", *Food Chemistry*, vol. 279, pp. 408–415, 2019.

[DGA 00] DGAl, "Direction générale de l'alimentation. Données d'exposition de la population française aux résidus de pesticides, plomb, cadmium, arsenic et radionucléides par la voie alimentaire", *Notre Alimentation*, vol. 24, pp. I–IV, 2000.

[DGC 00] DGCCRF, Bilan de surveillance de la contamination radioactive pour l'année 2000, 2000. Available at: http://www.minefi.gouv.fr/fonds_documentaire/dgccrf/01presentation/activites/labos/2000/enqradioactivite.htm.

[ELS 20] ELSAMAN R., ALI G.A.M., UOSIF MAM. et al., "Transfer factor of natural radionuclides from clay loam soil to sesame and Cowpea: radiological hazards", *International Journal of Radiation Research*, vol. 18, no. 1, January 2020.

[FES 07] FESENKO S., ISAMOV N., HOWARD B.J. et al., "Review of Russian language studies on radionuclide behaviour in agricultural animals: Part 1. Gut absorption", *Journal of Environmental Radioactivity*, vol. 98, pp. 85–103, 2007.

[FES 09] FESENKO S., HOWARD B.J., ISAMOV N. et al., "Review of Russian-language studies on radionuclide behaviour in agricultural animals: Part 4. Transfer to poultry", *Journal of Environmental Radioactivity*, vol. 100, pp. 815–822, 2009.

[GRE 01] GREEN N., "The effect of storage and processing on radionuclide content of fruit", *Journal of Environmental Radioactivity*, vol. 52, pp. 281–290, 2001.

[HAM 98] HAMEL S.C., BUCKLEY B., LIOY P.J., "Bioaccessibility of metals in soils for different liquid to solid ratios in synthetic gastric fluid", *Environmental Science & Technology*, vol. 32, pp. 358–362, 1998.

[HOW 09a] HOWARD B.J., BERESFORD N.A., BARNETT C.L. et al., "Radionuclide transfer to animal products: revised recommended transfer coefficient values", *Journal of Environmental Radioactivity*, vol. 100, pp. 263–273, 2009.

[HOW 09b] HOWARD B.J., BERESFORD N.A., BARNETT C.L. et al., "Quantifying the transfer of radionuclides to food products from domestic farm animals", *Journal of Environmental Radioactivity*, vol. 100, pp. 767–773, 2009.

[HOW 09c] HOWARD B.J., BERESFORD N.A., BARNETT C.L. et al., "Gastrointestinal fractional absorption of radionuclides in adult ruminants", *Journal of Environmental Radioactivity*, vol. 100, pp. 1069–1078, 2009.

[IAE 94] IAEA, Handbook of parameter values for the prediction of the radionuclide transfer in temperate environments, IAEA Technical report Series no. 364, IAEA, Vienna, 1994.

[ICR 01] ICRU, "ICRU Report No. 65, Quantities, units and terms in radioecology", *Journal of ICRU*, vol. 1, no. 2, pp. 1–44, 2001.

[ICR 16] ICRP, "ICRP Publication 134, Occupational intakes of radionuclides: Part 2", *Annals of the ICRP*, vol. 45, nos. 3-4, pp. 1–352, 2016.

[ICR 17] ICRP, "ICRP Publication 137, Occupational intakes of radionuclides: Part 3", *Annals of the ICRP*, vol. 46, nos. 3-4, pp. 1–487, 2017.

[ICR 19] ICRP, "ICRP Publication 141, Occupational intakes of radionuclides: Part 4", *Annals of the ICRP*, vol. 48, nos. 2–3, pp. 1–514, 2019.

[IRS 21] IRSN, Exposition de la population française aux rayonnements ionisants. Bilan 2014-2019, IRSN report 2021-00108, IRSN, Fontenay-aux-Roses, 2021.

[LET 07] LETTNER H., HUBMER A., BOSSEW P. et al., "$^{137}$Cs and $^{90}$Sr transfer to milk in Austrian alpine agriculture", *Journal of Environmental Radioactivity*, vol. 98, pp. 69–84, 2007.

[MAL 04] MALEK M.A., NAKAHARA N., NAKAMURA R., "Removal of $^{137}$Cs in japanese catfish during preparation for consumption", *Journal of Radiation Research*, vol. 45, pp. 309–317, 2004.

[MAY 96] MAYES R.W., BERESFORD N.A., HOWARD B.J., "Use of the true absorption coefficient as a measure of bioavailability of radiocaesium in ruminants", *Radiation and Environmental Biophysics*, vol. 35, pp. 101–109, 1996.

[MIT 01] MITCHELL N.G., "Models for radionuclide transfer to fruits and data requirements", *Journal of Environmental Radioactivity*, vol. 52, pp. 291–307, 2001.

[MOS 21] MOSTAFA Y., MOSTAFA A., KADHIM N.F. et al., "The plant transfer factor of natural radionuclides and the soil radiation hazard of some crops", *Environmental Monitoring and Assessment*, vol. 193, p. 320, 2021.

[MUS 84] MUSSALO-RAUHAMAA H., JAAKKOLA T., MIETTINEN J.K. et al., "Plutonium in Finnish Lapps. An estimate of the gastrointestinal absorption of plutonium by man based on a comparison of the plutonium content of Lapps and southern Finns", *Health Physics*, vol. 46, pp. 549–559, 1984.

[OCH 20] OCHEJE J.A., TYOVENDA A.A., "Determination of the transfer factor and dose rate of radionuclides in some selected crops in Kogi state, Nigeria", *IOSR Journal of Applied Physics (IOSR-JAP)*, vol. 12, no. 3, Ser. I, pp. 7–12, 2020.

[OLU 18] OLUYIDE S.O., TCHOKOSSA P., AKINYOSE F.C. et al., "Assessment of radioactivity levels and transfer factor of natural radionuclides around iron and steel smelting company located in Fashina Village, Ile-Ife, Osun State, Nigeria", *Facta Universitatis, Series: Working and Living Environmental Protection*, vol. 15, no. 3, pp. 241–256, 2018.

[ONO 19] ONONUGBO C.P., AZIKIWE O., AVWIRI G.O., "Uptake and distribution of natural radionuclides in Cassava crops from Nigerian Government Farms", *Journal of Scientific Research & Reports*, vol. 23, no. 5, pp. 1–15, 2019.

[OUL 01] OULD-DADA Z, FAIRLIE I., READ C., "Transfer of radioactivity to fruit: significant radionuclides and speciation", *Journal of Environmental Radioactivity*, vol. 52, pp. 159–174, 2001.

[PAA 20] PAATERO J., SALMINEN-PAATERO S., "Transfer of transuranium elements along the food chain lichen-reindeer-man. A review of investigations in Finnish Lapland", *Journal of Environmental Radioactivity*, vol. 212, p. 106126, 2020.

[PAN 19] PANG C., WANG W., TUO F. et al., "Determinations of $^{210}$Pb contents in food, diet, environmental samples and estimations of internal dose due to daily intakes", *Journal of Environmental Radioactivity*, vol. 203, pp. 107–111, 2019.

[RAI 11] RAINBOW P.S., LUOMA S.N., WANG W.X., "Trophically available metal. A variable feast", *Environmental Pollution*, vol. 159, pp. 2347–2349, 2011.

[REI 91] REINFELDER J.R., FISHER N.S., "The assimilation of elements ingested by marine copepods", *Science*, vol. 251, pp. 794–796, 1991.

[RUB 99] RUBY M.V., SCHOOF R., BRATTIN W. et al., "Advances in evaluating the oral bioavailability of inorganics in soil for use in human health risk assessment", *Environmental Science & Technology*, vol. 33 pp. 3697–3705, 1999.

[SHE 06] SHEPPARD S.C., SHEPPARD M.I., TAIT J.C., SANIPELLI B.L., "Revision and meta-analysis of selected biosphere parameter values for chlorine, iodine, neptunium, radium, radon and uranium", *J. Environ. Radioactivity*, vol. 89, pp. 115–137, 2006.

[SHE 10] SHEPPARD S.C., LONG J.M., SANIPELLI B., "Verification of radionuclide transfer factors to domestic-animal food products, using indigenous elements and with emphasis on iodine", *Journal of Environmental Radioactivity*, vol. 101, pp. 895–901, 2010.

[STE 01] STEWART A., BRUDENELL A., COLLINS C.D., "Deposition of gaseous radionuclides to fruit", *Journal of Environmental Radioactivity*, vol. 52, pp. 175–189, 2001.

[STE 06] SHEPPARD S.C., SHEPPARD M.I., TAIT J.C. et al., "Revision and meta-analysis of selected biosphere parameter values for chlorine, iodine, neptunium, radium, radon and uranium", *Journal of Environmental Radioactivity*, vol. 89, pp. 115–137, 2006.

[STR 92] STRAND P., SELNÆS T.D., BØE E. et al., "Chernobyl fallout: internal doses to the Norwegian population and the effect of dietary advice", *Health Physics*, vol. 63, pp. 385–392, 1992.

[THO 03] THORNE M.C., "Estimation of animal transfer factors for radioactive isotopes of iodine, technetium, selenium and uranium", *Journal of Environmental Radioactivity*, vol. 70, pp. 3–20, 2003.

[UCH 09] UCHIDA S., TAGAMI K., SHANG Z.R. et al., "Uptake of radionuclides and stable elements from paddy soil to rice: a review", *Journal of Environmental Radioactivity*, vol. 100, pp. 739–745, 2009.

[UE 21] UE, "Règlement d'exécution (UE) 2021/1533 de la commission du 17 septembre 2021 imposant des conditions particulières à l'importation de denrées alimentaires et d'aliments pour animaux originaires ou expédiés du Japon à la suite de l'accident survenu à la centrale nucléaire de Fukushima et abrogeant le règlement d'exécution (UE) 2016/6", *Journal officiel de l'Union européenne*, vol. L 330, pp. 72–85, 20 September 2021.

[UNS 00] UNSCEAR, Sources and Effects of Ionizing Radiation, Annex D Medical radiation exposures, United Nations, New York, 2000.

[WAL 00] WAISBERG M., BLACK W.D., WAISBERG C.M. et al., "The effect of pH, time and dietary source of cadmium on the bioaccessibility and adsorption of cadmium to/from lettuce (*Lactuca sativa* L. cv. *Ostinata*)", *Food and Chemical Toxicology*, vol. 42, pp. 835–842, 2000.

[WAL 03a] WALLACE W.G., LEE B.G., LUOMA S.N., "Subcellular compartmentalization of Cd and Zn in two bivalves. I. Significance of metal-sensitive fractions (MSF) and biologically detoxified metal (BDM)", *Marine Ecology Progress Series*, vol. 249, pp. 183–197, 2003.

[WAL03b] WALLACE W.G., LUOMA S.N., "Subcellular compartmentalization of Cd and Zn in two bivalves. II. The significance of trophically available metal (TAM)", *Marine Ecology Progress Series*, vol. 257, pp. 125–137, 2003.

[WAR 00] WAISBERG M., BLACK W.D., WAISBERG C.M., HALE B., "The effect of pH, time and dietary source of cadmium on the bioaccessibility and adsorption of cadmium to/from lettuce (*Lactuca sativa L. cv. Ostinata*)", *Food and Chemical Toxicology*, vol. 42, pp. 835–842, 2000.

[YAD 17] YADAV P., GARG V.K., SINGH B. et al., "Transfer factors and effective dose evaluation due to natural radioactivity in staple food grains from the vicinity of proposed nuclear power plant", *Expo Health*, doi: 10.1007/s12403-017-0243-0, 2017.

# 5

# Human Bioaccumulation of Radionuclides

## 5.1. Human bioaccumulation of radionuclides

The bioaccumulation of radionuclides follows the same mechanisms as those for the bioaccumulation of stable isotopes of the same metallic element. These mechanisms are well known for vertebrates, particularly mammals, the class to which humans belong.

### 5.1.1. *Transmembrane passage*

Uptake of pollutants occurs through various processes [AMI 17]. When a pollutant enters an organism, whatever its route of entry (oral, pulmonary, cutaneous), it first reaches the general circulation (blood). Then, from this internal fluid, it is distributed to the various tissues and organs, before eventually being excreted.

Movement within the body also involves the passage across multiple membrane barriers. Transport across membranes takes many forms (Figure 5.1). Membranes are generally assemblies of phospholipids and proteins in molecular bilayers, with the hydrophilic protein poles facing outwards and the hydrophobic poles facing inwards. Only lipophilic molecules can therefore cross the membrane by simple diffusion, according to the concentration gradient, from the more concentrated to the more diluted medium. Hydrophilic molecules have to pass through pores or channels, or enlist the help of various types of transporters.

Filtration (Figure 5.2) enables the passage of neutral, ionized or non-ionized water-soluble molecules, and to a lesser extent liposoluble and insoluble molecules, depending in particular on their size (Figure 5.2). Transport can be passive, i.e. without energy expenditure, but with facilitated diffusion. Facilitation (Figure 5.1) is due to two types of transport proteins: pore-forming protein channels and transport proteins (transporters).

Radionuclides are unstable metals and metalloids. From the point of view of passage through biological membranes, their behavior will be identical to that of their stable counterparts. In mammals, as in other organisms, individuals can control the internal concentration of radionuclides, if these are isotopes of an essential stable element, either by limiting the influx, increasing the efflux or neutralizing the toxicity of the latter, although in this case radiotoxicity will always be present.

As radionuclides are elements, they are mainly affected by filtration, facilitated diffusion and transporters. The problem may be different when the radionuclide is incorporated into an organic molecule, in which case passive diffusion is conceivable.

Biological membranes are lipid bilayers with a hydrophobic core that prevents the passage of radionuclides. To cross the lipid barrier, radionuclides must enter and leave the membrane via dedicated proteins called transporters [CAT 09].

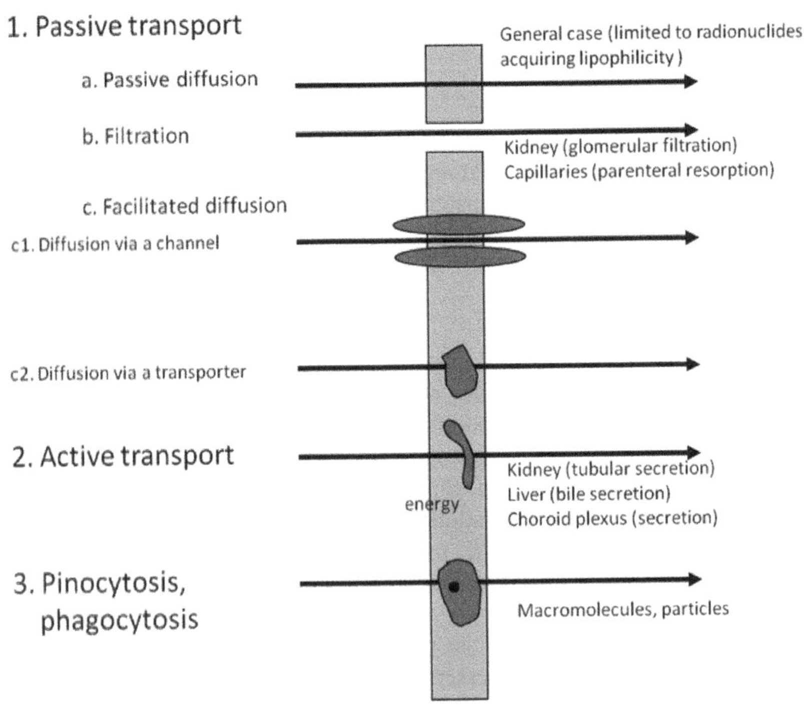


**Figure 5.1.** *Radionuclide transmembrane pathways*

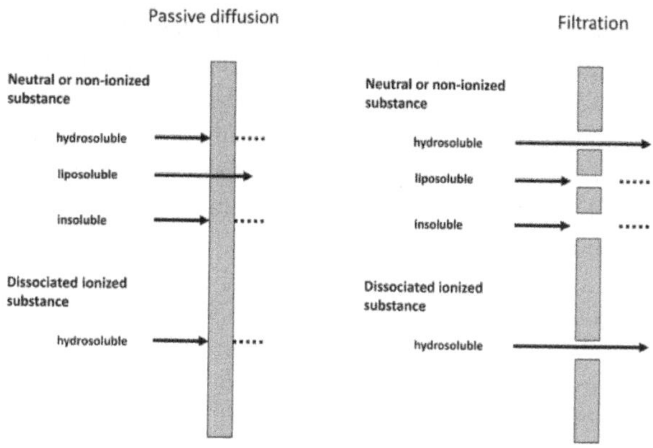


**Figure 5.2.** *Influence of molecule solubility on two cases of passive transport (Figure 5.1), simple diffusion and filtration*

Radionuclides such as metals can use membrane transporters to cross the lipid barrier. These transporters are dedicated proteins embedded in the membrane. There are three main types of transporters [AMI 17]. These include ZIP (Zinc–Iron Permease family), OPT (Oligo Peptide Transporters), magnesium transporters and Nramp (Natural resistance-associated macrophage proteins), CDF (Cation Diffusion Facilitator), co-transporters, ion channels, $P_{1B}$-ATPase-type transporters and ATP-binding cassette transporters [CAT 09].

In the case of radionuclides, part of the transmembrane passage is facilitated by transport through ion channels [BLU 01]. Ionic competition between isotopes of the same element, or between isotopes of chemically related elements (such as cesium and potassium (alkaline metals), or strontium and calcium (alkaline-earth metals)), explains much of the uptake and subsequent bioaccumulation of radionuclides [BLU 01].

### 5.1.2. *Epithelial and cerebral barriers*

Entry of a radionuclide into an organism can occur by three main routes: passage via food and water through the intestinal epithelium, passage of aerosols or radioactive particles through the pulmonary epithelium and percutaneous passage. The characteristics of both the biological barrier and the radionuclide determine the radioactive contaminant's rate of transfer in the body, its kinetics of entry, retention and elimination.

Transfers across the epithelial (skin and digestive tract) and cerebral (blood–brain and blood–liquid) barriers will vary greatly depending on the radionuclide. Percutaneous passage may be greatly facilitated if the skin is damaged.

Skin passage is relatively low for uranium (0.4%), cesium (1.3%) and cadmium (0.6%). Intestinal uptake of uranium in mammals ranges from 0.06 to 2.8%, and in humans from 1.4 to 5%. The intestinal passage of cadmium is similar (5%), but that of plutonium is a hundred times lower than that of uranium. On the contrary, cesium intestinal uptake is 80%, and tritium uptake in the form of tritiated water is rapid and complete [DUB 09b]. One-third of internal contamination is due to percutaneous exposure and two-thirds to respiratory exposure.

Percutaneous penetration is highly dependent on skin condition, and can be significant in the case of injury [NCR 07]. Uranium penetration is 0.4% in vivo in rats. The permeability coefficient of tritiated water is approximately $4.50.10^{-3}$ $cm.h^{-1}$. Skin absorption coefficients are $10^{-4}$, $2.7.10^{-4}$ and $10^{-3}$ $cm.h^{-1}$, respectively, for $^{60}Co$, Cs and $CdCl_2$ [DUB 09b].

Intestinal absorption of uranium is fairly well known. The fraction absorbed varies from 1.4% to 5%. ICRP publications 100 and 137 [ICR 06, ICR 17] give $f_A$ (gastrointestinal transfer coefficient) values of 0.002 for insoluble forms of uranium ($UO_2$, U $O_{38}$, etc.) and 0.02 for soluble forms. The absorption of plutonium is generally a hundred times lower than that of uranium, and $f_A$ is equal to between $3.10^{-8}$ and $1.10^{-3}$ [ICR 19]. For cobalt-60, intestinal absorption in the form of chloride varies from 2 to 20%. These values will vary according to the food containing the cobalt and the individual's nutritional status (20–95%). The ICRP adopts a value of 0.1 for the gastrointestinal transfer coefficient $f_A$ of cobalt for all forms, except insoluble oxide forms, where the value adopted is 0.05 [ICR 16a]. Cesium-137 is strongly and rapidly absorbed from the intestinal lumen (80% in one hour). The ICRP recommends an $f_A$ of 1 for cesium. The same applies to tritium in the form of tritiated water. Cadmium absorption, on the contrary, is relatively low (5%). Absorption values for children are considerably higher than those quoted above for adults [DUB 09b].

The blood–brain barrier (BBB) is not absolutely impervious to the passage of metals including radionuclides. Such passage has been observed for uranium [LEM 03]. The blood–liquid barrier formed by the choroid plexus also allows the passage of various metals and their radionuclides. This is particularly true when these are in organic form, such as methylmercury [DUB 09b]. The effect of plutonium has not yet been demonstrated, although exposure to this radionuclide has been shown to cause brain tumors in rats [SAN 92].

### 5.1.3. *Body distribution*

After passing through membrane barriers, the radionuclide is distributed throughout the body, reaching the sites of its biological or toxic action. In vertebrates, this cellular and tissue distribution takes place mainly via the blood. This liquid carries the radionuclide to all organs (Figure 5.3).

#### 5.1.3.1. *Blood transport*

Blood transport of the radionuclide takes two forms: a free form, corresponding to the water-soluble fraction of the pollutant dissolved in the aqueous phase of the blood, and a bound form, resulting from the binding of the non-water-soluble fraction to certain plasma constituents. The free form mainly concerns water-soluble chemical pollutants, i.e. neutral or ionizable substances (ionized or not), i.e. the ionized forms of liposoluble acidic or basic molecules. The bound form mainly concerns liposoluble or insoluble fractions in the blood, which can only be transported when attached to certain blood constituents, proteins and/or cells. In vertebrates, the proteins involved in this binding are mainly albumin, and secondarily globulins and fibrinogen. For pollutants such as arsenic and lead, and radionuclides of cesium, it is the figurative elements of the blood, such as red blood cells, that ensure transport. Carrier-pollutant bonds are usually low-energy (hydrogen bonds, hydrophobic bonds, electrostatic bonds, etc.). They are therefore immediately reversible and non-specific, since several pollutants are carried by the same type of transporter.

The balance in blood between free and bound forms varies from 0 to 100%. This balance depends on a number of factors, such as the nature of the xenobiotic, saturation of transport protein binding sites, animal species, competitive phenomena and pathological factors.

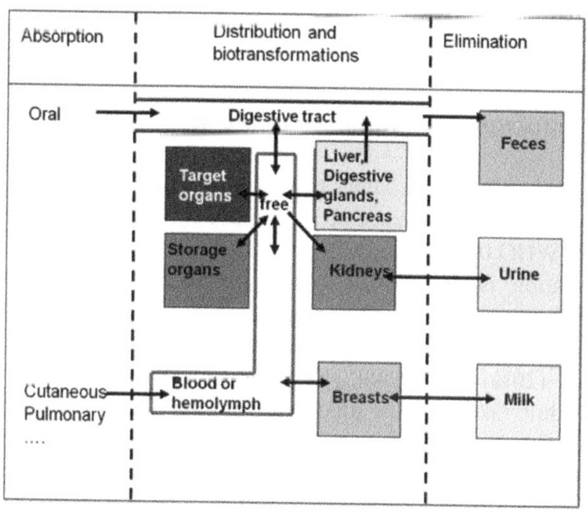


**Figure 5.3.** *General diagram of radionuclide fate in humans*

#### 5.1.3.2. *Cell and tissue distribution*

From the bloodstream, pollutants are distributed in their free form to the various organs. This distribution, or organotropism, is governed by the laws of membrane passage. It mainly takes place via passive diffusion and filtration through the basal laminae of cells. Numerous physico-chemical factors are involved in cell diffusion. These are mainly liposolubility, ionization and molecule size. Tissue distribution is conditioned by factors such as tissue blood flow, tissue-specific factors and organ particularities. These particularities are due to the roles of these organs, such as pollutant entry organs (lungs), metabolizing organs (liver, pancreas) or excretory organs (kidneys, etc.).

The most remarkable cases of organotropism in humans concern iodine, which bioaccumulates preferentially in the thyroid gland, alkaline earths (such as strontium isotopes), which accumulate in bone tissue, and alkalines (such as cesium isotopes), which are found mainly in muscle mass.

### 5.1.4. *Molecular mechanisms*

Radionuclides cause numerous disturbances at the molecular and cellular levels, as all the components of a cell are potential targets for radioactive contaminants. These include the disruption of redox homeostasis, DNA damage, interactions with elements of interest (elements whose radioactive isotopes are of particular interest from a radiation protection point of view) and proteins, the detoxification mechanisms involved and their energetic cost [DUB 09a]. This aspect will be developed in Chapter 8.

### 5.1.5. *Bioaccumulation of natural radionuclides*

Bioaccumulation of uranium comes mainly from food and drinking water[1]. Daily intake through ingestion is estimated at 1–4 $\mu g.d^{-1}$, a thousand times higher than the amount inhaled [WHO 04]. The main contributors are plants grown on uranium-rich soils. Contamination is mainly root-based, and plants whose tubers, such as potatoes, are consumed are a non-negligible source [COL 01]. The amount of uranium in a human is approximately 40 μg, distributed mainly in the muscles (40%), skeleton (20%) and blood (10%) [WHO 04]. According to ICRP 23 [ICR 75], an adult takes in 1.9 $\mu g.d^{-1}$ of uranium via the digestive tract and 0.007 $\mu g.d^{-1}$ via the respiratory tract. They excrete 0.05–0.5 $\mu g.d^{-1}$ via urine and 1.4–1.8 $\mu g.d^{-1}$ via feces. For this reason, biological monitoring is carried out on feces.

1 Before it was treated, Badoit water was the richest in uranium in France (1–2 $Bq.L^{-1}$).

Thorium is relatively abundant, but its very low concentration in the biosphere makes its contribution to natural human exposure negligible. Daily absorption due to natural thorium has been estimated, for an adult, at $3.10^{-5}$ Bq by inhalation and $4.10^{3}$ Bq by ingestion, giving an annual effective dose of 3 μSv [GAM 97].

Polonium-210 is found mainly in marine foods. Concentrations of polonium-210 in crustaceans are 50,000 times higher than in seawater, 10,000 times higher in mollusks and 2,000 times higher in fish. The edible parts of mussels can contain 10–120 $Bq.kg^{-1}$ of fresh weight polonium-210. However, its concentration in tobacco is not negligible (0.01 $Bq.g^{-1}$ of tobacco). $^{210}Po$ contamination is significant in certain foods such as sardines ($3.10^{4}$ $Bq.kg^{-1}$) [CAR 11].

Some regions of the world have a high level of natural radioactivity. This is the case in Ramsar (Iran), where the inhabitants are subject to high levels of natural irradiation, particularly from radium-226. This is verified by measuring tooth contamination, which is 0.18 and 0.32 $mBq.g^{-1}$, respectively, in controls and inhabitants of Ramsar [AGH 06]. The same applies to $^{210}Po$, the absorption of which into the blood through the digestive system is 30–40%. The highest concentrations of $^{210}Po$ and $^{210}Pb$ were found in the skeleton and hair. Average $^{210}Po$ concentrations in Ramsar residents were almost three times higher than in control groups in 2000 [SAM 05].

The concentrations of various radionuclides from the uranium and thorium families in human tissues are shown in Table 5.1.

| Radionuclide | Liver | Muscle, etc. | Bones | Lung | Kidney |
|---|---|---|---|---|---|
| $^{210}Pb$ | 400 | 100 | 3,000 | 200 | 200 |
| $^{210}Po$ | 600 | 100 | 2,400 | 200 | 600 |
| $^{226}Ra$ | 4.1 | 4.1 | 260 | 4.1 | 4.1 |
| $^{228}Ra$ | 3 | 2 | 100 | 20 | 2 |
| 230Th | 9 | 1 | 2–70 | 20 | 5 |
| $^{232}Th$ | 3 | 1 | 6–24 | 20 | 3 |
| $^{238}U$ | 3 | 5 | 100 | 20 | 30 |

**Table 5.1.** *Reference values for uranium and thorium series radionuclide concentrations in human tissue (mBq.kg-1) (modified from [UNS 08])*

According to UNSCEAR [UNS 08], $^{210}Po$ is distributed as follows in the human body (expressed in Bq concentration, with dose in mSv per year in brackets): 16 (0.38) in the skeleton, 3.4 (0.06) in the muscles, 1.4 (0.4) in the liver, 0.3 (0.16) in the lungs, 0.23 (0.01) in the spinal cord, 0.2 (0.33) in the kidneys, 0.17 (0.02) in the blood and 0.04 (0.11) in the spleen.

For $^{210}Po$, hepatic and pulmonary values between 1977 and 1979 were 3.2 and 0.3 $Bq.kg^{-1}$ dw, respectively, in Laplanders and 0.6 and 0.1 $Bq.kg^{-1}$ dw in Southern Finns [MUS 84].

According to the SFRP [SFR 12], annual intakes of the main radionuclides are approximately 36,000, 800 and 40 $Bq.year^{-1}$, respectively, for potassium-40, rubidium-87 and polonium-210 (1,400 for the latter radionuclide in Arctic populations). For cosmogenic radionuclides, the average annual intake is 20,000, 1,000, 500 and 50 $Bq.year^{-1}$, respectively, for carbon-14, beryllium-7, tritium and sodium-22.

Five natural radionuclides permeate the human body: potassium-40, carbon-14, rubidium-87, tritium and radium-226, with an average of 4,500, 3,700, 450, 40 and 13 Bq, respectively, for a 20–25 year-old man[2] [GRA 06, SFR 12]. As a result, the radioactivity of a 70 kg man is approximately 8,700 Bq (Figure 5.4). The concentration of radium in the human body varies from 0.05 to $0.32.10^{-9}$ g [SPI 56].

**Figure 5.4.** *Radionuclide bioaccumulation in the human body*

## 5.1.6. *Bioaccumulation of artificial radionuclides*

Bioaccumulation of artificial radionuclides will obviously be highly variable, depending on the content of these various radionuclides in the anthroposphere.

2 These values vary with three parameters: gender, weight and age. For men, the maximum K content is reached at 20–25 years of age (approximately equal to 2 $g.kg^{-1}$) and decreases with age (1.4 $g.kg^{-1}$). For women, at 20–25 years of age, the maximum K content is 1.6 $g.kg^{-1}$.

The main radionuclides involved in a reactor accident are iodine radionuclides, cesium-134 and 137, strontium-90 and plutonium radionuclides (Figure 5.4)[3].

The main parameters determining radionuclide contamination for members of the public have been the subject of ICRP publications 56, 67, 69 and 71 [ICR 89, ICR 93, ICR 95a, ICR 95b]. The ICRP has recently updated the main biokinetic and internal exposure parameters for a large number of radionuclides for nuclear workers. ICRP definitions of the reference adult male and reference adult female have been summarized in publications 89 and 110 [ICR 02, ICR 09].

Publication 145 [ICR 20b] describes the construction of mesh-type reference computational phantoms (MRCPs) for adults, which are the modeling counterparts of voxel-type reference computational phantoms. However, even though they provide anatomically more realistic representations of internal anatomy than the old stylized phantoms, voxel phantoms have their limitations, mainly due to voxel resolution, particularly with regard to small tissue structures (e.g. the lens of the eye) and very thin tissue layers (e.g. the stem cell layers of the stomach wall mucosa and intestinal epithelium). There are a series of 10 phantoms to simulate children of both sexes at birth (newborn), aged 1, 5, 10 and 15 years. The publication presents the development of age-specific absorbed fractions (SAFs) for internally emitted photons, electrons, alpha particles and neutrons, in a manner similar to the SAF values for adults [ICR 20a].

Dose coefficients for assessing internal exposure to radionuclides after inhalation and ingestion require the use of biokinetic models (Publication 130, [ICR 15]), radionuclide decay pattern data (Publication 107, [ICR 08]) and SAF values in a calculation phantom [ICR 16b].

A series of five reports replacing publications 30 and 68 have been and will be published by the ICRP to provide revised dose coefficients for occupational intakes of radionuclides by inhalation and ingestion. The first report [ICR 15] summarizes the concepts and methods adopted by the ICRP. The second report [ICR 16a] provides data for the elements hydrogen (H), carbon (C), phosphorus (P), sulfur (S), calcium (Ca), iron (Fe), cobalt (Co), zinc (Zn), strontium (Sr), yttrium (Y), zirconium (Zr), niobium (Nb), molybdenum (Mo) and technetium (Tc). The third publication in the series provides data for the elements ruthenium (Ru), antimony (Sb), tellurium (Te), iodine (I), cesium (Cs), barium (Ba), iridium (Ir), lead (Pb), bismuth (Bi), polonium (Po), radon (Rn), radium (Ra), thorium (Th) and uranium (U) [ICR 17]. The fourth publication contains data on the elements lanthanum (La),

3 The proportion between the two cesium isotopes depends on the time the fuel is in the reactor. The longer the time, the higher the activity of $^{134}$Cs, as it is an activation product, whereas $^{137}$Cs is a fission product. At Chernobyl, cesium-134 activity was twice that of cesium-137. At Fukushima, the opposite was true: 134 was twice as low as 137.

cerium (Ce), praseodymium (Pr), neodymium (Nd), promethium (Pm), samarium (Sm), europium (Eu), gadolinium (Gd), terbium (Tb), dysprosium (Dy), holmium (Ho), erbium (Er), thulium (Tm), ytterbium (Yb), lutetium (Lu), actinium (Ac), protactinium (Pa), neptunium (Np), plutonium (Pu), americium (Am), curium (Cm), berkelium (Bk), californium (Cf), einsteinium (Es) and fermium (Fm) [ICR 19]. Finally, the fifth publication summarizes data for the elements beryllium, fluorine, sodium, magnesium, aluminum, silicon, chlorine, potassium, scandium, titanium, vanadium, chromium, manganese, nickel, copper, gallium, germanium, arsenic, selenium, bromine, rubidium, rhodium, palladium, silver, cadmium, indium, tin, hafnium, tantalum, tungsten, rhenium, osmium, platinum, gold, mercury, thallium, astatine and francium [ICR 22].

#### 5.1.6.1. *Radioactive iodine*

An IPSN synthesis [IPS 01a] for iodine-129 indicates that this dietary radionuclide is rapidly absorbed (90% in 1 hour) from the stomach and small intestine into the blood plasma, and that inhaled iodine is rapidly transferred in its entirety to the blood. The dietary absorption factors ($f_A$) are 1, 0.2 and 0.01, respectively, for F, M and S for workers [ICR 17]. For members of the public, the biokinetic model for iodine is described in ICRP Publication 56 [ICR 89]. It assumes that 30% of the iodine fraction present in the blood accumulates in the thyroid and that the remaining 70% is eliminated directly in the urine. The biological half-life in blood is 0.25 days. Iodine radionuclides incorporated into thyroid hormones leave the thyroid with a half-life of approximately 80 days, and enter other tissues where they are retained with a half-life of 12 days. Of the iodine radionuclides leaving the thyroid, 80% enter the bloodstream and 20% are excreted in the feces in organic form. Those in the blood can then either be reabsorbed by the thyroid, or excreted in the urine.

For an adult European with normal thyroid function, the fraction captured by the thyroid is between 20% and 25%. Thyroid dysfunction may modify this value. Hypothyroid adults have low thyroid incorporation. Hyperthyroid adults, on the contrary, have a reduced biological iodine half-life in the thyroid. The effective (physical and biological) half-life of iodine-129 is 120 days for the whole body. In workers, the absorption ($f_A$) of F forms is 1 [ICR 17].

#### 5.1.6.2. *Cesium-137*

Cesium behaves like its chemical analog potassium. Gastrointestinal absorption is high (approximately 98%), but may be lower from foods such as game (78%) [HEN 89]. Similarly, mother-to-fetus transfer is very high, while skin absorption is very low. Cesium distribution in the body is relatively homogeneous, but due to muscle mass, it is found in large quantities in the muscles (80%). In children contaminated by the Chernobyl accident, cesium bioaccumulated more than in adults, preferentially in the thyroid gland, adrenals, pancreas and kidneys [BAN 03]. The presence of cesium in the thyroid is due to the fact that cesium and iodine migrate very easily into the fuel (outline), reaching the cladding and forming cesium iodide [LEF 86, ZER 15]. Indeed, once the nuclear fuel cladding is destroyed during reprocessing, cesium iodide salts are released into the environment and are available for absorption by human thyroids.

As a result of cesium-137 contamination of lichens and reindeer eating these plants, the average muscle burden of Lapps consuming reindeer meat rose between 1963 and 1965 from 18.5 to 33.3 kBq for men and from 8.9 to 16.1 kBq for women [LID 67], with peak values in 1965 of 64–75 kBq. A second peak occurred following the Chernobyl accident, leading to new contamination problems. "Pragmatically", the intervention level for reindeer meat was raised from 600 to 6000 Bq $kg^{-1}$.

The concentration of $^{137}Cs$ and $^{40}K$ was determined in the breast milk of Italian women. Abballe et al. [ABB 08] consider that the health risk for breast-fed infants is not significant. However, this study dates from 1998 to 2000, more than 10 years after the Chernobyl accident. It shows that the persistence of cesium mobility is limited.

$^{137}Cs$ content in the liver and lungs of Lapps averaged 79.9 and 43.7 $Bq.kg^{-1}$ dw, respectively, between 1977 and 1979. In southern Finland, the $^{137}Cs$ content of liver and lungs was less than 1.9 $Bq.kg^{-1}$ [MUS 84].

In its stable form, cesium ($^{133}Cs$) is retained only in very small quantities by humans. The ICRP estimates the mass of stable cesium present in an adult at 10 µg, compared with stable potassium masses of 140 g in men and 90 g in women [IPS 01d]. As body weight increases, the biological half-life of cesium increases with age (20 days for children and 150 days for adults). It is higher in men (150 days) than in women (100 days). For workers, the ICRP assumes 110 days [ICR 17]. A linear relationship links the biological half-life of cesium to total potassium and muscle mass. Urinary excretion of cesium outweighs fecal excretion. 80% of cesium is stored in skeletal muscle mass [ICR 17].

#### 5.1.6.3. *Strontium-90*

The kinetic model for strontium is described in ICRP Publication 67 [ICR 93]. Strontium is mainly deposited in the two main bone structures (trabecular or cancellous and cortical or compact), where it is retained. It is also retained in the liver and other soft tissues.

Kinetic studies show the qualitative similarity of calcium and strontium distributions. However, organisms discriminate quantitatively between the two metabolic processes, to the advantage of calcium [IPS 01b].

The food chain is the main route of strontium intake by humans, and according to Galle [GAL 97], inhalation represents only a minor contribution.

Strontium is mainly deposited in trabecular and cortical (or compact) bone, where it is retained. It is also retained in liver and other soft tissues. Kinetic studies show that the distributions of calcium and strontium are qualitatively similar. Quantitatively, however, the body discriminates between the two metabolic processes, to the advantage of calcium. The food chain is the main route of strontium intake by humans, with inhalation representing only a minor contribution [GAL 97]. Its biological half-life in human is about 10 years. Absorption factors ($f_A$) are 0.25, 0.05 and 0.0025 for F, M and S, respectively [ICR 16a].

#### 5.1.6.4. *Cobalt*

Absorption of cobalt by the digestive tract varies between 20% and 95% of the quantity ingested, depending on the age of the individual. The preferred site of retention of assimilated cobalt is the liver, although significant bone retention may occur. Of the total cobalt assimilated, half is directly eliminated within 0.5 days, 5% reaches the liver, and the balance (45%) is homogeneously distributed in the tissues. Of the cobalt transferred to tissues, 60% is eliminated within six days, 20% within 60 days, and the remaining 20% within 800 days. Cobalt is eliminated mainly in the urine. Absorption factors ($f_A$) are 0.1, 0.02 and 0.001 for F, M and S, respectively [ICR 16a].

#### 5.1.6.5. *Ruthenium-106*

Historically, one of the first concerns about anthropogenic irradiation of a civilian population arose in 1959 in the Irish Sea, with the contamination of seaweed by ruthenium-106 from liquid discharges from the Windscale (now Sellafield) reprocessing plant. The plant discharged $74.10^{12}$ Bq per month of $^{106}$ Ru, resulting in a concentration of $3.10^{-3}$ $Bq.cm^{-3}$ in seawater. The seaweed *Porphyra* concentrates more than 1,800 times the concentration of $^{106}$Ru present in seawater, i.e. a concentration of 5,550 $Bq.kg^{-1}$. The Welsh traditionally used this seaweed to make laver bread, consuming an average of 16 g a day. The concentration of $^{106}$Ru in the laver bread was 2,590 $Bq.kg^{-1}$, resulting in a radiation dose of 7 mSv per year. The gastrointestinal absorption coefficients used by the ICRP are 0.05, 0.01 and $5.10^{-4}$ for types F, M and S, respectively [ICR 17].

#### 5.1.6.6. *Phosphorus-32*

Another early example can be found in the United States, where the first plutonium-bearing reactors discharged cooling water directly into the Columbia River, in particular phosphorus-32 which was discharged at a rate of 555 $Bq.L^{-1}$ before dilution by the waters of the Columbia River. Fish were contaminated with this $^{32}P$ at a rate of 555 $Bq.kg^{-1}$ in winter and 185,000 $Bq.kg^{-1}$ in summer, i.e. an average of 9,250 $Bq.kg^{-1}$ of fish flesh. The concentration factor varied from 1,900 to 4,200 [WAT 70]. Heavy fish consumers annually ate 200 fish-based meals, i.e. approximately 40 kg per year, with an average $^{32}P$ concentration for the most commonly consumed species of 1,480 $Bq.kg^{-1}$. This resulted in an irradiation dose of 3 mSv to the nervous system. The ICRP's gastrointestinal absorption coefficient is 0.8 for all phosphorus compounds [ICR 16a].

#### 5.1.6.7. *Iridium*

Iridium absorbed from the digestive tract is transferred 20% to the liver, 4% to the kidneys, 2% to the spleen, 54% uniformly to all other organs, and the remaining 20% is eliminated in the excreta. Iridium deposited in an organ or tissue is eliminated according to two kinetics: a fast one for 20% with a biological half-life of eight days and a slow one for 80% with a biological half-life of 200 days [COO 82]. The ICRP uses absorption factors of 0.01 and 0.002 for F and M, respectively [ICR 17].

#### 5.1.6.8. *Uranium*

Uranium is poorly absorbed and eliminated mainly via the feces. After systemic dispersion in the blood, soluble uranium salts remain in the kidneys (rat) for around 15 days before being eliminated by the urinary route. Bones are the preferred retention organs for uranium, with a biological half-life of 300–5,000 days [WRE 85]. An $f_A$ of 0.02 for soluble forms and 0.002 for insoluble forms is generally recommended [ICR 17].

#### 5.1.6.9. *Plutonium*

For the general public, the main routes of entry for plutonium are inhalation and ingestion. For workers, injuries can be a significant route of entry.

The fate of inhaled plutonium is estimated using the ICRP Publication 66 respiratory model [ICR 95a], which divides the human respiratory system into six zones (from the nose to the lymph nodes). The activity median aerodynamic diameter (AMAD) of aerosol particles is a key parameter in the deposition model [MÉT 10].

At the height of the nuclear testing era (1960s), dietary intake was more prevalent than inhalation: 2,400 and 280 mBq, respectively, for populations in the northern hemisphere, and 60 and 7 mBq 10 years later [ANS 09].

Our knowledge of the fate of plutonium in humans comes from human experiments on the systemic distribution of plutonium carried out by Wright Langham from 1945 to 1947 on 18 hospitalized people in what was presumed to be the terminal phase, as well as monitoring urinary elimination. In addition, we have post-mortem analyses of almost 2,000 people contaminated by radioactive fallout from atmospheric nuclear testing. There was no difference between men and women; the liver and skeleton were the major retention organs, and elimination was slow, taking several years [MÉT 10]. In the United States, there is a registry of people contaminated during their working lives (United States Transuranium and Uranium Registries, USTUR) (see: www.ustur.wsu.edu) [MÉT 10].

Concentrations of plutonium-239, plutonium-240, strontium-90 and total alpha emitters were measured in children's teeth from all over Great Britain and Ireland in 1995/1996. Concentrations of plutonium, but not strontium-90 or total alpha emitters, decreased with increasing distance from the Sellafield nuclear fuel reprocessing plant. However, the absolute concentrations of plutonium measured ($5 \pm 4$ mBq.kg$^{-1}$ ash weight) were so low that O'Donnell et al. [ODO 97] considered the radiological hazard insignificant.

The average $^{239,240}$Pu content in bones (ribs), liver and lungs of Lapps was 1.7, 15.5 and 1.2 mBq.kg$^{-1}$ dw, respectively. The average plutonium body burden of Lapps was 41 mBq between 1977 and 1979 [MUS 84].

#### 5.1.6.10. *Americium*

Americium is a moderately transferable compound, deposited mainly in the skeleton, liver and gonads of all species. As such, its behavior is similar to that of other transuranic elements. However, it differs from plutonium in that it has a shorter organ retention time and is less toxic, with the exception of americium-242m and 243, which are just as harmful as plutonium. The ICRP assumes a biological half-life of 20 years for humans, but extrapolates this from animal experiments. The absorption coefficient ($f_A$) used is $5.10^{-4}$ [ICR 19].

#### 5.1.6.11. *Neptunium*

Of the various possible ways of incorporating neptunium, inhalation is the most common mode of contamination for nuclear workers. The quantity of material absorbed after inhalation depends on various factors (particle size, specific surface area, physico-chemical form, etc.). According to the ICRP [ICR 95b], neptunium is classified as category M or S, depending on its physico-chemical form. Ingestion is the most common route of exposure for members of the public. Gastrointestinal absorption of neptunium is between 0.1% and 1% of the ingested mass. A transfer coefficient $f_1$ of $5.10^{-4}$ for humans is proposed by the ICRP [ICR 89]. However, absorption depends on numerous physiological (species, age of the animal, trophic status) and physical parameters (physico-chemical form of the ingested compound, mass and nature of the diet, etc.). In particular, absorption is maximal in newborns [PAQ 94]. The ICRP model [ICR 89] assumes that 50–70% of the neptunium absorbed by an adult is deposited in the skeleton, 5–10% in the liver, 1–4% in the kidneys and 3% in other soft tissues. The main route of elimination for this radionuclide is via the urine. Neptunium is present in large quantities in free form in the blood, and a third of this plasma concentration is eliminated within 24 hours [THO 82].

#### 5.1.6.12. *Thorium*

Thorium can enter the body through ingestion or dust inhalation, or more rarely through injury. Absorbed into the bloodstream, thorium is mainly deposited in the liver and skeleton. The latter is predominant, since 60% of the thorium inhaled or ingested by an individual is bound to bone. In children, thorium uptake by the body is four to eight times greater than in adults, resulting in greater bone uptake [ICR 93]. The absorption coefficient ($f_A$) used is $5.10^{-4}$ [ICR 19].

## 5.2. Protecting people from radionuclides

Various solutions have been devised to protect humans from the harmful effects of radionuclides, particularly when they enter the body. Historically, the search for radioprotectors began very early on. The distribution of potassium iodide (KI) tablets was recommended in the event of a nuclear accident. Similarly, the containment of people living in the vicinity of an accident was considered and used during the most recent nuclear accidents (Chernobyl and Fukushima).

### 5.2.1. *The search for chemical "radioprotectors"*

As early as 1945, the US Army began studies to find a chemical "radioprotector" (or chelator) that would prevent radionuclides from being bioaccumulated by military personnel operating in a radioactively contaminated atmosphere and soil, either to defend or conquer this polluted territory. Up to 1970, over 12,000 chemical compounds were tested. Similar studies were carried out by other nuclear-armed armies.

In France, research began in 1962 at the Centre de recherche du service de santé des armées (CRSSA). A good radioprotector immediately prevents lethality and ensures the soldier's survival in a highly hostile environment. The radioprotector must provide a good dose reduction factor (DRF). This is the ratio between the total irradiation doses for 50% of the population treated with the radioprotector and the control population ($DL_{50}$ in 30 days for small mammals). The ideal radioprotector would have a DRF factor equal to 2, be active at sub-toxic doses and have a prolonged action of at least six hours [FAT 81]. The main known radioprotectors are aminothiols or derivatives, as well as synthesized thiophosphates [WAS 76, PHI 77]. Despite the discovery of effective radioprotective compounds, the human application of chemical radioprotection remains difficult. The main reasons for this are the loss of activity at low administration doses and the short duration of action of these compounds.

### 5.2.2. *The use of potassium iodide tablets in the event of a nuclear accident*

The administration of iodine tablets is recommended by the IAEA in the event of a nuclear accident as soon as possible after the first radioactive release, or as a preventive measure shortly before the accident, as the effect is transient[4]. However, in the case of the last two major nuclear accidents, this distribution proved disastrous. In Fukushima, for example, the administration of the iodine tablets needed to prevent thyroid cancer proved highly complex. The Japanese authorities failed to manage their distribution properly, and the population did not know if and when to ingest them, which, combined with a breakdown in communications and a loss of confidence in the authorities, led to a chaotic situation. Prophylaxis based on KI did not work at all [BOI 12].

What is the real toxicity of KI? There is evidence of serious dysfunction following iodine intake in newborns and the elderly [SPA 12]. In the perinatal rat, the prophylactic use of stable iodine may have two combined actions: an interesting inhibitory effect on thyrocyte growth, and a toxic action resulting, depending on the stage of maturation, in either hyperfunction or follicular destruction, but without constituted hypothyroidism [BOU 00]. In humans, two categories in addition to children require special attention: people allergic to iodine and adults over 40, where iodine intake is not recommended. For pregnant women, on the contrary, there would be no problem [CSS 15, AMI 20]. The potential adverse effects of iodine thyroid blocking (ITB) have not been systematically studied to date [SPA 12]. This is one of the reasons why the IRSN launched the PRIODAC project in 2014. The aims of this project are 1) to assess the side effects associated with repeated intakes of KI, 2) to gain a better understanding of the molecular mechanisms regulating

---

4 Ideally, stable iodine should be taken before discharge. After six hours, taking stable iodine has virtually no effect, as iodine passes very quickly into the bloodstream and the incorporated radioactive iodine has already settled in the thyroid.

iodine metabolism, 3) to revise the French regulatory MA (market authorization) of 65 mg KI tablets and 4) to develop new recommendations linked to the administration of KI towards greater international harmonization [BEN 18].

### 5.2.3. *Short and extended containment*

In the event of a nuclear accident, containment facilities in the homes of local residents are not equipped for this function, and containment time is necessarily short. On the contrary, for longer confinement periods, it is necessary to use nuclear shelters, if they are equipped with air filtration. Containment requires all air intakes to be caulked, all ventilation in homes to be completely eliminated and all traffic to be banned. Problems became insoluble after 10 days, in particular due to shortages of water and medication [COH 22]. Furthermore, confinement caused a significant proportion of medical staff to flee the disaster zone [BOI 12].

### 5.2.4. *Radiation protection is all about prevention*

While the administration of iodine tablets can be beneficial for children, we must not lose sight of the fact that a nuclear accident releases large quantities of a wide variety of radionuclides into the environment, and not just radioactive iodine isotopes.

## 5.3. Therapeutic removal of radionuclides from humans

There is currently no effective therapeutic treatment to induce the removal of uranium from humans [PAQ 09].

For soluble plutonium, there is an effective treatment based on diethylene triamine penta-acetic acid (DTPA) [ANS 09]. This large molecule acts as a chelator to which the radionuclide binds. The biological half-life of the radionuclide is thus reduced to that of the chelator, which is very short. The same applies to several other radionuclides, such as americium, cobalt and thorium. The recommended treatment for contamination is 0.5 g DTPA, renewable if necessary. This treatment accelerates the natural elimination of these radionuclides, whatever the route of incorporation or initial chemical form. The earlier they are administered, the more effective their excretion, but this depends on the organ in question. DTPA is easily excreted from blood and extracellular fluids and prevents deposition in target organs [BHA 95]. In the absence of well-established and proven therapeutics, this treatment is also recommended in cases of iridium contamination, since DTPA complexes this radionuclide and induces its urinary elimination.

In the case of cobalt, it may be possible to administer cobalt gluconate (Cobalt oligosol) sublingually [OPR 97].

The treatment of cesium consists of the ingestion of Prussian Blue (iron ferrocyanide), which insolubilizes cesium in the intestinal lumen and induces its excretion. Other products (ammonium or potassium iron hexacyanoferrate) have been tested and appear to be more effective [NEI 91]. Pectin, a natural substance, has proved to be a good chelator for cesium. This product is used by the association Chernobyl Children International, founded by Vassili Nesterenko. It is still used to treat children in Belarus, in particular, who still live in contaminated areas. In the event of skin contamination, cesium is difficult to eliminate, but its biological half-life on the skin is short, at around four days.

In the event of radioactive iodine contamination, a special protective measure is to administer stable iodine orally, to prevent the fixation of radioactive iodine by iodine saturation of the thyroid gland. Stable iodine is administered via 130 mg KI tablets, i.e. 100 mg stable iodine (2 x 65 mg tablets). This administration must be carried out as soon as possible, and no later than a few hours after contamination. According to the French interministerial circular of March 10, 2000 on the revision of special response plans for basic nuclear infrastructures, it is recommended for a calculated thyroid dose greater than or equal to 100 mSv. The dosage is one tablet for adults, children over 12 and pregnant women, half a tablet for children aged 3–12 and a quarter tablet for children under 3 years of age. In the event of a major release of radioactive iodine, the other emergency measure is to evacuate the population from the area through which the cloud passes.

For neptunium, treatment involves minimizing body diffusion of the radionuclide from its route of entry, and then promoting its excretion by natural means. DTPA is not recommended, as it promotes skeletal deposition. The use of 3,4,3-LIPOHO would be more effective, but has not yet been approved for marketing [PAQ 94]. Natural excretion of neptunium is low, and elimination kinetics depend on the organ and mode of administration [PAQ 94]. The ICRP [ICR 93] proposes a behavior of neptunium similar to that of plutonium, but results from animal experiments tend to show different results [IPS 01c].

Ansoborlo et al. [ANS 06] review current knowledge of actinide speciation in human biological media. These parameters are essential for a good understanding of the retention and excretion of these radionuclides, and for the design of removal therapies.

In the case of internal strontium contamination, treatment consists of reducing intake by ingesting sodium alginate and stimulating excretion. This can be achieved by isotope dilution with calcium gluconate or strontium lactate, or by acidification with ammonium chloride, which will promote strontium excretion [OPR 97].

## 5.4. Conclusions

Humans are exposed to natural contamination of their environment by radionuclides of cosmic or telluric origin. The level of radioactivity in the environment has been highly variable over time. For example, when life first appeared on Earth, natural radioactivity from uranium was twice as high as it is today. Similarly, following the numerous American and Soviet tests of atomic bombs in the atmosphere in the early 1960s, environmental contamination, especially in the northern hemisphere, peaked in 1963 for various artificial radionuclides such as cesium-137, strontium-90 and various plutoniums.

Exposure to natural radioactivity is permanent. Spread out over days, months and years, it occurs at a low dose rate. It is said to be chronic. Our body's cells probably repair its effects better than those of equivalent irradiation concentrated over a short period of time. A person's exposure to this natural radioactivity varies enormously from one individual to another, depending not only on their tastes for travel or mountaineering, or habits when it comes to airing the house, but also on where they live. In certain regions, such as Ramsar in Iran, soil contamination is much higher than the global average.

People's eating habits will have a considerable impact on their natural radionuclide contamination. This is particularly true for some coastal dwellers, who generally consume far more seafood than mainlanders. If their shellfish consumption is high, their polonium-210 contamination will be high.

The use of nuclear energy leads to anthropogenic contamination of the environment, and consequently of human beings, with various artificial radionuclides and with the additional natural contamination of some natural radionuclides. This increase in contamination during routine use of nuclear energy is small compared with natural contamination. However, in the event of accidents at any stage of the nuclear fuel cycle, this additional contamination, both natural and artificial, can be significantly higher than natural contamination (see Chapter 12).

The exposure pathways for this artificial contamination are external and internal. External exposure is mainly due to contamination of the atmosphere and soil. The internal routes are mainly inhalation of contaminated air and ingestion of artificially radioactive food and drinking water.

The transfer coefficients of radionuclides from a foodstuff to humans will vary greatly depending on the nature of the radionuclide and, in particular, its chemical speciation. Culinary processing of a food may have a significant effect on the quantitative aspect of its transfer to humans.

The world's regions are variously contaminated by radionuclides, either naturally or artificially as a result of industrial or military activities, for example. As a result, human exposure to natural and artificial radionuclides varies widely, depending on people's habits and where they live.

In some cases, the radionuclide that has entered an individual's body can be more or less rapidly and completely excreted. This is highly dependent on the chemical nature of the radionuclide.

The incorporation kinetics, tissue distribution (or organotropism) and elimination rate (biological half-life) of a radionuclide accumulating in an individual will also be highly dependent on the nature of the radionuclide.

Various regulatory values for the concentration of certain radionuclides in foodstuffs have been laid down in several countries and in the European Union, particularly following the Chernobyl accident. The same process was repeated in Japan after the Fukushima accident.

## 5.5. References

[ABB 08] ABBALLE A., BALLARD T.J., DELLATTE E. et al., "Persistent environmental contaminants in human milk: Concentrations and time trends in Italy", *Chemosphere*, vol. 73, pp. S220–S227, 2008.

[AGH 06] AGHAMIRI S.M.R., GHORBANI Z., DARAFSHEH A. et al., "$^{226}$Ra concentration in the teeth of habitants of areas with high level of natural radioactivity in Ramsar", *Journal of Environmental Radioactivity*, vol. 89, pp. 212–218, 2006.

[AMI 17] AMIARD J.C., *Les risques chimiques environnementaux. Méthodes d'évaluation et impacts sur les organismes*, 2nd edition, Lavoisier, Tec&Doc, Paris, 2017.

[AMI 20] AMIARD J.C., *Nuclear Accidents: Prevention and Management of an Accidental Crisis*, vol. 3, ISTE, London, and John Wiley & Sons, New York, 2020.

[ANS 06] ANSOBORLO E., PRAT O., MOISY P. et al., "Actinide speciation in relation to biological processes", *Biochimie*, vol. 88, pp. 1605–1618, 2006.

[ANS 09] ANSOBORLO E., PAQUET F., DEN AUWER C. et al., "Plutonium", in MÉNAGER M.T., GARNIER-LAPLACE J., GOYFFON M. (eds), *Toxicologie nucléaire environnementale et humaine*, Lavoisier, Paris, pp. 445–467, 2009.

[BAN 03] BANDAZHEVSY Y.I., "Chronic Cs-137 incorporation in children's organs", *Swiss Medical Weekly*, vol. 133, pp. 488–490, 2003.

[BEN 18] BENDERITTER M., POURCHER T., MARTIN J.C. et al., "Do multiple administrations of stable iodine protect population chronically exposed to radioactive iodine: what is PRIODAC research program (2014–22) teaching us?", *Radiation Protection Dosimetry*, pp. 1–13, 2018.

[BHA 95] BHATTACHARYYA M.H.., BREITENSTEIN B.D.., MÉTIVIER H. et al., *Traitement de la contamination accidentelle des travailleurs*, Editions de Physique, Paris, 1995.

[BLU 01] BLUST R., "Radionuclide Accumulation in Freshwater Organisms", in VAN DER STRICHT E. and KIRCHMANN R. (eds), *Radioecology. Radioactivity and Ecosystems*, UIR, Liège, pp. 275–302, 2001.

[BOI 12] BOILLEY D., "Les plans d'urgence et d'évacuation", *Enseignements de Fukushima*, Greenpeace France, Paris, pp. 15–27, March 2012.

[BOU 00] BOURAHLA K., Toxicité chimique et radiologique des isotopes de l'iode étude expérimentale chez le rat en période périnatale, Thesis, Paris XII, 2000.

[CAR 11] CARVALHO F.P., "Polonium ($^{210}$Po) and lead ($^{210}$Pb) in marine organisms and their transfer in marine food chains", *Journal of Environmental Radioactivity*, vol. 102, pp. 462–472, 2011.

[CAT 09] CATTY P., VERRET F., LECOMTE-PRADINES C. et al., "Transport membranaire des métaux et métalloïdes", in Ménager M.T., Garnier-Laplace J., Goyffon M. (eds), *Toxicologie nucléaire environnementale et humaine*, pp. 65–82, Lavoisier, Paris, 2009.

[COH 22] COHEN F., BEHAR A., "La trilogie des " protecteurs " en cas d'attaque nucléaire, et leurs limites", *Médecine & guerre nucléaire*, vol. 37, no. 2, pp. 12–14, 2022.

[COL 01] COLLE C., GARNIER-LAPLACE J., ROUSSEL-DEBEL S. et al., "Comportement de l'uranium dans l'environnement", in MÉTIVIER H. (ed.), *L'uranium, de l'environnement à l'homme*, EDP Sciences, Les Ulis, 2001.

[COO 82] COOL D.A., "Committed dose equivalents from an accidental inhalation of insoluble 192Ir", *Health Physics*, vol. 42, pp. 380–386, 1982.

[CSS 15] CSS (CONSEIL SUPÉRIEUR DE LA SANTÉ), Publication du Conseil Supérieur de la Santé No. 9275. Accidents nucléaires, environnement et santé à l'ère post-Fukushima ; Partim : Protection de la thyroïde, 4 March 2015.

[DUB 09a] DUBLINEAU I., MÉSANGER M.T., "Mécanismes d'interactions aux niveaux moléculaire et cellulaire. Introduction", in MÉNAGER M.T., GARNIER-LAPLACE J., GOYFFON M. (eds), *Toxicologie nucléaire environnementale et humaine*, Lavoisier, Paris, pp. 155–156, 2009.

[DUB 09b] DUBLINEAU I., PETITOT F., LESTAEVEL P. et al., "Transfert à travers les barrières épithéliales et cérébrales", in MÉNAGER M.T., GARNIER-LAPLACE J., GOYFFON M. (eds), *Toxicologie nucléaire environnementale et humaine*, pp. 133–152, Lavoisier, Paris, 2009.

[FAT 81] FATOME M., "La radioprotection chimique", *Radioprotection*, vol. 16, no. 2, pp. 113–124, 1981.

[GAL 97] GALLE P., *Toxiques nucléaires*, 2nd edition, Masson, Paris, 1997.

[GAM 97] GAMBINI D.J., GRANIER R., *Manuel pratique de radioprotection*, Lavoisier, Tec&Doc, Paris, 1997.

[GRA 06] GRASTILLEUR C., "La surveillance des radionucléides dans les aliments par la direction générale de l'alimentation", *Bulletin épidémiologique de l'AFSSA*, vol. 23, pp. 6–7, 2006.

[HEN 89] HENRICHS H., PARETZKE H.G., VOIGT G. et al., "Measurements of Cs absorption and retention in man", *Health Physics*, vol. 57, pp. 571–578, 1989.

[ICR 75] ICRP, Report on the task group on reference man, ICRP Publication 23, Pergamon Press, Oxford, 1975.

[ICR 89] ICRP, "ICRP Publication 56, Age-dependent doses to members of the public from intake of radionuclides: Part 1", *Annals of the ICRP*, vol. 20, no. 2, pp. 1–122, 1989.

[ICR 91] ICRP, "ICRP Publication 67, Age-dependent doses to members of the public from intake of radionuclide: Part 2: Ingestion Dose Coefficients", *Annals of the ICRP*, vol. 23, nos. 3–4, pp. 1–167, 1991.

[ICR 93] ICRP, "ICRP Publication 67, Age-dependent doses to members of the public from intake of radionuclides. Part 2: Ingestion dose coefficients", *Annals of the ICRP*, vol. 23, pp. 3–4, 1993.

[ICR 95a] ICRP, "ICRP Publication 69, Age-dependent doses to members of the public from intake of radionuclides: Part 3: Ingestion dose coefficients", *Annals of the ICRP*, vol. 25, no. 1, pp. 1–74, 1995.

[ICR 95b] ICRP, "ICRP Publication 71, Age-dependent doses to members of the public from intake of radionuclides: Part 4: Inhalation dose coefficients", *Annals of the ICRP*, vol. 25, no. 3, pp. 1–405, 1995.

[ICR 02] ICRP, "ICRP Publication 89, Basic anatomical and physiological data for use in radiological protection reference values", *Annals of the ICRP*, vol. 32, pp. 3–4, 2002.

[ICR 06] ICRP, "ICRP Publication 100, Human alimentary tract model for radiological protection", *Annals of the ICRP*, vol. 36, pp. 1–164, 2006.

[ICR 08] ICRP, "ICRP Publication 107, Nuclear decay data for dosimetric calculations", *Annals of the ICRP*, vol. 38, no. 3, 2008.

[ICR 09] ICRP, "ICRP Publication 110, Adult reference computational phantoms", *Annals of the ICRP*, vol. 39, no. 2, 2009.

[ICR 15] ICRP, "ICRP Publication 130, Occupational intakes of radionuclides: Part 1", *Annals of the ICRP*, vol. 44, no. 2, pp. 3–188, 2015.

[ICR 16a] ICRP, "ICRP Publication 134, Occupational Intakes of radionuclides: Part 2", *Annals of the ICRP*, vol. 45, nos. 3-4, pp. 1–352, 2016.

[ICR 16b] ICRP, "ICRP Publication 133, The ICRP computational framework for internal dose assessment for reference adults: specific absorbed fractions", *Annals of the ICRP*, vol. 45, no. 2, pp. 1–74, 2016.

[ICR 17] ICRP, "ICRP Publication 137, Occupational intakes of radionuclides: Part 3", *Annals of the ICRP*, vol. 46, nos. 3–4, pp. 1–487, 2017.

[ICR 19] ICRP, "ICRP Publication 141, Occupational intakes of radionuclides: Part 4", *Annals of the ICRP*, vol. 48, nos. 2–3, pp. 1–514, 2019.

[ICR 20a] ICRP, "ICRP Publication 143, Paediatric reference computational phantoms", *Annals of the ICRP*, vol. 49, no. 1, pp. 1–293, 2020.

[ICR 20b] ICRP, "ICRP Publication 145, Adult mesh-type reference computational phantoms", *Annals of the ICRP*, vol. 49, no. 3, pp. 1–201, 2020.

[ICR 22] ICRP, "ICRP Publication 151, Occupational intakes of radionuclides: Part 5. ICRP", *Annals of the ICRP*, vol. 51, nos. 1–2, 2022.

[IPS 01a] IPSN, Fiche radionucléide. Iode 129. Aspects sanitaires, 2001.

[IPS 01b] IPSN, Fiche radionucléide. Strontium 90 et yttrium 90. Aspects sanitaires, 2001.

[IPS 01c] IPSN, Fiche radionucléide. Neptunium 237. Aspects sanitaires, 2001.

[IPS 01d] IPSN, Fiche radionucléide. Césium 137 + Barium 137m. Aspects sanitaires, 2001.

[LEF 86] LEFÈVRE J., *Les déchets nucléaires*, Eyrolles, Paris, 1986.

[LEM 03] LEMERCIER V., MILLOT X., ANSOBORLO E., MÉNÉTRIER F., FLÜRY-HÉRARD A., ROUSSELLE C., SCHERRMANN J.M.., "Study of uranium transfer across the blood-brain barrier", *Radiat Prot Dosimetry*, vol. 105, pp. 243–245, 2003.

[LID 67] LIDEN K., GUSTAFSSON M., "Relationships and seasonal variation of Cs 137 in lichen reindeer and man in northern Sweden 1961-1965", in ABERG B., HUNGATE F.P. (eds), *Radiological Concentration Processes*, Pergamon Press, Oxford, pp. 193–208, 1967.

[MÉT 10] MÉTIVIER H., *Plutonium. Mythes et réalités*, EDP Sciences, Les Ulis, 2010.

[MUS 84] MUSSALO-RAUHAMAA H., JAAKKOLA T., MIETTINEN J.K. et al., "Plutonium in Finnish Lapps. An estimate of the gastrointestinal absorption of plutonium by man based on a comparison of the plutonium content of Lapps and southern Finns", *Health Physics*, vol. 46, pp. 549–559, 1984.

[NCR 07] NCRP, Development of a biokinetic model for radionuclide contaminated wounds and procedures for their assessment, dosimetry and treatment, National Council on Radiation Protection and Measurements, Report 156, Bethesda, MD, 2007.

[NEI 91] NEILSON P., DRESOW B., FISHER R. et al., "Inhibition of intestinal absorption and decorporation of radiocaesium in humans by hexacyanferrates", *International Journal of Radiation Applications and Instrumentation*, vol. 18, pp. 821–826, 1991.

[ODO 97] O'DONNELL R.G., MITCHELL P.I., PRIESTB N.D. et al., "Variations in the concentration of plutonium, strontium-90 and total alpha-emitters in human teeth collected within the British Isles", *Science of the Total Environment*, vol. 201, pp. 235–243, 1997.

[OPR 97] OPRI, Intervention médicale en cas d'accident radiologique, report, 1997.

[PAQ 94] PAQUET F., MÉTIVIER H., MASSE R., "Le neptunium : pourquoi de nouvelles études ?", *Radioprotection*, vol. 29, pp. 387–396, 1994.

[PAQ 09] PAQUET F., ADAM-GUILLEMIN C., ANSOBORLO E. et al., "Uranium", in MÉNAGER M.T., GARNIER-LAPLACE J., GOYFFON M. (eds), *Toxicologie nucléaire environnementale et humaine*, Lavoisier, Paris, pp. 411–443, 2009.

[PHI 77] PHILLIPS T.L. "Chemical modification of radiation effects", *Cancer*, vol. 39, pp. 987–999, 1977.

[SAM 05] SAMAVATA H., SEAWARD M.R.D., AGHAMIRI S.M.R. et al., "$^{210}$Po and $^{210}$Pb content in environmental and human body samples in the Ramsar area, Iran", *International Congress Series*, vol. 1276, pp. 225–226, 2005.

[SAN 92] SANDERS C.L., DAGLE G.E., MAHAFFEY J.A., "Incidence of brain tumors in rats exposed to an aerosol of $^{239}PuO_2$", *International Journal of Radiation Biology*, vol. 62, pp. 97–102, 1992.

[SFR 12] SFRP, La radioactivité naturelle en dix épisodes, available at: http://www.sfrp.asso.fr/spip.php?rubrique28, 2012.

[SPA 12] SPALLEK L., KRILLE L., REINERS C. et al., "Adverse effects of iodine thyroid blocking: A systematic review", *Radiation Protection Dosimetry*, vol. 150, no. 3, pp. 267–277, 2012.

[SPI 56] SPIERS F.W., "Radioactivity in man and his environment. Presidential address", *The British Journal of Radiology*, vol. 29, no. 344, pp. 409–417, 1956.

[THO 82] THOMPSON R., "Neptunium – the neglected actinide: a review of the biological and environmental literature", *Radiation Research*, vol. 90, pp. 1–32, 1982.

[UNS 08] UNSCEAR, Sources and Effects of Ionizing Radiation, United Nations, New York, 2008.

[WAS 76] WASHBURN J.M., "Prediction of the effective radioprotective dose of WR 2721 in human through an intraspecies tissue distribution study", *Radiation Research*, vol. 66, pp. 100–105, 1976.

[WAT 70] WATSON D.G., CUSHING C.E., COUTANT C.C. et al., Radioecological studies on the Columbia river. Part I, BNWL 1377, 1970.

[WHO 04] WHO, Uranium in drinking-water, Background document for development of WHO guidelines for drinking-water quality, 2004.

[WRE 85] WRENN M.E., DURBIN P.W., HOWARD B. et al., "Metabolism of ingested uranium and radium", *Health Physics*, vol. 48, pp. 601–663, 1985.

[ZER 15] ZERBIB J.C., "Tchernobyl, effets sanitaires et environnementaux", *Les cahiers de Global Chance*, vol. 37, pp. 42–62, 2015.

# 6

# Estimates of Human Radiation Doses

## 6.1. Introduction: from exposure to dose

In assessing the risk associated with the presence of radionuclides in the human environment [ADA 09, AMI 17], after the initial stages (hazard assessment, identification of exposure routes and quantification of radioactive contamination of the anthroposphere and humans), the dose received by humans as a result of the various exposures must be evaluated as accurately as possible.

The human exposure dose is calculated in three stages. Once the absorbed dose has been quantified, the equivalent dose is calculated, followed by the effective dose (Figure 6.1).

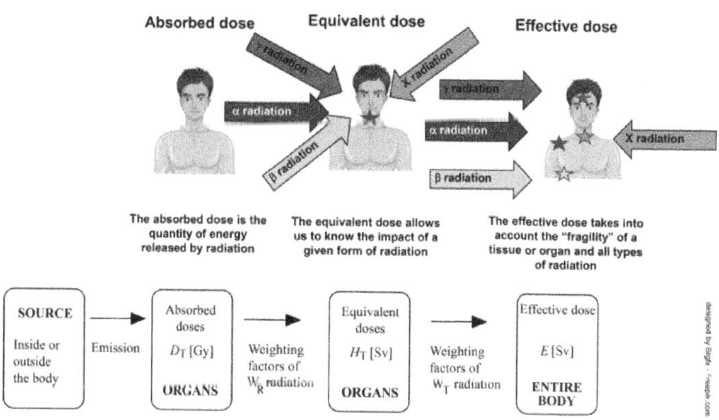


**Figure 6.1.** *Steps in calculating effective dose*

This chapter describes the approach used to calculate these three doses.

## 6.2. Expressing exposure

The radiation inducing one unit of electrostatic charge (esu) in a one-kilogram sample of dry air at normal pressure and temperature is expressed in coulomb per kilogram ($C.kg^{-1}$). This unit replaces the roentgen (R), which considered a volume of one cubic centimeter. In a standard atmosphere, the equivalence is 1 R $\approx$ 0.258 $mC.kg^{-1}$ and 1 $mC.kg^{-1} \approx$ 3.88 R [BIA 11].

### 6.2.1. *Absorbed dose*

The absorbed dose from ionizing radiation is officially expressed in Gray (Gy), the international system's standardized unit equivalent to one joule per kilogram ($J.kg^{-1}$).

However, in the past, they were calculated using the rad (abbreviation for radiation absorbed dose) as a unit. One rad corresponds to 100 $erg.g^{-1}$, i.e. 100 erg of energy absorbed by one gram of a given biological tissue. One erg is equal to 1 ten-millionth of a joule. When time is a factor, the expression is made in dose rate (or dose per unit of time). An example of a dose rate unit is the millirad per hour ($mrad.h^{-1}$)

The old unit, the roentgen (R), measured the amount of ionization in the air caused by the radioactive decay of atomic nuclei. In non-bone biological tissue, one roentgen is roughly equivalent to 0.93 rad. In air, one roentgen is equal to 0.87 rad. Dials calibrated in $mR.h^{-1}$ indicate milliroentgen per hour.

In humans, unlike other living beings, the estimation of radiation dose is not limited to the absorbed dose. In fact, two important factors – the energy of the radiation and the type of biological tissue – have a strong influence on radiosensitivity.

### 6.2.2. *Equivalent dose*

The nature and energy of ionizing radiation is estimated by linear energy transfer, and this is taken into account to calculate the equivalent dose.

#### 6.2.2.1. *Relative biological effectiveness (RBE)*

Relative biological effectiveness (RBE) is a measure used to compare the biological effect of two types of radiation. To characterize RBE, the reference generally used is X-radiation (X-photon) for a linear energy transfer (LET) of

3 KeV.μm$^{-1}$. RBE varies with a number of factors, including the organ exposed and the age of exposure. The choice of the RBE factor used is a tricky one, particularly for alpha emitters and certain betas. There is currently no consensus on the subject, and this remains a delicate point for doses delivered by radionuclides with a short linear path. In the literature, RBE values vary widely. This depends on the criterion taken into account (cell death, genetic damage, etc.), the tissue examined, the dose delivered (external or internal), the type of exposure (acute or chronic), and the energy of the alpha or beta particle.

Generally, an RBE of 20 is recommended for stochastic effects in humans, and an RBE of 5 for deterministic effects in non-human populations [UNS 96].

Some RBEs can be very high. For example, for mouse sperm irradiated with $^{210}$Po, an RBE of 245 was obtained by Rao et al. [RAO 91]. For hematopoietic tissue irradiated with $^{239}$Pu, an RBE ranging from 150 to 360 has been proposed by various authors [JIA 94, LOR 96]. In addition, alpha particles can trigger genomic instability that can be transmitted to offspring [LIT 98, WRI 98].

Similarly, in the case of tritium, a low-energy beta emitter that was until recently considered to be not very radiotoxic, more and more scientists believe that the RBE of 1 is unsuitable and that an RBE of 2 or even 5 should be used [GAZ 10].

#### 6.2.2.2. *Linear energy transfer (LET)*

Linear energy transfer (LET) is a quantity that describes the energy transferred by an ionizing particle through matter, per unit distance. It varies according to the nature and energy of the ionizing radiation. LET is typically used to quantify the effect of ionizing radiation on biological matrices. It has an important relationship with stopping power (average energy loss of the particle per distance traveled).

#### 6.2.2.3. *The weighting factor or consideration of RBE and LET*

The quality factor or weighting factor (Q or $W_R$) to take account of the type of radiation and its energy varies from 1 to 20 (Table 6.1). For neutrons, the ICRP (International Commission on Radiological Protection) authorizes the use of a continuous relationship law instead of discrete values.

The weighting factor is used to calculate the equivalent dose ($H_T$), whose unit is the joule per kilogram and whose name is the sievert (Sv). Formerly, the unit was the rem (1 rem = 1 rad x Q). The terminology “equivalent” comes from the fact that we calculate a dose equivalent to the dose delivered by $\gamma$ and X-rays (or $W_R$ = 1), which serve as a reference, to produce the same stochastic biological effect.

| Type of radiation | Weighting factor (WR) |
|---|---|
| Alpha | 20 |
| Beta | 1 |
| Gamma and X-rays | 1 |
| Neutrons (various energies) | 5–20 |
| Protons (various energies) | 1–5 |

**Table 6.1.** *Weighting factors ($W_R$) (according to [ICR 91])*

### 6.2.3. *Effective dose*

The second factor that determines the severity of the harmfulness of ionizing radiation in humans is the nature of the biological tissue receiving the radiation.

#### 6.2.3.1. *Tissue weighting factor*

The relationship between the probability of occurrence of stochastic effects and the equivalent dose is not the same from one irradiated organ or tissue to another, depending on their radiosensitivity. Different tissues or organs have different radiosensitivities. To take account of this, a tissue weighting factor is used. A limited number of dose conversion factors (DCF or $W_T$) are used for regulatory purposes to convert the energy released. This tissue weighting factor varies from 0.01 to 0.12 depending on the organ (Table 6.2).

| Tissue | Weighting factor (WT) |
|---|---|
| Colon | 0.12 |
| Stomach | 0.12 |
| Bone marrow | 0.12 |
| Spleen | 0.12 |
| Breast | 0.12 (0.05) |
| Gonads | 0.08 (0.20) |
| Other organs not referenced here | 0.08 |
| Bladder | 0.05 |
| Liver | 0.05 |
| Esophagus | 0.05 |
| Thyroid | 0.05 |
| Surface bone | 0.01 |
| Skin | 0.01 (0) |
| Salivary gland | 0.01 (0) |
| Brain | 0.01 (0) |
| Therefore, for the whole body a total of: | 1.00 |

**Table 6.2.** *Tissue weighting factor ($W_T$) from [ICR 07] (ICRP Publication 60 values in brackets ([ICR 91])*

#### 6.2.3.2. *Effective dose calculation*

Effective dose (E) is obtained after double weighting: effective dose = $W_R$ x $W_T$ x absorbed dose. Like the equivalent dose, it is expressed in sieverts (formerly rem, Röntgen equivalent man, 1 Sv = 100 rem). The fact that effective dose has the same unit (the sievert) as equivalent dose is a cause of confusion. Moreover, effective dose has its limits, as the tissue weighting factors used to define it are not constant and vary more or less for each tissue as a function of absorbed dose, as well as dose rate.

Adding up all the effective doses for each organ gives the whole-body effective dose (Figure 6.1).

For external irradiation exposures, the ICRP presents radionuclide-specific organ and effective dose rate coefficients for members of the public resulting from environmental external exposures to both photon and electron radionuclide emissions. They are calculated using computer phantoms representing the ICRP's newborn, 1-, 5-, 10-, 15-year-old and adult men and women reference. The data show that, in general, the smaller the body mass of the phantom, the higher the organ and effective dose.

The data show that the smaller body mass of young children will result in higher dose rate coefficients due to smaller masses of overlying tissue shielding internal organ doses and greater proximity to the source in the case of soil contamination. However, age-related differences in effective dose rate coefficients are generally not significant for important radionuclides. Therefore, the difference in effective dose between an adult and an infant is 60–140% at a photon energy of 0.05 MeV, while it is less than 70% above a photon energy of 0.10 MeV [ICR 20].

### 6.2.4. **Committed dose due to internal contamination**

The committed dose from a single intake of a given radionuclide is the total dose received by the individual concerned throughout their lifetime. It therefore depends on the effective half-life. Thus, for iodines-131, 123 and 129 with effective periods of 7.6 days, 13 hours and $16.10^6$ years respectively, the committed dose will be 13, 0.13 and 30 Gy.

In the case of internal radioactive contamination, assessing the radiation dose is more difficult than in the case of external contamination. Unlike external irradiation, where the dose is delivered in its entirety at a given moment, the internal dose builds up over time, and will vary according to the organs and tissues irradiated. This dose is known as the committed dose, and can only be an estimated and calculated quantity. Committed doses are estimated for a lifetime.

Two types of biomathematical models are used to assess the activity that has remained in contaminated organs and tissues: biokinetic metabolic models and biophysical models. Biokinetic models assess the fate of the radionuclide or stable isotopes of the same chemical element in a standard human. These models are therefore representative of an anatomo-physiological model of the average human. Biophysical models are dosimetric models of radiation energy absorption in organs and tissues. They involve simulations of radiation–biological matter interaction in a virtual human context.

Assessing the committed dose therefore appears to be a multidisciplinary process, requiring numerous steps and corresponding tools. An IAEA guide [AIZ 04] provides valuable information. Real progress has been made in estimating internal dose and reducing uncertainties in dosimetric interpretation [DAV 10]. Similarly, the INRS has published a guide to the models to be used by default in the event of internal exposure of workers. These models can be found in ICRP publications 30 (gastrointestinal and systemic), 53, 80 and 106 (radiopharmaceuticals), 66 (respiratory), 67, 69 and 71 (systemics), 100 (food) and in NCRP report 156 for a wound [INR 12].

The ICRP [ICR 16] provides dose coefficients for assessing internal exposure to radionuclides. These are radiological protection quantities giving either the equivalent dose to the organ, or the effective dose per radionuclide intake after ingestion or inhalation (publication 130) [ICR 15]. It also lists radionuclide decay data (energies and yields of emitted radiation), in its publication 107 [ICR 08] and specific absorbed fraction (SAF) values.

#### 6.2.4.1. *Biokinetic models*

Biokinetic models give a mathematical description of the radionuclide's pathway, according to its physico-chemical form, in relation to the relevant sector of the organism. Two anatomo-physiological sectors are particularly important: entry points and the systemic sector (blood, extracellular fluid and organs).

This is defined as the quantity of radionuclide in becquerels (Bq) taken up by the organism from the external environment, in the form of a chemical compound. It is the physico-chemical characteristics of the compound that play a role here, and not the nature of the radionuclide. This incorporation can take place via two main natural routes: inhalation and ingestion, and artificially via injection.

Incorporation by inhalation (INH) requires a model of the respiratory system in which aerosol deposition in different regions is quantified as a function of particle size. The amount deposited differs from the amount inhaled by the existence of an exhaled quantity. Deposition is followed by purification, which takes place with decreasing exponential kinetics at all respiratory stages, via two processes. The first is mechanical, with particles passing through the digestive tract via the bronchial ciliary apparatus. The second is biochemical, involving dissolution in bronchial mucus and alveolar fluid (surfactant). Part of the deposit passes into the systemic sector. The capacity of the deposit to migrate to this sector is known as transferability, and is quantified in three classes: slow, medium and fast by the value of the associated biological period.

In the case of incorporation by ingestion (ING), the digestive model adopts a segmentation of the gastrointestinal tract, with transfer kinetics between segments and systemic absorption adapted to the duration of digestive transit.

The incorporation by injection (INJ) corresponds to the case of percutaneous blood penetration or penetration following injury. The variability of such penetration makes modeling very difficult, as local kinetics are not easily accessible. When only part of the contaminant diffuses into the systemic sector, the problem of a significant skin dose arises.

Retention by the entry routes of the digestive, respiratory systems, etc., is the fraction of the deposited quantity present at any time t. This retention is modeled differently in the case of the respiratory system, where it is modeled by region and defined in relation to incorporation I and for the tracheobronchial (TB) region.

The systemic sector corresponds to the absorption (passage of the radionuclide through the blood and extracellular fluids) and distribution of the radionuclide in the organs and tissues according to the metabolism specific to the corresponding stable element.

Retention modeling loses the generic character it had in the gateway sector, and representations are made in different compartments depending on the organs or part of the organs concerned, and linked by transfer coefficients. Retention may be diffuse (as in the case of tritium and cesium) or localized (as in the case of plutonium in bone and liver). Models distinguish between two types of elimination: rapid and slow exponential.

Effective elimination is the fraction of the quantity present at a given time t in an organ or tissue represented by a compartment, which leaves it by combining biological and radioactive decay, per unit of time. It is obtained mathematically by deriving the retention equation. Elimination, through its return to the extracellular and vascular compartments, is followed by excretion, but also for certain contaminants by partial recycling (as in the case of iodine), which complicates modeling.

Excretion is the final metabolic step. It comprises two pathways, the urinary and the metabolic (hepatic) fecal pathway.

For systemic retention, it is sometimes necessary to take into account the kinetics of the entry points. This is particularly true of inhalation, where for certain forms of radionuclides, the respiratory tract becomes a reservoir delivering its activity to the blood with very long biological and effective periods. This is the case for uranium and plutonium in the form of poorly transferable oxides.

#### 6.2.4.2. *Biophysical models*

These are models of energy absorption in organs and tissues. These models take into account source-organ and target-organ pairs.

These organ pairs are reduced to simple geometric solids and arranged in a virtual human form, for which radiation–matter interactions within and between these volumes are simulated. For penetrating radiation (X- or gamma photons), the source tissue of the target tissue is the contaminated neighboring tissue, and itself if it has concentrated the radionuclide.

Although already highly developed, the assessment of committed dose is constantly evolving. While the biophysical part is fairly stable, the metabolic part is subject to regular updates of biokinetic parameters based on recent data and incorporation calculation software that has been developed.

#### 6.2.4.3. *From committed dose to effective dose*

As with external contamination, the same tissue and energy weighting factors are applied to calculate the effective dose from the committed dose (dose assessed over a lifetime). The assessment takes into account how the radioactive substance is ingested and how it passes through the body.

As with other doses, committed doses are calculated for both ingestion and inhalation exposure pathways. Similarly, equivalent committed doses and effective committed doses are calculated. The final dose to a tissue (or organ), or to the organism as a whole, is estimated on the scale of a human lifetime. The calculation is based on 50 years for adults and 70 years for children, to take account of the gradual elimination of radioactive substances, or the non-elimination of some of them.

Alpha emitters, such as uranium or plutonium and actinides generated by reactors, are far more dangerous than beta emitters such as fission products. Similarly, inhalation of alpha emitters (which bind to the lungs) is dozens of times more dangerous than ingestion.

Radioactive isotopes have the same metabolism as stable isotopes of the same element, or as chemical elements with similar properties. For example, radium binds to bone tissue like calcium, and iodine-131 binds to the thyroid like normal iodine. As this gland plays a major role in growth, children and adolescents are much more sensitive than adults to the ingestion of radioactive iodine. The variation with age is significant. An infant will be around 10 times more sensitive to iodine-131 than an adult.

## 6.3. Limits to the estimation of effective doses

The limitations of effective doses include the fact that we have to model this estimate, that radionuclide microdistribution is not taken into account, that some radionuclides transmute, thus changing their chemical nature, that interactions may exist between radiotoxicity and toxicity, and that there is considerable individual variability. Furthermore, the fact that effective dose has the same unit (the sievert) as equivalent dose is a cause of confusion.

### 6.3.1. *Modeling*

Direct measurement of radiation dose is limited. The intrinsic difficulty lies in the fact that it is physically impossible to measure the dose received by each individual for all the sources to which they are subjected. Doses must therefore be reconstructed from ambient measurements and exposure models. For example, to model, the GRNC's specialized working group (GT4) used the ICRP database [ICR 91] (Figure 6.2). All modeling has its limits of representativeness.

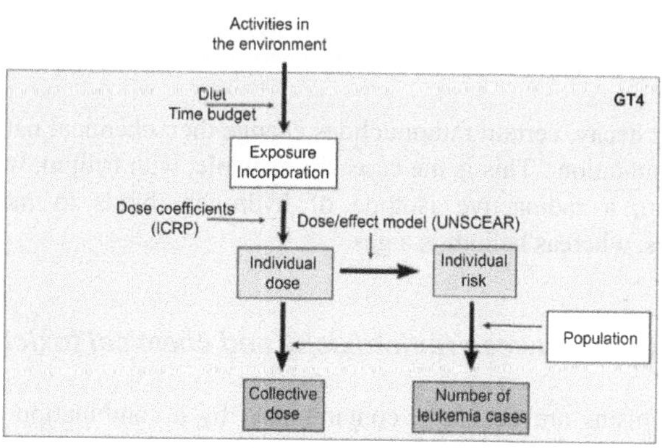


**Figure 6.2.** *Estimating the risk of leukemia in young people according to the methodology adopted by GRNC WG4 (modified from [GRN 99])*

### 6.3.2. *Microdistribution*

The ICRP's approach is macro-dosimetric, i.e. contamination at the organ or tissue level is assumed to be homogeneous. Unfortunately, this is not always the case. There is frequently a heterogeneous distribution at the sub-cellular level, with certain cellular organelles or macromolecules accumulating certain radionuclides more strongly than others. This is known as microdistribution.

This is particularly important for alpha emitters, which have a short pathway through living matter, so their distribution will have a considerable impact on the harmful effects they cause. Therefore, an alpha emitter located on a biological membrane and the same one bound to DNA will not cause the same type of harm.

This is also true for electron emitters (Auger and beta emitters). While various toxicological and epidemiological studies have been carried out on alpha emitters, there are enormous gaps when it comes to Auger and beta emitters. Electron emitters have natural origins, such as potassium-40, lead-210, radium-228, thorium-231 and thorium-234 for Auger emitters, and carbon-14 and tritium for beta emitters. They also have industrial origins, such as iron-55, cobalt-57, nickel-59, zinc-65, krypton-83m, molybdenum-94 and xenon-133 for Auger emitters, and iodine-131, strontium-90 and yttrium-90 for beta emitters. These electron emitters are widely used in nuclear medicine, either for therapeutic or diagnostic purposes, along with other Auger emitters (iodines-123 and 125, indium-111, platinum-193m and 195m, bromines-77 and 80m) or specifically manufactured beta emitters (samarium-153). Good dose and risk estimation presupposes precise cellular localization and adequate toxicological knowledge (particularly for beta emitters) of these radionuclides [PAQ 10].

### 6.3.3. *Transmutation*

During their decay, certain radionuclides change their chemical nature, a process known as transmutation.. This is the case, for example, with tritium, which becomes helium. Tritium, a radioactive isotope of hydrogen, binds to most biological macromolecules, whereas helium is a gas.

### 6.3.4. *Interactions between radiotoxicity and chemical toxicity*

Living organisms are generally contaminated by a combination of stable and radioactive pollutants. While the estimation of effective dose takes into account all the irradiation caused by the various radionuclides, and therefore theoretically all radiotoxicity, it does not take into account the chemical toxicity of radionuclides and any other stable chemical pollutants. The effect of irradiation is likely to be more harmful for individuals in less-than-optimal health.

### 6.3.5. *Individual variability*

Radiosensitivity varies according to sex and age, and also, for the same category of the population, according to the individual. However, all reasoning is based on a theoretical average individual and on three fundamental concepts: i) the relationship between dose and cell damage is linear, ii) all individuals are considered to be identical, and iii) the response of each individual to cell damage is determined by chance. Therefore, all individuals exposed to the same dose of product should have the same risk of developing cancer.

As in the case of chemical risk estimation [AMI 17], the uncertainty linked to inter-individual variability ($UF_H$) should be applied to account for a population that may be more radiosensitive than the general population.

Individual variability is of great importance in radiotherapeutic applications, so we have been trying to decipher the underlying mechanisms. Screening for certain gene mutations is not enough to describe all human radiosensitivity. Approaches involving the examination of specific biological functions, such as DNA repair, appear more promising for predicting superacute reactions following radiotherapy [GRA 11].

## 6.4. Dose rates

Exposure doses irradiate individuals instantaneously or spread out over the long term. This is reflected in the dose rate. This rate is expressed in units of dose per unit of time. This is the case, for example, for absorbed doses in $Gy.s^{-1}$ (and their sub-units $\mu Gy.year^{-1}$) and for equivalent doses in $Sv.h^{-1}$ (and their sub-units $mSv.year^{-1}$ or in $\mu Sv.h^{-1}$).

## 6.5. Occupational doses

Occupational exposure to ionizing radiation can arise from activities using radiation or radioactive substances in industry, medicine, teaching and research, and can also occur when workers are exposed to natural sources of radiation.

The worldwide annual number of workers exposed to natural and artificial sources of ionizing radiation was estimated by UNSCEAR [UNS 22] at around

24 million for the period 2010–2014. Around 52% of these were employed in sectors involving exposure to natural sources of radiation (12.6 million) and around 48% were employed in sectors involving exposure to artificial sources of radiation (11.4 million). For exposure to natural sources of radiation over the period 2010–2014, mining and processing of coal and minerals other than coal and uranium accounted for 94% of the annual number of workers. Around 11.8 million people were employed in mining operations (70% in coal mines and 30% in other mining operations, excluding uranium mines).

The estimated annual number of monitored workers exposed to artificial sources worldwide rose to over 11.4 million in 2010–2014, compared with around 10 million in the period 1995–1999. The medical sector dominated the workforce exposed to artificial sources, accounting for around 80% of the total workforce and around 75% of the collective dose [UNS 22].

## 6.6. Collective doses

All the quantities described above are individual doses and concern only an individual with their habits and life history. It may be interesting to know the exposure of a group or human population. In this case, it is possible to calculate the collective equivalent dose and collective effective dose. The calculation is simple, and involves multiplying the average dose (equivalent or effective) by the number of individuals in the group or population concerned. The unit for these collective quantities is the man-sievert, or man.Sv. These quantities are used in particular to manage groups of workers so as to reduce the collective dose.

If the reasoning does not concern a single generation, but several generations, then the quantity used is the dose commitment. This quantity is calculated as the integral over an infinite time of the dose rate per head. This quantity will be used, for example, to estimate the irradiation dose sustained in the case of long-term waste storage.

These two values are used primarily for radioactive risk management. They are the only values that can be used to judge implementation of the optimization principle ALARA ("as low as reasonably achievable"). The optimization principle is a pragmatic approach to acting responsibly, consensually and fairly in a context of risk uncertainty. Apart from radioactive risk, this principle is also used for some chemical pollutants such as dioxins and biotoxins [AMI 17].

## 6.7. References

[ADA 09] ADAM-GUILLERMIN C., FRELON S., LEBARON-JACOBS L. et al., “Concepts de base pour une évaluation dosimétrique pertinente des expositions aux radionucléides”, in MÉNAGER M.T., GARNIER-LAPLACE J., GOYFFON M. (eds), *Toxicologie nucléaire environnementale et humaine*, Lavoisier, Tec&Doc, Paris, pp. 47–62, 2009.

[AIE 04] AIEA, *Évaluation de l'exposition professionnelle due à l'incorporation de radionucléides*, Guide de sûreté RS-G-1.2. STI/PUB/1077, IAEA, Vienna, 2004.

[AMI 17] AMIARD J.C., *Les risques chimiques environnementaux. Méthodes d'évaluation et impacts sur les organismes*, 2nd edition, Lavoisier, Tec&Doc, Paris, 2017.

[BIA 11] BIAU A., “Dosimétrie individuelle, grandeurs et unités”, *Radioprotection*, vol. 46, pp. S15–S21, 2011.

[DAV 10] DAVESNE E., Optimisation des programmes de surveillance de la contamination interne par l'étude des incertitudes liées à l'évaluation dosimétrique, DRPH/SDI/2010-11 report. ISRN/IRSN-2010/135, Thesis, Université de Paris Sud Orsay, 2010.

[GAZ 10] GAZAL S., AMIARD J.C., *Le tritium. Actualité d'aujourd'hui et de demain*, Lavoisier, Tec&Doc, Paris, 2010.

[GRN 99] GRNC (GROUPE RADIOÉCOLOGIE NORD COTENTIN), Estimation des niveaux d'exposition aux rayonnements ionisants et des risques de leucémies associés de populations du Nord-Cotentin, Summary report, July 1999.

[GRA 11] GRANZOTTO A., JOUBERT A., VIAU M. et al., “Réponse individuelle aux radiations ionisantes : quel(s) test(s) prédictif(s) choisir ?”, *Comptes rendus biologies*, vol. 334, pp. 140–157, 2011.

[ICR 91] ICRP, “1990 recommendations of the International Commission on Radiological Protection”, *Annals of the ICRP*, vol. 21, nos. 1–3, 1991.

[ICR 07] ICRP, “The 2007 recommendations of the International Commission on Radiological Protection”, *Annals of the ICRP*, vol. 37, nos. 2–4, pp. 1–332, 2007.

[ICR 08] ICRP, “Nuclear decay data for dosimetric calculations”, *Annals of the ICRP*, vol. 38, no. 3, 2008.

[ICR 15] ICRP, “Occupational intakes of radionuclides: Part 1”, *Annals of the ICRP*, vol. 44, no. 2, pp. 3–188, 2015.

[ICR 16] ICRP, “The ICRP computational framework for internal dose assessment for reference adults: specific absorbed fractions”, *Annals of the ICRP*, vol. 45, no. 2, pp. 1–74, 2016.

[ICR 20] ICRP, “Dose coefficients for external exposures to environmental sources”, *Annals of the ICRP*, vol. 49, no. 2, pp. 1–145, 2020.

[INR 12] INRS, "Surveillance médico-professionnelle de l'exposition interne aux radionucléides en installations nucléaires de base. Extraits des recommandations élaborées selon le guide méthodologique '*Recommandations pour la pratique clinique*'", *Références en santé au travail*, no. 130, pp. 55–70, 2012.

[JIA 94] JIANG T.N., LORD B.I., HENDRY J.H., "Alpha particles are extremely damaging to developing hemopoiesis compared to gamma irradiation", *Radiation Research*, vol. 137, pp. 380–384, 1994.

[LIT 98] LITTLE J.B., "Radiation-induced genomic instability", *International Journal of Radiation Biology*, vol. 74, pp. 663–671, 1998.

[LOR 96] LORD B.I., MASON T.M., "On the relative biological effectiveness of alpha-particle irradiation with respect to hemopoietic tissue", *Radiation Research*, vol. 145, pp. 510–518, 1996.

[PAQ 10] PAQUET F., BARBEY P., BARDIES M. et al., L'évaluation et la gestion des risqué associés aux expositions aux radionucléides émetteurs Auger et bêta. Avis et propositions de pistes de recherche, DS 2010-03 report, IRSN, 2010.

[RAO 91] RAO D.V., NARRA V.R., HOWELL R.W. et al., "Induction of sperm head abnormalities by incorporated radionuclides: Dependence on subcellular distribution, type of radiation, dose rate, and presence of radioprotectors", *Radiation Research*, vol. 125, pp. 89–97, 1991.

[UNS 96] UNSCEAR, Sources and Effects of Ionizing Radiation, United Nations, New York, 1996.

[UNS 22] UNSCEAR, Evaluation of occupational exposure to ionizing radiation, Sources, Effects and Risks of Ionizing Radiation, vol. IV, Annex D, United Nations, New York, 2022.

[WRI 98] WRIGHT E.G., "Radiation-induced genomic instability in haemopoetic cells", *International Journal of Radiation Biology*, vol. 74, pp. 681–687, 1998.

# 7

# Estimating Human Radiation Doses

## 7.1. Natural external radiation doses

Members of the public are constantly exposed to ionizing radiation emitted by natural and artificial radioactive sources. These exposures vary from person to person, depending on where they live (geolocation, building materials and construction methods), their lifestyle (diet, sedentary lifestyle) and their professional activity (air transport, mining, spas, etc.). We will examine each of these sources, their origins, modes of action and associated exposure levels.

### 7.1.1. *Telluric exposure*

The Earth was originally highly radioactive, but little by little, the short- and medium-lived isotopes that originally made it up have disappeared. Only radioelements with half-lives measured in billions of years remain: potassium-40 ($^{40}K$) with a half-life of 1.2504 ± 0.0030.$10^9$ years, uranium-235 ($^{235}U$) with a half-life of 0.704 ± 0.001.$10^9$ years, uranium-238 ($^{238}U$) with a half-life of 4.468 ± 0.005.$10^9$ years and thorium-232 ($^{232}Th$) with a half-life of 14.02 ±0.06.$10^9$ years.

$^{40}K$ is a beta-emitting radionuclide that emits energetic gamma radiation (1.461 MeV emitted at 10.55%), which plays an important role in both external and internal human exposure. While $^{40}K$ has no radioactive parentage[1], $^{235}U$ and $^{238}U$, as well as $^{232}Th$, are at the head of a natural radioactive family (see Figure 7.1); the $^{238}U$ parentage chain [IRS 14] whose radionuclides, along with those of $^{40}K$ and $^{14}C$, play a major role in the exposure of the world's population to ionizing radiation.

---

1 There are eight other natural radioelements, present in the Earth's crust, which do not give radioactive lineages and whose periods are greater than 10 billion years ($^{50}V$, $^{87}Rb$, $^{115}In$, $^{138}La$, $^{144}Nd$, $^{147}Sm$, $^{176}Lu$ and $^{187}Re$). However, these isotopes are beta or alpha emitters ($^{144}Ne$ and $^{147}Sm$) and therefore play no role in the external exposure of living beings.

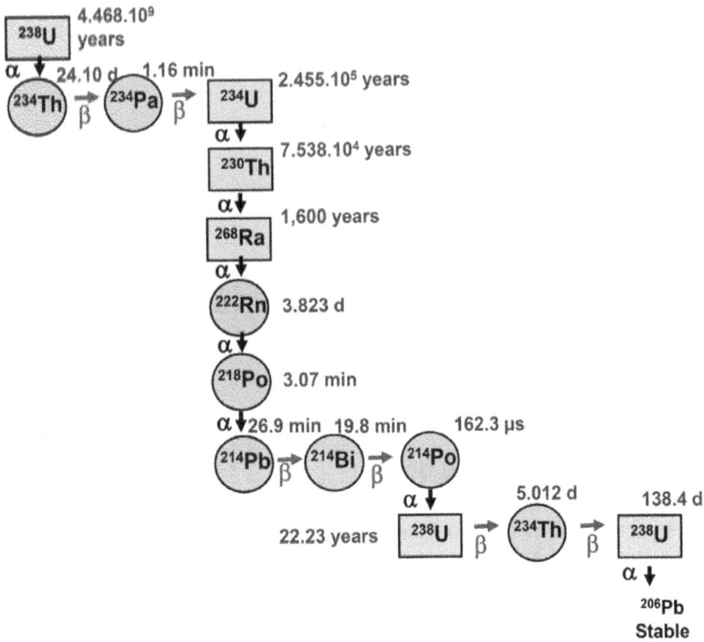


**Figure 7.1.** *Uranium-238 parentage chain. Physical half-lives taken from [CEA 14]*

Telluric exposure is caused by gamma radiation from $^{40}K$ and the descendants of the three radioactive chains. Granitic zones, which contain potassium (in the form of $K_2O$) and uranium, deliver the highest dose rates.

The $^{238}U$ radioactive family, the $^{232}Th$ family and $^{40}K$ each contribute around *one-third* of the total external exposure suffered on a standard soil, typically between 10 and 200 nanograys per hour (nGy.h$^{-1}$), or 0.09–1.8 mGy.yr$^{-1}$ [RAN 99].

In a more recent publication [IRS 21c], the IRSN was able to map telluric gamma dose rates in France, based on data from radio-photo-luminescent dosimeters (RPLs) distributed in 17,404 localities.

For the general population, the average effective dose due to telluric exposure is **0.63 mSv.year$^{-1}$**, with a range of average rates varying from 0.36 mSv.year$^{-1}$ to 1.1 mSv.year$^{-1}$.

### 7.1.2. *Cosmic exposure*

On August 7, 1912, Austrian physicist Viktor Franz Hess (1883–1964) discovered cosmic radiation by measuring the rate of air ionization (number of ions created per second and per $cm^3$ of air) as a function of altitude. A year later, he made seven ascents to an altitude of 5,350 m. He observed that between 4,000 and 5,000 m, ionization almost doubled in intensity. As there was little difference between day and night[2], Hess concluded that the radiation was of cosmic origin, i.e. coming from outside our solar system. It was the American physicist Robert Millikan (1868–1953), who proposed (1927) to call these rays, cosmic rays [FER 06].

Another English physicist, Arthur Compton (1892–1962), demonstrated that cosmic rays vary according to the latitude at which they are measured (weaker at the equator than at the poles), due to the influence of the Earth's magnetic field. They propagate at speeds close to that of light.

In 1938, French astronomer Pierre Auger (1898–1993) discovered that the penetration of particles into the Earth's atmosphere, which collide with the oxygen and nitrogen in the air[3], triggers large "sheaves of particles", which are secondary particles in contrast to primary particles of cosmic origin, which are mostly atomic nuclei, positively charged heavy ions. Virtually all chemical elements are represented, from hydrogen to iron [FER 06]. It is these heavy ions that deliver high effective doses.

Exposure to cosmic radiation at sea level delivers an average effective dose per capita of 0.32 $mSv.year^{-1}$ (0.037 $\mu Sv.h^{-1}$). For an individual, the dose varies according to the time spent indoors (which provides protection by a factor of 0.8) and outdoors. Depending on geographical location and altitude (as altitude increases, the protective layer of air decreases, increasing the dose), the average effective dose per inhabitant due to cosmic radiation in France varies from 0.3 to 1.1 $mSv.year^{-1}$ at ground level. Each flight within mainland France increases this dose by 0.7–4 µSv. For Mainland France–Overseas routes, the dose is incremented by 0.023–0.103 mSv per flight (Paris–Papeete) [IRS 21b].

Worldwide, the estimated number of people employed in civil aviation, mainly exposed to cosmic radiation in 2020, was approximately 0.7 million. The annual

2 Hess ruled out the possibility that the ionizations observed were due to the sun by flying at night and even during a solar eclipse [KOT 13].

3 It is these secondary particles that produce carbon-14, the radioactive isotope of the long-lived element carbon (5,700 years), by interacting with air nitrogen (mainly with the nitrogen-14 isotope, which is the majority in the element nitrogen N), following the reaction $^{14}N$ (n, p) 14 C. Other modes of formation by neutron activation exist for oxygen and nitrogen isotopes in air: $^{16}O$, $^{17}O$, $^{15}N$ and $^{13}N$ [DEL 85].

collective effective dose from natural sources was approximately 24,300 man.Sv (12,800 man.Sv from coal mining and processing, 9,500 man.Sv from mining and processing of other minerals, and approximately 2,000 man.Sv from exposure of flight personnel to cosmic radiation) [UNS 22b].

## 7.2. Natural internal radiation doses

Soil-borne thorium, uranium and their descendants, as well as potassium-40, are absorbed by plants and then by animals. This makes them the primary source of natural radioactivity for human internal contamination via the food chain. For the respiratory pathway, the main contaminant is radon-222, a radioactive noble gas directly descended from radium-226 in the uranium-238 chain (Figure 7.1).

### 7.2.1. *Population exposure to radon*

Radon is an inert, radioactive gas produced by the decay of uranium and thorium chains. Along with its descendants, radon-222 is considered to be the main source of exposure for the world's population. In the three radioactive descendant chains of $^{235}U$, $^{238}U$ and $^{232}Th$, we encounter three radon isotopes in a gaseous state, mixed with air and therefore inhalable: radon-219 ($^{219}Rn$) with $^{235}U$, radon-222 ($^{222}Rn$) with $^{238}U$ and radon-220 ($^{220}Rn$) with $^{232}Th$. However, only radon-222, with a half-life of 3.88 days, plays a significant role in internal lung exposure. The half-lives of $^{219}Rn$ (3.98 s) and $^{220}Rn$ (55.8 s) are too short to cause significant exposure of the pulmonary alveoli.

Radon-222, directly produced by radium-226[4], will decay into seven radionuclides, solid-phase descendants, before completing the decay chain with a stable isotope: lead-206. These isotopes will diffuse into the atmosphere, binding or not, to inactive aerosols in the air, or deposit on the ground or walls. These descendants are solid particles whose behavior differs from that of radon. They are characterized by a high diffusion coefficient in the air and constitute the free fraction, which tends to attach to aerosols and form the attached fraction.

The dose to the lung will depend on the rate of inhaled descendants. Since not all the descendants are present in the air, the ratio of the potential energy of the mixture to that of the equilibrium state with radon-222 (19 MeV$\alpha$ produced by the decay of poloniums 218, 214 and 210) is called the equilibrium factor. This equilibrium factor varies from 0 to 1, and the dose increases linearly with it. It decreases as the ambient air renewal rate increases. The average value of this factor varies between

4 Radium-226 is an alpha-emitting radionuclide with a long half-life: 1,600 ± 7 years. On a human scale, it is a virtually constant source of radon-222.

0.3 and 0.6 in uranium mines. In French mines, a factor of 0.22 has been used to calculate miners' exposure [VAC 08].

Among the seven radioactive progeny of radon-222 are three alpha emitters, isotopes of polonium ($^{218}$Po, $^{214}$Po and $^{210}$Po). The first two have short half-lives (3.05 min and 164 μs), but $^{214}$Po gives rise to the relatively long-lived $^{210}$Pb (22.23 years), a low-energy beta and X emitter. It is the last isotope, $^{210}$Po (138.38 d), which is potentially the most damaging of the seven radionuclides, in terms of dose.

When you inhale air containing radon and its descendants, most of the radon will be re-exhaled, but it is the solid descendants that will settle on the lining of the pulmonary alveoli, delivering significant doses to the repository.

A fraction of radon is dissolved in the blood. The contribution of radon gas itself is of the order of only 2–5% of the effective dose [IRS 18b].

#### 7.2.1.1. *Radon exposure of the French population*

Figure 7.2 shows the location and extent of granitic zones in France, as well as the variation in mass activity of these rocks in milligrams of uranium per kilogram of rock [IEL 17], given that one milligram of uranium[5] has an activity of 26 becquerels.

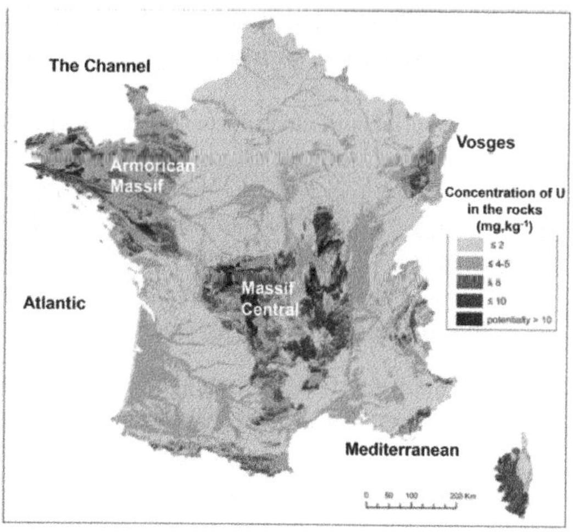


**Figure 7.2.** *Map of the distribution of uranium content in mg.kg$^{-1}$ of rock in the main geological units of mainland France (modified from [IEL 17])*

5 The activity of one gram of natural uranium is distributed as follows: $^{238}$U = 0.99275 g activity = 12,346 Bq, $^{23}$U = 0.00719 g activity = 575 Bq and $^{234}$U = 0.000057 g activity = 13,110 Bq for a total of 26,031 Bq.g$^{-1}$ from Unat [IRS 01].

This map, the result of joint research by the CNRS and BRGM [IEL 17], *has provided the* IRSN with a map of radon emission potential of geological origin in France. These data on mass activity in $U_{nat}$, which make it possible to assess the telluric dose, are also used to study and model radon transfer from rocks and soils to the surface, and more particularly at the soil-building interface. Data on radon activity in France are used for epidemiological studies on the health risk associated with exposure to natural radioactivity, and also for regulatory provisions related to the three radon-potential zones[6] that have been defined in France[7].

The measurement campaign, organized between 1982 and 2003 by the French Ministry of Health and IRSN in over 10,000 buildings throughout mainland France, estimated the average radon concentration in dwellings at 90 $Bq.m^{-3}$ for the whole of France[8] (the median[9] is 50 $Bq.m^{-3}$). The measurements (12,641) showed significant disparities (from 1 to 11) from one department to another and, within a department, from one building to another. For example, the average was only 24 $Bq.m^{-3}$ in Paris, but 264 $Bq.m^{-3}$ in Lozère [IRS 17a]. Among the highest average radon levels, 9% of measurements were above 200 $Bq.m^{-3}$, 2.3% above 400 $Bq.m^{-3}$ and 0.5% above 1,000 $Bq.m^{-3}$ [IRS 12].

However, in the average value of 90 $Bq.m^{-3}$, each measurement point in a home counts as just one unit. If we take into account the number of people living in each dwelling on a regular basis or for part of the year, we obtain an average radon concentration, adjusted for dwelling type and seasonality, and weighted by dwelling density (based on INSEE data for each department), which is equal to 68 $Bq.m^{-3}$.

This is the value that best reflects the average exposure of the French population [IRS 21b]. It is lower than the arithmetic mean, as the most densely populated departments generally have lower averages, due to a greater number of multistory buildings (the influence of radon emitted by the ground is then very much reduced, but there is still radon released by building materials).

#### 7.2.1.2. *Calculation of the lung dose delivered by radon*

Initially, to calculate the lung dose due to radon, the ICRP favored a dosimetric approach. In this case, radon dose is calculated by measuring the average activity of

6 Three exposure risk zones are defined in article R. 1333-29 of the French Public Health Code: zone 1, municipalities with low radon potential; zone 2, municipalities with low radon potential but where particular geological factors may facilitate the transfer of radon to buildings; and zone 3, municipalities with significant radon potential.

7 Order of June 27, 2018 delimiting areas with radon potential in French territory.

8 This average value is higher than in the United Kingdom (20 $Bq.m^{-3}$) and lower than in Sweden (108 $Bq.m^{-3}$), where there are large quantities of uranium- and thorium-rich granites and pegmatites, as well as vast expanses of uranium-rich aluminous shales [MJÖ 03].

9 The "median value" is such that 50% of observed results are above and 50% below it.

radon in the air inhaled during the day, and relating the becquerels inhaled to the resulting dose. The ICRP dose coefficients (expressed in Sv.Bq$^{-1}$) thus make it easy to calculate an effective dose (in Sv) from cumulative exposure data (in Bq).

The dosimetric approach is based on a human respiratory model developed by the ICRP and adopted in 1994 (ICRP 66) [ICR 94]. This model makes it possible to assess the fraction of inhaled radioactive aerosols deposited in the bronchi and pulmonary alveoli, according to their aerodynamic properties. The smaller the diameter, the greater the number of aerosols entering the bronchioles. Once deposited, radioactive particles are removed from the lung by mechanical transport[10] to the alimentary tract and by absorption into the bloodstream. The energy delivered to lung cells is calculated on the basis of their position relative to the radionuclide deposits and the pathway of the alpha particles emitted. In calculating the effective dose, a weighting factor $w_R = 20$ is used to calculate the dose resulting from alpha radiation, and a tissue weighting factor $w_T = 0.12$ is used to calculate the dose to the lung, in relation to that to the whole body.

However, in 1993, the ICRP showed that the dosimetric approach underestimated dose assessment. The Commission now favors the exposure–risk relationship derived from epidemiological studies of several cohorts of uranium miners. These relationships are statistically robust[11] and make it possible to estimate risk from radon concentration data (Bq.m$^{-3}$) without having to assess the effective dose (Sv).

The new dose coefficient for radon, which the ICRP proposes to use for radon exposure in the home (ICRP 137) [ICR 17], leads to the following relationship: a dose of 10 mSv.year$^{-1}$ is reached for a radon activity volume of 215 Bq.m$^{-3}$.

With this new dose relationship, the weighted average exposure of 68 Bq.m$^{-3}$ corresponds to an annual dose of 3.16 mSv.year$^{-1}$.

This new assessment significantly increases the estimated annual dose to the public, and demonstrates the predominant role played by radon-222 in public exposure in France. The IRSN's previous radon dose assessment, using the old ICRP dose coefficients, was 1.43 mSv.year$^{-1}$ for a total dose of 4.5 mSv.year$^{-1}$.

---

10 The bronchopulmonary trachea, in particular, is lined with ciliated cells, which beat in coordinated waves, sending aerosols trapped in mucus back to the aerodigestive tract. Arriving at the upper part of the trachea, these particles tip over into the digestive system. Those that pass through the intestinal barrier enter the circulating bloodstream, while others remain in the digestive system until they are excreted.

11 In 2008, the results of epidemiological studies on 11 cohorts were available, with follow-ups ranging from 10 to 30 years and involving 1,294 to 50,001 minors [VAC 08].

Using ICRP 137 coefficients [IRS 17a, b], the IRSN estimates the average annual dose to the public at 6.5 mSv.year-1, including 3.5 mSv.year-1 for radon, which alone accounts for 54% of the annual dose.

The effective dose due to radon can vary, depending on the municipality, from 0.75 to 47 mSv.year$^{-1}$ [IRS 21d]. These dose values apply to adults only. For children, the values vary with each age group, as respiratory rates vary with age and are lower than for adults.

#### 7.2.1.3. *The health impact of radon*

Of the seven radon-222 descendants, three are alpha emitters, which deliver their energy to the epithelium of the respiratory tract[12] and four are beta-gamma emitters, which irradiate the basal layer beneath the mucociliated epithelium. Radon descendants can thus induce cancer in these irradiated tissues. Since 1988, radon-222 and its decay products have been classified by the IARC as carcinogenic to humans[13]. After tobacco, it is the second cause of lung cancer.

The quantitative study of the health risk associated with radon in housing in mainland France was carried out by the IRSN and Santé Publique France in 2018 [AJR 18]. It showed that of the 30,000 annual deaths from lung cancer in France over the study period (2008–2012)[14], approximately 3,000 deaths were attributable to internal exposure due to radon.

Of the 3,000 fatal cancers attributable to radon, 75% occur in active smokers, 20% in ex-smokers and 5% in non-smokers. Approximately 33% of lung cancer deaths attributable to radon are due to exposure levels above 100 Bq.m$^{-3}$.

The results of numerous epidemiological studies, 14 of which estimate an exposure–risk relationship, are concordant and show that the risk is proportional to radon exposure [VAC 08], and that it is significant for continuous domestic exposure over 30 years, from radon concentrations above approximately 200 Bq.m$^{-3}$.

12 This thin tissue forms the superficial layer of the mucosa. It comprises several cell types (ciliated, mucous, serous, basal, Clara) that line the pulmonary branches. The cells become non-ciliated in the respiratory bronchioles.

13 Volume 43, *Man-made Mineral Fibers, and Radon*, 1988, Volume 78, *Ionizing Radiation*, Part 2: *Some Internally Deposited Radionuclides*, 2001, and Volume 100D, *Radiation*, 2012.

14 In 2012, lung cancer incidence was 39,495 cases (28,211 males and 11,284 females) and mortality 29,949 (21,326 males and 8,623 females) [BIN 13].

In some regions, radon exposure in homes[15] can reach levels close to those observed in uranium mines in France [VAC 08].

Since the 1990s, several authors have suggested that radon inhalation may cause irradiation of the blood-forming bone marrow. A critical analysis of all available epidemiological results does not support the hypothesis of a causal relationship between radon exposure and the risk of leukemia [LAU 01]. Analysis of mortality among French uranium miners also revealed no excess mortality from leukemia, but a significant excess of mortality from central nervous system tumors. No increase was significantly associated with the length of employment, nor with exposure to radon, uranium dust or gamma radiation. This excess has never been observed in other cohort analyses [VAC 08].

### 7.2.2. *Internal exposure to natural radionuclides*

Potassium-40: although the content of the isotope $^{40}$K (1.25.10$^9$ years) in the element potassium is very low (0.0117%), this radionuclide is the most abundant in the environment. Its activity is measured in hundreds of Bq.kg$^{-1}_{sec}$ in continental and marine soils and sediments. A major element for living organisms, potassium (and with it $^{40}$K) is strongly transferred to plants and animals, which explains the activities of the order of hundreds of Bq.kg$^{-1}$ in fresh foodstuffs.

Our body's potassium content varies according to age and sex. Maximum potassium levels are reached around the age of 20, at 2g.kg$^{-1}$ for men and 1.6 g.kg$^{-1}$ for women. It then decreases slowly (to 1.6 and 1.2 g.kg$^{-1}$). With a specific activity of 30 Bq.g$^{-1}$ of $^{40}$K per gram of potassium, the radioactive load of men and women is equal to 60 Bq.kg$^{-1}$ and 48 Bq.kg$^{-1}$, respectively, at the age of 20.

For an adult, the daily intake of potassium is 3.3 g.d$^{-1}$ or 100 Bq.d$^{-1}$ (ICRP 23) [ICR 75], which results in an average body burden renewal of 42 days for a standard man and 29 days for a woman. The average daily intake of $^{40}$K for a vegetarian diet is approximately 300 Bq.d$^{-1}$ [PAU 97].

Carbon-14: the $^{14}$C content (5,568 years) of the element carbon is extremely low (1.2.10$^{-12}$). There are a thousand billion times fewer radioactive $^{14}$C atoms than stable $^{12}$C atoms (98.89%). The proportion of carbon-13 is 1.108% [DÉL 85]. The production of carbon-14 by cosmic rays leads to an activity of this isotope close to 226 Bq.kg$^{-1}$ of stable carbon in all components of the terrestrial environment.

15 In France, approximately 1.25 million individual dwellings exceed 200 Bq.m$^{-3}$, 300,000 are above 400 Bq.m$^{-3}$ and 60,000 exceed 1,000 Bq.m$^{-3}$ [AUR 03], i.e. average levels higher than those found in French uranium mines.

However, since 1945, the numerous aerial nuclear tests[16] have increased the content of this radioactive isotope in the element carbon, mainly in the northern hemisphere, notably in 1961–1962. In 1961, the USSR tested the world's largest tested nuclear weapon, the "Tsar Bomba", with an energy of over 55 megatons (approximately 3,700 times that of Hiroshima), and in 1962, 178 aerial tests were carried out by the United States and the USSR. Carbon mass activity, at $^{14}C$, had risen to 420 $Bq.kg^{-1}$ of carbon in 1963–1965, following the aerial nuclear tests, and then declined slowly to return to the cosmogenic level of 225 Bq/kg of C over the period 2018–2020 [IRS 21b].

The total mass of carbon in the human body is approximately 0.2 $kg.kg^{-1}$, i.e. with 226 Bq of $^{14}C$ per kg of carbon, a load of 45 $Bq.kg^{-1}$. The daily intake of carbon into the body (0.35 $kg.d^{-1}$ for a 70 kg man) leads to the renewal of the body's carbon load in 40 days [PAU 97].

Polonium-210: radon-222 decay in the atmosphere produces a source of polonium-210 aerosols, the intensity of which varies according to geographical location and the presence of mining activities. Once on the ground, $^{210}Po$ migrates very little and remains in the first few centimeters of soil. It is a radionuclide that will contaminate plants. $^{210}Po$ is transferred to animals mainly by ingestion, with relatively high transfer factors. These environmental and animal contaminations mean that $^{210}Po$ is likely to induce significant human exposure compared with other natural or artificial radionuclides.

Consumption of fish and especially seafood, which are significantly richer in $^{210}Po$ than fruit and vegetables, results in a higher effective dose when consumed regularly. Tobacco, which also contains $^{210}Po$, represents an additional internal exposure for smokers[17] [IRS 16a].

Internal contamination: based on measurements in foodstuffs and drinking water, and on the results of surveys of French eating habits, the IRSN has estimated the dose received by a French resident at 0.55 mSv per year. Nearly a quarter of this dose is due to shellfish consumption. Also, depending on lifestyle habits, in particular the consumption of fish and seafood or smoking tobacco, the internal dose will vary from 0.4 to 3.1 $mSv.year^{-1}$ [IRS 21b].

16 More than 2,400 nuclear tests, including 543 atmospheric tests, were carried out by the United States, Russia, Great Britain, France and China between 1945 and 1980 [IRS 13]. Almost a third of these tests (32.8%) were carried out in 1961–1962 by the United States and the USSR, before they signed the Partial Test Ban Treaty on August 5, 1963, which came into force on October 10, 1963 and obliges member states not to carry out nuclear explosions in the atmosphere, outer space or underwater.

17 Tobacco leaves contain a few tens of $Bq.kg^{-1}$ of lead-210 and polonium-210. The annual dose from one cigarette a day is estimated at 10 μSv [IRS 21b].

### 7.2.3. *Radon exposure of workers*

A wide variety of occupational activities expose workers to radon, such as uranium miners (as well as those in other mines), work in quarries or basements, in agricultural caves (cheese cellars, wine cellars, mushroom beds) and in spas. Radon dose is linked to the worker's respiratory rate. For heavy physical activity, the ICRP assumes an average respiratory flow of 1.2 $m^3.h^{-1}$ and for sedentary activity 0.86 $m^3.h^{-1}$.

Uranium miners are subject to three distinct radiological exposures: internal exposure due to inhalation of radon-222 and its descendants, inhalation of uranium ore dust and external exposure, is practically 360°, from gammas emitted by uranium-bearing rock. External exposure is measured in millisieverts (mSv), internal exposure to radioactive dust in $Bq.m^{-3}$ and radon exposure in special units called working level months (WLM). A miner's total exposure is the sum of these three components.

The alpha-emitting descendants of radon-222 are polonium-218 (Eα = 6.001 MeV emitted at 99.98%), polonium-214 (Eα = 7.687 MeV emitted at 99.99%) and polonium-210 (Eα = 5.304 MeV emitted at 99.99%). If all three are present (equilibrium = 1), they deliver a total of 18.973 MeV per radon decay to lung tissue. The special unit chosen for exposure is the working level (WL), which corresponds to exposure to a radon concentration of 3,700 $Bq.m^{-3}$ (0.1 $\mu Ci.m^{-3}$) at equilibrium with its short-lived descendants.

If a miner is exposed for one month (170 h) to 1 WL, they will have had an exposure of 1 WLM. In epidemiological studies, the exposure indicator used is the number of WLM accumulated during the study period.

## 7.3. Estimates of artificial radiation doses to humans

In addition to public exposure from cosmic radiation and radionuclides of cosmic and terrestrial origin, there are also exposures from various sources of artificial radiation, such as the after-effects of nuclear testing (mainly the 543 atmospheric tests) and the accidental destruction of reactor No. 4 at the Chernobyl power plant[18], as well as internal and external medical exposures for diagnostic purposes.

18 The accident occurred on Saturday, April 26, 1986 at 1:23 a.m.

### 7.3.1. *External and internal medical exposures*

For 2002, 2012 and 2017, IRSN has carried out a review of the population's exposure to ionizing radiation, due to diagnostic medical imaging examinations (conventional, dental and diagnostic interventional radiology, computed tomography (CT) scan and nuclear medicine). After an increase of almost 90% between 2002 and 2012 in the average annual effective dose per patient, the last two reports show similar dose values: 1.56 mSv in 2012 and 1.53 mSv in 2017.

In 2017, 45% of the French population was exposed to ionizing radiation for medical diagnostic purposes. This resulted in external exposures (conventional radiography, mammography, CT scan, interventional radiology for specific diagnostic purposes) or internal exposures (artificial imaging after the administration of a radionuclide). The dose received can vary from approximately 0.1 mSv for a chest X-ray to 15 mSv for an abdominal-pelvic CT scan [IRS 21b].

CT scan, the biggest contributor to dose, accounts for just under 13% of diagnostic procedures, but contributes almost 75% of the collective effective dose (1.14 mSv). The slight drop observed in 2017 (0.03 mSv) is essentially linked to conventional radiology, which is the second-largest contributor to collective effective dose (down from 0.28 to 0.18 mSv), even though it accounts for the majority of procedures performed (55%). From 2012 to 2017, the average annual number of diagnostic medical imaging procedures performed in France rose from 1,247 to 1,187 per 1,000 beneficiaries.

Nuclear medicine saw its contribution rise from 0.12 mSv to 0.17 mSv in 2017, an increase of almost 44%, equivalent to the increase in procedure frequencies. It is the third largest contributor [IRS 20a].

Worldwide, medical exposure remains by far the main source of anthropogenic exposure of the general population to ionizing radiation. Some 4.2 billion medical radiological examinations are carried out every year. The total annual collective effective dose to the world's population of 7.3 billion people is estimated at 4.2 million Sv. UNSCEAR has estimated that 6.2 million radiotherapy treatments are carried out each year, including approximately 5.8 million external beam treatments and 0.4 million brachytherapy treatments. In addition, an estimated 1.4 million radionuclide therapy treatments are carried out each year. CT contributes the most (62.6%) to the overall collective dose, but accounts for only approximately 10% of all examinations. The estimated annual effective dose per capita from medical radiological examinations has decreased slightly from the Committee's previous assessment in 2008, from 0.65 to 0.57 mSv [UNS 22a].

### 7.3.2. *Exposure due to atmospheric nuclear blasts*

The production of carbon-14 by cosmic rays in the air leads to an activity of this isotope of approximately 226 $Bq.kg^{-1}$ of stable carbon in all components of the Earth's environment. However, the numerous nuclear tests carried out in the atmosphere have increased the content of this radioactive isotope in the element carbon. More than 2,400 nuclear tests, including 543 atmospheric blasts, were carried out by the United States, Russia, Great Britain, France and China[19] between 1945 and 1980 [REN 08]. These aerial blasts of atomic or thermonuclear bombs were responsible for the formation of a large number of radionuclides produced by nuclear fission, as well as activation products of the various materials making up the devices. Many of these will find their way into the food chain. As the majority of launches (United States and Russia) were carried out in the northern hemisphere, this is also where cumulative deposition is greatest [AMI 23].

The large number of aerial blasts, more than 90% of which were carried out in 1961–1962, resulted in the highest level of radioactive fallout in France in 1963. The average annual dose received by the French population that year was estimated at 0.32 mSv.

Today, average doses from nuclear testing are due to three radionuclides: carbon-14 ($^{14}C$), strontium-90 ($^{90}Sr$) and cesium-137 ($^{137}Cs$). The dose increment caused by $^{14}C$ from nuclear testing in the 2020s is only a few percent of the dose due to the natural activity of the incorporated $^{14}C$.

The dose from $^{137}Cs$ cannot be distinguished from the after-effects of Chernobyl fallout. In areas particularly affected by fallout from the Chernobyl accident, the contribution of cesium-137 represents only 2–15% of the total dose rate, with a dose equivalent rate ranging from 2 to 7 $nSv.h^{-1}$ [IRS 21b]. The effective dose associated with the intake of $^{90}Sr$ by an adult is very low, estimated at 1.3 $\mu Sv.year^{-1}$ [IRS 20a].

### 7.3.3. *Artificial exposure through food and drink*

As a result of fallout from nuclear blasts around the world and the Chernobyl accident, France was heterogeneously contaminated by $^{137}Cs$, $^{90}Sr$, plutonium and americium-241 (produced by the decay of $^{241}Pu$).

$^{137}Cs$ induces both external and internal exposure through the ingestion of foodstuffs. On the basis of the $^{137}Cs$ content of various foodstuffs and beverages, the IRSN has calculated the doses associated with ingestion of this radionuclide for an adult, taking into account most types of foodstuff [IRS 21c]. The dominant foodstuffs

19 The last atmospheric blast was carried out by China in October 1980 [IRS 13].

in terms of dose are wild mushrooms (0.5–3 μSv.year$^{-1}$) and game meat (0.01–0.25 μSv.year$^{-1}$).

The total effective dose linked to the incorporation of $^{137}$Cs by adult ingestion was 1 μSv.year$^{-1}$ in 2020. This dose was 30 times lower than that estimated for 1964, and almost 300 times lower than that estimated for 1986 in France.

The mass activities of plutonium-239 and 240 in foodstuffs are extremely low, and the order of magnitude of the effective dose is 0.05 μSv.year$^{-1}$ [IRS 21b].

### 7.3.4. *Exposure to seafood products*

The presence of man-made radionuclides in the marine environment has many and varied origins. Discharges of these radionuclides into the sea, whether deliberate (atomic bomb testing, discharges from nuclear reactors and reprocessing plants) or unintentional (linked to incidents or accidents), are dispersed by currents[20] [AMI 23]. The radionuclides formed attach themselves to particles suspended in the water, which gradually settle to the bottom and end up stored in the sediment up to several centimeters deep.

Marine organisms can, through physiological mechanisms such as water filtration, accumulate certain radionuclides with a concentration factor of five to several tens of thousands, compared to seawater.

If we consider the case of $^{210}$Po ($^{238}$U chain), while the mass activity of fresh fruit, vegetables, fruit and roots is of the order of 120 mBq.kg$^{-1}$, large fish are at 2 Bq.kg$^{-1}$, small fish at 10 Bq.kg$^{-1}$, crustaceans at 18 Bq.kg$^{-1}$ and shellfish at 27 Bq.kg$^{-1}$ (i.e. 225 times the content of fruit and vegetables) [IRS 20a].

The average effective dose to a member of the French public, mainly from $^{210}$Po in foodstuffs, is 0.32 mSv.year$^{-1}$, corresponding to an average consumption of 25 kg of fish per year and approximately 10 kg of shellfish per year[21]. The variability in annual dose is mainly due to shellfish consumption, with a low value of 0.19 mSv (no seafood consumption), while a high value of 2.32 mSv corresponds to shellfish consumption of over 40 kg per year [IRS 20a].

20 Some radionuclides, such as tritium and carbon-14, have both natural and man-made origins [AMI 23].

21 The average, minimum and maximum values for the annual quantities consumed of different food categories (kg.year$^{-1}$), used by the IRSN for dose calculations, are derived from the results of 11 food surveys conducted in France from 1991 to 2015 [IRS 20a].

### 7.3.5. *Occupational exposure*

In France, the total number of people occupationally exposed to ionizing radiation is close to 390 000 for the last five IRSN annual publications (2017–2021). The four main activities exposed (94.6%) are medical and veterinary (59.7%), nuclear (27.7%), non-nuclear industry (4.3%), and research and teaching (2.8%).

Table 7.1 shows trends in the number of people exposed, and in individual and collective doses. The majority of workers exposed to natural ionizing radiation (included in nuclear activity) are civil aviation personnel.

| | All exposed workers | | | All exposed civilian flight crews | | |
|---|---|---|---|---|---|---|
| **Year** | **Numbers monitored** | **Collective dose (H.Sv)** | **Average dose ($mSv.year^{-1}$)** | **Numbers monitored** | **Collective dose (man.Sv)** | **Average dose ($mSv.year^{-1}$)** |
| **2015** | 372,881 | 104.4 | 0.98 | 19,612 | 38.8 | 1.98 |
| **2016** | 378,304 | 107.5 | 0.96 | 20,671 | 41.9 | 2.02 |
| **2017** | 384,198 | 100.6 | 1.03 | 22,422 | 47.1 | 2.10 |
| **2018** | 390,363 | 104.1 | 1.12 | 23,356 | 48.7 | 2.1 |
| **2019** | **395,040** | **112.3** | **1.20** | **24,429** | **53.44** | **2.20** |
| **2020** | 387,452 | 72.5 | 0.78 | 21,949 | 22.38 | 1.01 |
| **2021** | 392,180 | 82.7 | 0.85 | 20,390 | 22.54 | 1.13 |

**Table 7.1.** *Comparison of changes in the external dosimetry of all workers exposed to ionizing radiation with that of flight personnel (based on [IRS 16a, IRS 17b, IRS 18a, IRS 19, IRS 20b, IRS 21a, IRS 22])*

The collective dose is the sum of individual doses and is expressed in man.sievert (man.Sv).

The years of the coronavirus pandemic (2020–2021) significantly reduced professional activities and, consequently, individual and collective doses. We will therefore consider data for 2019 unaffected by the sharp reductions in activity.

Although flight crews account for only 5.8% of exposed workers, their collective dose is 47.6% of that of the 395,040 exposed workers as a whole (16 times higher). Aircraft crews, who are exposed to natural cosmic radiation, are therefore the most exposed occupational group in France among workers exposed to ionizing radiation.

This is not an intuitive result, for although France has the second largest fleet of nuclear reactors in the world, and is the most nuclearized country in the world, if the number of its reactors is related to the population[22], it is natural *cosmic* radiation that causes the highest average individual doses among all workers exposed to ionizing radiation.

Worldwide, the average annual collective effective dose for nuclear workers was approximately 5,500 man.Sv (approximately 480 man.Sv to nuclear fuel cycle workers, approximately 4,500 man.Sv to medical workers and approximately 440 man.Sv to industrial workers). The annual effective dose was estimated at approximately 2 mSv for workers exposed to natural sources and 0.5 mSv for workers exposed to man-made sources. The world average annual collective effective dose is approximately 30,000 man.Sv [UNS 22b].

## 7.4. Some cases of human exposure

The first nuclear air strikes were accompanied by several exposures to radioactive aerosols, carried out without the public's knowledge. Internal contamination experiments, notably with plutonium, were also carried out without the knowledge of several thousand people, who knew absolutely nothing about radioactive materials and the risks associated with their incorporation into their bodies.

### 7.4.1. *Guinea pigs*

In 1993, an investigation by American journalist Eileen Welsome revealed to the general public that, in the 1940s–50s, individuals had been contaminated, without their knowledge, with plutonium, in order to study the effects of radiation. Other experiments (1960–1972) involved whole-body irradiation with X-rays, at doses ranging from 50 to 200 mSv per day, up to a total dose of 3 sieverts[23] [BEL 95].

These were planned experiments, conducted according to protocols, which were carried out by American doctors on several thousand adults, as well as on disabled children (with calcium and radioactive iodine) in a spirit strongly reminiscent of those conducted by Nazi doctors in concentration camps.

The first known injection took place on May 14, 1945. A 58-year-old house painter named Albert Stevens, suffering from a stomach tumor, was injected

22 0.84 reactors per million inhabitants in France, 0.42 in Japan (before the Fukushima accident), 0.27 in Russia, 0.24 in the United States and 0.035 in China.

23 The lethal dose at 50% is of the order of 4 sieverts.

intravenously with 3.5 μCi of plutonium-238 and 0.046 μCi of plutonium-239[24]. Three days later, his spleen, 9th rib, part of his pancreas, the left lobe of his liver, two-thirds of his stomach, part of his peritoneum, several lymph nodes and subcutaneous tissue were removed. The victim survived to then die in 1966 at the age of 79. A report by the US Department of Energy, published in February 1995, puts the number of victims similar to Albert Stevens at 9,000, spread over 154 planned experiments [BEL 95].

Inhabited areas that have been deliberately, or accidentally, contaminated by fallout from air strikes have also been uncovered through declassification of "Confidential Defense" documents, notably in the United States and France. For example, the inhabitants of Rongelap, an atoll in the Marshall Islands, discovered in declassified documents that they had not been the victims of a whim of the weather, but had been deliberately left on their island in the foreseeable path of the radioactive cloud from the Castel Bravo blast[25], "for study purposes" [ANO 21].

On March 1, 1954, 51 hours after the blast, the inhabitants of Rongelap saw the arrival of an American army ship. The soldiers evacuated the population, who abandoned all their clothes and were given a bar of soap. They then underwent a series of examinations by American doctors. The studies were numerous and meticulous. Shortly afterwards, many of the inhabitants suffered the first deterministic effects of radiation (skin burns, hair loss, diarrhea) caused by the high radioactive load of the whitish dust that fell to the ground after the blast. Every year, they were re-examined, as had previously been "conceived and planned" in the 1953 report entitled "Project 4.1". The blast heavily contaminated 250 atoll and island residents, 28 US servicemen and 24 sailors from a Japanese tuna fishing boat[26]. A cohort of 2,273 inhabitants of 12 Marshall Islands atolls[27] was followed to observe the prevalence of thyroid nodules caused by radioactive iodine produced by the Bikini nuclear blast [AMI 19].

---

24 These activities, expressed in terms of the units of the time, correspond to 129.5 kBq of plutonium-238 and 1.7 kBq of plutonium-239, which delivered a committed dose of over 30 mSv (dose calculated assuming ingestion).

25 "Castle Bravo" is the name of the powerful H (hydrogen) bomb tested by the United States on March 1, 1954, which released an energy of 15 megatons, i.e. around a thousand times the power of the atomic bomb dropped on Hiroshima on August 6, 1945 (16 ± 2 kt). The explosion, 7 meters from the surface of Bikini Atoll, created a crater some 2 km in diameter and 70 m deep.

26 This fine, whitish dust of burnt coral also fell on a Japanese fishing boat (the tuna fishing vessel *Daigo Fukuryu Maru*, "Lucky Dragon 5"). The 24 contaminated crew members returned to Japan on March 14. They suffered from nausea, headaches, burning, eye pain and bleeding gums. They were admitted to Tokyo hospitals with acute radiation syndrome. On September 23, 1954, a 40-year-old radio-operator died [AMI 19].

27 The country is made up of 30 atolls and approximately 1,100 islands.

On February 13, 1960, France carried out its first nuclear atmospheric test at Hamoudia, some 50 km from Reggane[28]. Long presented as “clean” by the French authorities, military documents dated 1961, declassified on April 4, 2013, have shown that the plume from the cloud and its radioactive fallout spread to the Maghreb, the Sahel and West Africa. Point zero of Gerboise Bleue was located in a desert considered uninhabited, whereas the number of inhabitants of the Reggane palm groves and the Touat valley is estimated at 40,000. This new delimitation has a direct impact on the compensation arrangements for victims of nuclear blasts, opened up by the Morin law (January 5, 2010) and its implementing decrees [TIL 14].

### 7.4.2. *Estimates of exposure around nuclear facilities*

Almost all nuclear facilities, such as reactors and spent fuel reprocessing plants, discharge radioactive liquid and atmospheric effluents, which are subject to authorization. The regulatory limits set for each radionuclide must comply with technical provisions (release methods, volumetric activity, temperature, etc.). For people living within a 10 km radius of a nuclear site, a dose of 1–10 μSv.year$^{-1}$ (or 0.001–0.01 mSv.year$^{-1}$) is used, depending on the site. Based on the population of the entire country, the average dose per inhabitant resulting from these industrial activities is 0.11 μSv.year$^{-1}$. All exposures from industrial and military sources have been rounded up to an average dose per inhabitant of 12 μSv.year$^{-1}$ (0.012 mSv.year$^{-1}$) [IRS 21c].

There may, however, be groups with higher exposure than the reference group, such as fishermen or farmers, who may consume their own produce. In the case of the La Hague plants, the *Groupe Radioécologie du Nord Cotentin* (GRNC) calculated, on the basis of an in-depth radioecological analysis, the average individual doses of several particular groups, notably those of fishermen from Les Huquets, which were 226 μSv in 1985 and 26 μSv in 1996, while those of a group of farmers (Pont-Durand), for the same years, were 53 and 59 μSv [GRN 99].

### 7.4.3. *Estimates of exposure due to fallout from Chernobyl and Fukushima*

From late April to early May 1986, the air masses contaminated by the Chernobyl accident that passed over France (first Corsica, then eastern France) produced radioactive deposits of varying activities, mainly iodine-131, cesium-137 and cesium-134. Two factors led to this variability: the gradual depletion of the

28 Code name: Blue Gerbil, power 70 kilotons, more than four times that of the bomb dropped on Hiroshima (16 ± 2 tons).

cloud's load during its journey, and the amplification of deposition in areas affected by rain, which washed the cloud away. Where rain fell, soil contamination was almost 100 times greater than that from dry fallout of radioactive aerosols. Dry deposition ranged from approximately 1,000 Bq.m$^{-2}$ in eastern France to approximately 100 Bq.m$^{-2}$ in western France, whereas rainfall in eastern France led to much greater and more heterogeneous radioactive deposition [REN 05].

Local contamination in the departments of Nord-Est, Franche-Comté, the Southern Alps and Corsica resulted in cesium-137 deposits in excess of 10,000 Bq.m$^{-2}$, and locally even 20,000 Bq.m$^{-2}$. On the east coast of Corsica, on the scale of a few tens of square kilometers, intense showers resulted in deposits exceeding 40,000 Bq.m$^{-2}$ [IRS 18c].

Average individual doses due to Chernobyl fallout and the after-effects of nuclear blasts will therefore vary according to place of residence: in 2020, a resident of the worst-affected areas of eastern France received an average dose of 46 microsieverts per year (μSv.year$^{-1}$) compared with 9.3 μSv.year$^{-1}$ for a person living elsewhere in France, i.e. an average dose per inhabitant of 12 μSv.year$^{-1}$ for the whole country [IRS 21b]. These doses are mainly due to external exposure from cesium-137 in the soil. The intake of cesium-137 through food ingestion is less than 1 μSv.year$^{-1}$. However, it may become non-negligible for people who eat a lot of wild mushrooms or game from the areas most affected by Chernobyl fallout [IRS 16b].

On March 11, 2011, 25 years after Chernobyl, the world's biggest civil nuclear disaster, whose releases spread far beyond the borders of Ukraine and nearby Belarus, a new nuclear cataclysm struck Japan. It was no longer a single reactor that exploded (as in Chernobyl), but three reactors that were destroyed in Fukushima: reactors 1, 2 and 3. Reactor 4 was shut down like the first three, but the core, removed from the vessel, was placed in the reactor's storage pool.

Mainland France was affected by air masses slightly contaminated by atmospheric releases from the Fukushima accident. Measurements revealed activity at $^{131}$I and $^{134}$Cs, as well as a slight increase in pre-existing levels of $^{137}$Cs. However, this influence was very slight and very limited in time. Maximum effective doses received by an adult were estimated at less than 2 μSv (0.002 mSv) for 2011. They have been negligible since [IRS 21b].

## 7.5. References

[AJR 18] AJROUCHE R., ROUDIER C., CLÉRO E. et al., "Quantitative health impact of indoor radon in France", *Radiation Environmental Biophysics*, vol. 57, no. 3, pp. 205–214, August 2018.

[AMI 23] AMIARD J.C., *Marine Radioecology*, vol. 6, ISTE, London, and John Wiley & Sons, New York, 2023.

[AMI 19] AMIARD J.C., *Military Nuclear Accidents: Environmental, Ecological, Health and Socio-economic Consequences*, vol. 1, ISTE, London, and John Wiley & Sons, New York, 2019.

[ANO 21] ANONYMOUS, "Cobayes humains", created 06-03-2008, updated 30-09-2021, Cobayes humains - PerleDeDiamant (blog4ever.com).

[ASN 18] ASN, L'ASN rend un avis favorable sur l'arrêté du 27 juin 2018 portant délimitation des zones à potentiel radon sur le territoire national, 23 July 2018.

[AUR 03] AURENGO A., "Le risque de cancer du poumon est-il surestimé ?", *La Revue de l'Autorité de sûreté nucléaire*, no. 153, pp. 52–54, June 2003.

[BEL 95] BELL R., "Les cobayes humains du plutonium", *La Recherche*, vol. 26, no. 275, pp. 384–393, April 1995.

[BIN 13] BINDER-FOUCARD F., BELOT A., DELAFOSSE P. et al., Estimation nationale de l'incidence et de la mortalité par cancer en France entre 1980 et 2012. Partie 1 – Tumeurs solides, Saint-Maurice (Fra): Institut de veille sanitaire, July 2013.

[CEA 14] CEA, Périodes radioactives. Table des valeurs recommandées, Technical note LIST/LNHB/2014/18, 2014.

[COP 04] COPPIN F., ROUSSEL-DEBET S., "Comportement du $^{210}$Po en milieu terrestre : revue bibliographique", *Radioprotection*, vol. 39, no. 1, pp. 39–58, 2004.

[DÉL 85] DÉLIBRIAS G., "Le carbone 14", in ROTH E., POTY B. (eds), *Méthodes de datation par les phénomènes nucléaires naturels : Application*, Masson, Paris, pp. 421–458, 1985.

[FER 06] FERNANDEZ B., *De l'atome au noyau. Une approche historique de la physique atomique et de la physique nucléaire*, Editions Ellipse, Paris, 2006.

[GRN 99] GRNC, Estimations des niveaux d'exposition aux rayonnements ionisants et des risques de leucémie associés des populations du Nord-Cotentin, Synthèse, IPSN, July 1999.

[ICR 17] ICRP 137, Occupational intakes of radionuclides: Part 3, SAGE Publications, Thousand Oaks, 2017.

[ICR 75] ICRP 23, Report of the Task Group on Reference Man, Pergamon Press, Oxford, 1975.

[ICR 93] ICRP 65, Protection Against Radon 222 at Home and at Work, September 1993.

[ICR 94] ICRP 66, "ICRP Publication 66, Human respiratory tract, modem for radiological protection", *Annals of the ICRP*, vol. 24, nos. 1–3, 492 p. Pergamon Press, Oxford, 1994.

[IEL 17] IELSCH G., CUNEY M., BUSCAIL F. et al., "Estimation and mapping of uranium content of geological units in France", *Journal of Environmental Radioactivity*, no. 166, pp. 210–219, January 2017.

[IRS 01] IRSN, Uranium naturel et environnement, 14 May 2001.

[IRS 12] IRSN, Le radon, December 2012.

[IRS 13] IRSN, Les retombées des essais aériens d'armes nucléaires sur le sol français, April 2013.

[IRS 14] IRSN, Origines de la radioactivité, April 2014.

[IRS 16a] IRSN, La radioprotection des travailleurs, Exposition professionnelle aux rayonnements ionisants en France : bilan 2015, June 2016.

[IRS 16b] IRSN, 1986-2016 : Tchernobyl, 30 ans après, May 2016.

[IRS 17a] IRSN, Le radon, November 2017.

[IRS 17b] IRSN, La radioprotection des travailleurs, Exposition professionnelle aux rayonnements ionisants en France : bilan 2016, June 2017.

[IRS 18a] IRSN, La radioprotection des travailleurs, Exposition professionnelle aux rayonnements ionisants en France : bilan 2017, June 2018.

[IRS 18b] IRSN, Nouveaux coefficients de dose pour le radon recommandés par la CIPR dans sa publication 137, report, 3 July 2018.

[IRS 18c] IRSN, Les retombées de Tchernobyl en France, 21 April 2018.

[IRS 19] IRSN, La radioprotection des travailleurs, Exposition professionnelle aux rayonnements ionisants en France : bilan 2018, September 2019.

[IRS 20a] IRSN, Exposition de la population aux rayonnements ionisants due aux actes d'imagerie médicale diagnostique réalisés en France en 2017, August 2020.

[IRS 20b] IRSN, La radioprotection des travailleurs, Exposition professionnelle aux rayonnements ionisants en France : bilan 2019, June 2020.

[IRS 21a] IRSN, La radioprotection des travailleurs Exposition professionnelle aux rayonnements ionisants en France : bilan 2020, June 2021.

[IRS 21b] IRSN, Exposition de la population française aux rayonnements ionisants, Bilan 2014-2019, 20 September 2021.

[IRS 21c] IRSN, Bilan de l'état radiologique de l'environnement français de 2018 à 2020, November 2021

[IRS 21d] IRSN, Campagne nationale de mesure du radon, 1 January 2021.

[IRS 22] IRSN, La radioprotection des travailleurs Exposition professionnelle aux rayonnements ionisants en France : bilan 2021, June 2022.

[KOT 13] KOTERA K., "Cent ans de rayons cosmiques", *Reflets de la physique*, no. 32, pp. 8–13, January 2013.

[LAU 01] LAURIER D., VALENTY M., TIRMANCHE M., "Radon exposure and the risk of leukemia: A review of the epidemiology study", *Health Physics*, vol. 84, no. 3, pp. 272–288, 2001.

[MJÖ 03] MJÖNES L. "Le radon en Suède : 50 années d'expérience", *La Revue de l'autorité de sûreté nucléaire,* pp 104–108, no. 153, June 2003.

[PAU 97] PAULIN R., "Radionucléides naturels", in GALLE P. (ed.), *Toxiques nucléaires*, pp. 3–22, Masson, Paris, 1997.

[RAN 99] RANNOU A., "Connaissance actuelle des sources d'irradiation naturelle", *Radioprotection*, vol. 34, no. 4, pp. 505–519, 1999.

[REN 05] RENAUD P., GARCIA-SANCHEZ L., MÉTIVIER J.M. et al., Interprétation phénoménologique et analytique de la relation pluie-dépôt utilisée pour la reconstitution des dépôts de $^{137}$Cs en France consécutifs à l'accident de Tchernobyl, DEI report 2005-03, 3 November 2005.

[REN 08] RENAUD P., Les retombées en France des essais nucléaires atmosphériques, IRSN, 2008.

[TIL 14] TILOUINE J., "Algérie : itinéraire d'un nuage radioactif", *Jeune Afrique*, 5 March 2014.

[UNS 22a] UNSCEAR, Sources, effects and risks of ionizing radiation, vol. I, Annex A, Evaluation of medical exposure to ionizing radiation, United Nations, New York, 2022.

[UNS 22b] UNSCEAR, Sources, effects and risks of ionizing radiation, vol. IV, Annex D, Evaluation of occupational exposure to ionizing radiation, United Nations, New York, 2022.

[VAC 08] VACQUIER B., Analyse de la mortalité dans la cohorte française des mineurs d'uranium, Thesis, Université Paris XI, 15 October 2008.

# 8

# The Biological Effects of Ionizing Radiation at the Molecular and Cellular Level

There are numerous effects of ionizing radiation (IR) at sub-cellular and cellular levels. The classic distinction is between genetic instability, bystander or neighborhood effects, abscopal or distant effects, clastogenic effects, transgenerational effects and effects on the immune system. Most of these phenomena are particularly observable at low doses (<100 mGy) and low dose rates (<100 $mGy.h^{-1}$). In the early days of research into the adverse effects of irradiation, the conventional paradigm adopted was that damage induction is targeted in cells affected by IR, and that there is transmission of damage to daughter cells. However, in addition to targeted effects, there are non-targeted and delayed effects of IR.

Since the 1990s, there has been a major shift in the field of radiobiology, from a DNA-centric view of how radiation damage occurs, to a much more biological view that appreciates the importance of macro- and micro-environments, hierarchical organization, underlying genetics, evolution, adaptation and signaling at all levels, from atoms to ecosystems [MOT 12].

To understand the effects of IR at sub-cellular and cellular levels, we need to understand not only the physico-chemical processes behind these biological effects, but also the biological structures and mechanisms of the basic unit of life.

## 8.1. Physico-chemical processes responsible for biological effects

As soon as IR was discovered in 1895, its dangers became apparent. In 1896, the first radiation-induced skin lesions were described, and in 1902, the first case of radiation-induced cancer was observed, followed by experimental induced cancers in 1910 and the description of genetic mutations in 1927. The effects of IR on an organism result from the transfer of energy to matter, a complex process.

### 8.1.1. *Physical interactions of ionizing radiation with living matter*

The energy released by atoms is propagated by electromagnetic waves (gamma or X-rays) or particles (neutrons, beta or alpha particles). This energy causes three types of physical interaction with living matter: ionization, excitation and heat transfer. Ionization occurs when the energy of radiation exceeds the binding energy of electrons, and an electron is torn from the electron cloud. Excitation occurs when the energy is not sufficient to pull an electron out, but can be sufficient to move an electron from the fundamental energy level to a higher energy level (less bonded layer), in which case the atom is said to be excited. Heat transfer occurs when the energy is not sufficient to excite the atom, but can increase its translational, rotational and vibrational kinetic energies. On average, three excitations and a higher number of heat transfers occur for one ionization.

### 8.1.2. *Physico-chemical reactions*

More precisely, these are radiochemical reactions resulting from physical interactions and occurring in very short timescales of the order of $10^{-12}$ seconds. They mainly involve the formation of free radicals, produced when one or more electrons on an atom's outer electron layer become single. This gives the radical entity a very high level of chemical reactivity, as free radicals tend to capture an electron to complete their electronic layer. This phenomenon is mainly observed with the electrons of water molecules, as the water content of living organisms is extremely high.

The reactions that follow the formation of free radicals occur between $10^{-7}$ and $10^{-3}$ seconds after the interaction of IR, and constitute the diffusional stage. These reactions depend on the linear energy transfer (LET) of the radiation. The higher the LET, the greater the number of ionizations, and the higher the probability of encounters between the free radicals formed. In the case of water, the number of molecules formed is high ($H_2$, $H_2O$, $H_2O_2$) with, in particular, hydrogen peroxide ($H_2O_2$) being very harmful to cells.

The reactions that follow free radical formation also depend on the presence of oxygen. In the absence of oxygen, the reaction stops by dimerization. In the presence of oxygen, reactions can lead to the formation of a hydroperoxide radical ($OOH^\bullet$), an electron-hungry oxidant or a superoxide radical ($O_2$-).

The reactions that follow the formation of free radicals also depend on the presence of organic molecules. A large number of reactions take place, leading to the formation of hydrogen peroxide, peroxide radicals ($RO^{\bullet}$, $ROO^{\bullet}$) and tetroxides (ROOOOH).

Free radicals, which are extremely toxic to cells, are inactivated by a large number of enzymatic systems, including superoxide dismutase, peroxidases and catalases.

## 8.2. The effects of ionizing radiation at the molecular level

Radiation energy is transferred directly or indirectly to the molecule. When energy is transferred directly to the molecule, it is ionized or excited, and the excess energy acquired is lost through fluorescence radiation or the breaking of chemical bonds.

In the case of *indirect* energy transfer, energy is transferred to molecules in the vicinity of the molecules concerned, which in turn cause damage to the first molecules. This is notably the case with free radicals formed during the radiolysis of water. The direct effect seems to be predominant.

All molecules can be directly or indirectly affected by IR (Table 8.1). However, the molecule disturbed by IR, which will have the greatest repercussions at cellular and tissue levels for the individual and their descendants, will be DNA.

| Nature of DNA damage | Average number of lesions per Gy per nucleus |
|---|---|
| Formation of intra- or inter-chain links | 30 |
| DNA double-strand break | 40 |
| Formation of DNA–protein bonds | 150 |
| Single-strand DNA break | 750 |
| Modification of sugar structure | 1,200 |
| Modification of base structure | 1,400 |

**Table 8.1.** *Average number of lesions of various types, induced by a 1 Gy dose absorbed by a nucleus*

### 8.2.1. *The effects of ionizing radiation on DNA*

#### 8.2.1.1. *DNA, the essential molecule of life*

Much of the impact of IR is a consequence of the deterioration of DNA (deoxyribonucleic acid) by ionization. Several aspects of genetics are useful for understanding the harmful effects of IR on living beings.

DNA is a molecule present in all living cells, containing all the information necessary for the development and functioning of an organism. It is also the medium of heredity, since it is transmitted during reproduction, whether completely or partially. It therefore carries genetic information and constitutes the genome of living organisms. DNA determines protein synthesis, via ribonucleic acid or RNA. In the eukaryotic cells that make up human cells, DNA is contained in the nucleus and a small proportion in the mitochondrial matrix and chloroplasts.

DNA contains one sugar (D-deoxyribose) and four nitrogenous bases associated in pairs (adenine and thymine, guanine and cytosine). The four nucleosides (A, T, G and C) combine with phosphoric acid to form nucleotides.

The genetic code of DNA is based on the succession of three bases for each amino acid. Some amino acids are coded by a single triplet, while others are coded by several triplets. There are 22 genetically encoded amino acids. Essential or indispensable amino acids are those that cannot be synthesized de novo by the body. In humans, eight amino acids are essential (tryptophan, lysine, methionine, phenylalanine, threonine, valine, leucine and isoleucine), to which cysteine, glycine and tyrosine must be added in certain cases.

#### 8.2.1.2. *DNA replication*

The primary objective of every life form is to transmit its genetic material, intact and unchanged, to the next generation. This must be achieved despite the constant attacks of endogenous and environmental agents on DNA [JAC 09].

The majority of higher organisms, including humans, are diploid organisms (haploid gametes and diploid zygotes), i.e. they possess double genetic information from each parent. A gene located in the same place, on the same chromosome, will therefore be represented by two alleles. If the alleles are identical, the organism is homozygous for that trait; otherwise, it is said to be heterozygous for that pair of alleles.

During meiosis, different haploid gametes are separated, each with two copies, i.e. two alleles, of each gene. As a result, for a heterozygous organism, four gametes are formed, two carrying the information of parent A and two carrying the information of parent B.

The principle of replication is to go from two to four strands of DNA. The two original strands separate, and each nucleotide that becomes free is joined by a complementary nucleotide supplied by the cytoplasm. Thanks to certain enzymes (DNA polymerase), these partners join to form a chain complementary to the other. In this way, each chain gives rise to a new double helix. Numerous enzymes are involved in replication (topoisomerase, helicase, DNA lipase, etc.).

#### 8.2.1.3. *DNA damage*

DNA can be damaged by a variety of factors, such as IR or oxidative stress [BAR 07]. DNA damage is caused by two main mechanisms: one-electron oxidation and hydroxyl radical-induced decomposition. The cell will respond to this stress with varying degrees of adaptation, and the consequences are shown in Figure 8.1. The consequences of DNA damage can have serious biological consequences, particularly during replication or transcription, so the quality of repairs is essential.

There are five types of DNA damage: single-strand breaks (SSBs), double-strand breaks (DSBs), base modification, bridging between surrounding proteins and DNA, and sugar alterations [LAU 05]. As living beings have always been subjected to irradiation, which was even more intense in the past[1], they have also developed highly effective repair mechanisms.

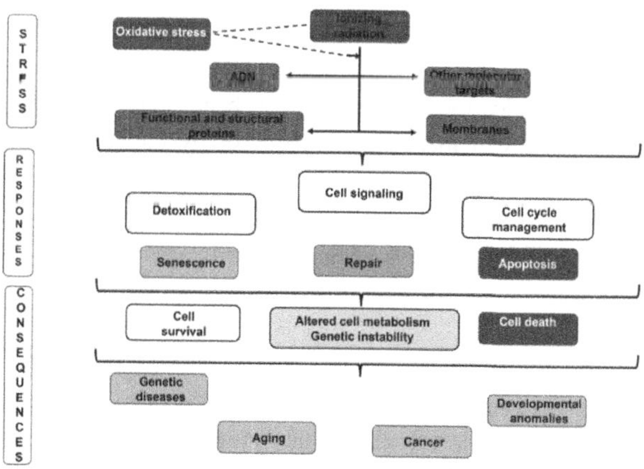


**Figure 8.1.** *Cellular responses and consequences of exposure to ionizing radiation (according to Bertrand et al., pp. 257–274 [MEN 09])*

1 4.5 billion years ago, the natural radioactivity of the Earth's crust was 5,850 $Bq.kg^{-1}$ and today only 1,330 $Bq.kg^{-1}$ [PRA 02].

### Single-strand breaks

For low-LET radiation, approximately 1,000 SSBs are produced per mammalian cell and per Gy. They result from the removal of a hydrogen atom from the sugar by the •OH radical, leading to the breaking of phosphate–sugar bonds. The dose–response relationship is linear, meaning that SSB formation increases with irradiation dose. The level of SSB is lower with higher LETs. SSBs are quickly repaired, within an hour of radiation exposure, and would have little impact on cell lethality.

### Double-strand breaks

These breaks are far more deleterious. For low-LET radiation, there are 40–100 per mammalian cell per Gy. They occur when the two chains break at sites that are fairly close to each other. Two mechanisms may explain their formation: either a •OH radical acts on the 2-deoxyribose and the radical is then transferred to the second strand, or several •OH radicals act on the DNA molecule at nearby sites. Repair of these breaks can be slow, which explains the radiosensitivity.

DNA DSBs are among the most significant genetic lesions introduced by IR that can initiate carcinogenesis. During a CT scan, the number of induced DSBs depends linearly on the dose–length product, a unit of radiodiagnostics that is proportional to both the local dose delivered and the length of the exposed body part. It is therefore possible to assess in vivo DSB induction and repair in individuals exposed to low doses of radiation [LOB 05].

### Modified bases

Some 30 base modification products have been identified by gas or liquid chromatography coupled with mass spectrometry or by nuclear magnetic resonance. The number of bases modified per eukaryotic cell per Gy is estimated at 2,000 at low LET. Modifications affect pyrimidine and purine bases. They mainly result from hydroxyl radical attack on the aromatic ring, but are also produced by direct ionization.

### DNA–protein bridges

After exposure to gamma radiation, bridging, in which the •OH radical may be involved, can occur between DNA and surrounding proteins (30 per cell per Gy), between the two strands or within the same strand. Tyrosine–thymine bridges, or bridges between different amino acids and thymine or cytosine, have been identified.

### Sugar alterations

The •OH radicals can lead, via the removal of a hydrogen atom, to the formation of a strand break by the release of 2-deoxyribose or by the alteration of the sugar that remains linked to the 3' or 5' end of the phosphodiester backbone, or to the formation of an abasic site, i.e. an altered sugar but linked by the 3' and 5' ends to the phosphodiester backbone. Various altered sugars have been identified.

Lesions at the molecular level can be quantified using three detection methods: tests to measure and characterize radiation-induced lesions (comet method, chromatography, molecular biology, pulsed-field electrophoresis), tests to determine the repair activity of these lesions and, finally, mutagenicity tests (Ames test, mutation of the HPRT (hypoxanthine–guanine phosphoribosyltransferase) gene). The main role of "omics" approaches is to understand gene function, but they also serve as biomarkers of deleterious effects and enable integrated analysis of the mechanisms of action of IR. These approaches are applied at different levels: messenger RNA (transcriptome), expressed proteins (proteome) and metabolites (metabolome) (Junot et al., pp. 235–256, in: [MEN 09]).

### 8.2.1.4. *DNA repair*

There are numerous repair mechanisms: repairs of mismatches (presence of pairs other than A.T or G.C), etc. Repairs of modified bases are carried out either by reversions, or by excision and replacement synthesis [FRI 06]. Nucleotide excision repair (NER) involves several enzymatic steps [HOE 01]. Damage is recognized by sensors, and signal transducers amplify the signal throughout the cell by activating effectors acting in their corresponding metabolic pathways. Transducers are activated by ATM or ATR, and mediators facilitate signal transmission (Figure 8.2).

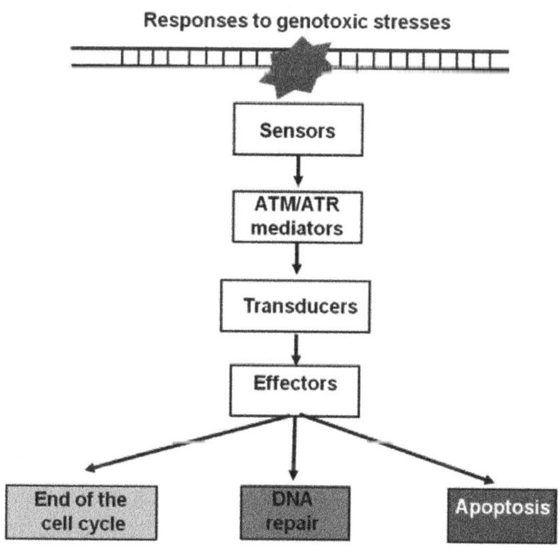


**Figure 8.2.** *Cellular response to genotoxic stress (according to Bertrand et al., pp. 257–274 [MEN 09])*

Base excision repair (BER) is probably the most common DNA repair mechanism in living organisms. The modified base is recognized by a specific DNA glycosylase, which excises it. The resulting AP site is cleaved by an AP endonuclease (APE), which generates a substrate for the synthesis and ligation steps, respectively, by a DNA polymerase β and a ligase 3.

In eukaryotes, DSBs are repaired in several ways: either by homologous recombination (HR), which restores the original sequence, or by other means, which may introduce potentially deleterious alterations. Alterations may be small in the case of non-homologous end joining (NHEJ), larger in the case of single-strand annealing (SSA) or loss of heterozygosity in the case of break-induced replication (BIR) ([JAC 02]; Radicella et al., pp. 175–194, in [MEN 09]).

The resulting lesions are usually removed by the BER pathway, resulting in short replacement fragments in one of the two DNA strands [LIN 00].

The main mechanisms involved in the repair of endogenous and IR-induced DNA damage are BER, HR and NHEJ [LAU 05, MON 05]. These mechanisms have been described by Averbeck [AVE 00], Hoeijmakers [HOE 01] and Rothkamm and Lobrich [ROT 03].

DNA DSBs are key events in the response to IR which, together with the individual genetic and epigenetic profile, can condition the fate of healthy tissue in an exposed individual. Over 70% of cells exposed to 5 Gy following radiotherapy have at least one focus persisting 24 hours after exposure [VAU 16]. Repair of this type of damage is therefore imperfect and slow.

### 8.2.2. *Epigenetics*

Epigenetics is a new discipline which studies the influence of the cellular or physiological environment on gene expression, and attempts to understand the mechanisms at work in this process. Indeed, the same gene can be read differently in different circumstances. The sequence of nucleotides (DNA), i.e. the genetic code, that makes up these genes remains unchanged. However, the proteins encoded by these genes may be produced at different times or in different places, depending on the epigenetic marks present on the genes. These marks result from a gene's environment. This environment includes the proteins that make up the chromosomes, the cell (nucleus, cytoplasm, cellular organelles), the tissue, the organ, the individual, etc., and all the conditions which affect individuals, such as diet, activities or location. This whole environment, from the molecular to the ecological level, is likely to modify gene expression (Figure 8.3).

Whether or not a cell can use the gene to make a given protein depends in particular on chemical modifications to the DNA, the gene's regulatory sequences and the proteins surrounding the DNA (histones). For example, enzymatic reactions such as methylation of regulatory sequences or histone deacetylation can render a gene "silent", i.e. prevent its expression. These enzymatic reactions themselves depend on the particular history of each cell and its environment.

This is why an identical genome present in the initial cell of an organism at the origin of the embryo will express itself differently, producing some 200 types of cell families in higher organisms. Similarly, a gene will not necessarily have identical expression depending on whether it has been transmitted by the father or the mother.

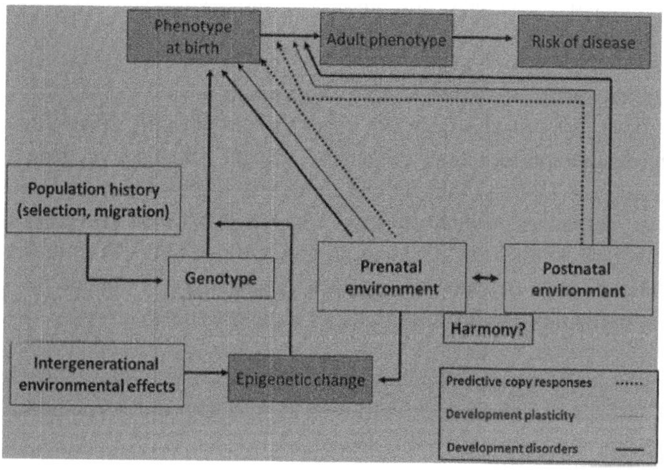


**Figure 8.3.** *Epigenetics. Nature of the adaptive predictive response (APR) determined by the predetermined and current postnatal environment*

Epigenetic changes include DNA methylation, histone modification and RNA-associated silencing [KOV 08]. Epigenetics is known to play an important role in many cellular processes, and epigenetic aberrations can lead to carcinogenesis. IR is involved in epigenetic phenomena.

Thus, fractional irradiation is able to induce epigenetic effects (variation in DNA methylation) in vivo in unirradiated cells (bystander effect, BE) in both spleen and skin, with distinct methylation changes in both cell types. In the spleen, DNA hypo-methylation has been observed alongside reduced expression of the methyl-binding protein MeCP2. Irradiation has also induced alterations in tissue-specific microRNAs in the skin and spleen [ILN 09].

Several authors [AYP 11a, AYP 11b] hypothesize that epigenetics, in particular DNA methylation and microRNA (miRNA) expression, could be the missing link in understanding radiation-induced genomic instability (RIGI). Indeed, epigenetic aberrations induced by low-LET irradiation are different from those induced by high-LET irradiation. This distinction is apparent in miRNA analyses in which only three miRNAs involved in two major pathways were altered after high-LET irradiations, while six miRNAs involved in five major pathways were altered after low-LET irradiations. The exact mechanisms of reduced radiosensitivity with high-LET radiation are unclear. For low-LET X-rays and carbon ions (13 keV.$\mu m^{-3}$), halogenated pyrimidine incorporation resulted in altered DNA repair kinetics, a higher initial number of DNA DSBs, more frequent chromosomal aberrations at the first post-irradiated metaphase and increased radiosensitivity with cell lethality [FUJ 13].

The current system of radiation protection assumes that exposure to low doses and/or low dose rates results in health risks that are linearly related to dose. The observation of radiation-induced epigenetic effects (changes in gene expression without altering the DNA sequence) and nonlinear responses, such as non-targeted and adaptive responses, which in turn can be controlled by gene expression networks, challenges this view of radiation protection, as the risk is underestimated. More realistic models of radiation-induced cancer should include an epigenetic contribution, particularly in the initiation and progression phases [BEL 20].

Some authors also believe that epigenetic effects are also relevant to the possible “beneficial” effects of low-dose and/or low-dose-rate exposures, including those due to natural background radiation [BEL 20].

The hereditary effects of low doses of radiation in early life can be either positively adaptive, or lead to increased disease formation, including cancer, diabetes and obesity. Bernal et al. [BER 13] found that mean DNA methylation was significantly increased in male offspring exposed to doses between 7 and 76 mGy, with maximal effects at 14 and 30 mGy ($P<0.01$). Thus, exposure to low doses of radiation during gestation induces epigenetic alterations leading to positive adaptive phenotypic changes. These changes are reversed by antioxidants, indicating that they are partly mediated by oxidative stress.

The heritability of epigenetic marks through meiotic cell divisions raises the question of the transmission of certain acquired characteristics through the generations [COR 20].

The literature on epigenetic phenomena such as DNA methylation, histone modifications, chromatin remodeling and non-coding RNA modulation in low-dose radiation responses is becoming abundant, and tentative progress has been made in understanding the epigenetic regulation of low-dose radiation responses. However, many questions remain to be answered [MA 10].

## 8.3. The effects of ionizing radiation at the sub-cellular level

### 8.3.1. *The main cell organelles*

Each cell is relatively autonomous, consisting of numerous organelles with various functions to keep it running, growing and multiplying. The main cellular organelles are those isolated within the cell by membranes, such as the nucleus. The latter contains the bulk of the cell's genetic material, particularly within the chromosomes; the mitochondria, the cell's energy factories; the endoplasmic reticulum (ER), rough and smooth; the lysosomes, whose digestive enzymes ensure the breakdown (hydrolysis) of used molecules and organelles; and the Golgi apparatus. They also include various large biomolecular complexes, such as the centriole, microtubules, cytoskeleton, flagella and nucleolus, as well as large RNA and protein complexes (ribosome, spliceosome, vault), large DNA and protein complexes (nucleosome), large protein complexes, etc.

### 8.3.2. *The organization of DNA into nucleotides or chromosomes*

The DNA double helix associates with histones (proteins), forming nucleotides which, by successive windings, form the chromosomes. During cell division, each DNA cell condenses into chromatids linked by a centromere. Chromosomes are generally short, with a variety of shapes, notably U- or V-shaped. They have two arms, sometimes of different lengths, separated by a primary constriction, the centromere. The centromere serves as a "guide" during meiosis. Chromatin is not inert, but rather a dynamic structure, carrying epigenetic information that modulates gene expression [COR 20].

### 8.3.3. *The effects of ionizing radiation on genomic instability and mutations*

Organisms are continually subjected to stresses such as IR, chemical contaminants and free radicals, which damage DNA and generate genetic instability. Genome instability manifests itself as chromosomal aberrations, genetic mutations, late cell death and aneuploidy. Genome stability is essential to counteract aging, cancer, neuronal and/or congenital disorders. Genetic instability is a characteristic of cancer cells and can be an initiating event in tumorigenesis and radiation-induced carcinogenesis. RIGI is observed in the progeny of irradiated cells that may or may not have been directly affected by a radiation energy deposition. Moreover, genetic instability can be maintained over cell generations (Bertrand et al., pp. 257–274, in [MEN 09]).

The challenge of genome integrity is immense, as the cell must prevent mutations and chromosomal aberrations. Thousands of lesions occur every day, mostly as a result of normal cellular metabolism and environmental genotoxins (tobacco, alcohol, infectious agents, IR, chemical toxins, etc.). Every cell has its own signaling and repair system [LIN 00, JAC 09].

### 8.3.4. *The appearance of mutations*

Mutations appear suddenly and randomly. They occur in all living organisms and in all cells. There are several classic types of mutation. Firstly, there are *somatic* mutations, which occur in organ stem cells, and *germline* mutations, which occur in gametes. Only germline mutations can be transmitted to offspring.

#### 8.3.4.1. *Frequency of mutations*

The mutation rate can be modified by the conditions surrounding the individual. The main parameters influencing the mutation rate are temperature, certain chemical substances, ultraviolet radiation and, of particular interest in this book, IR.

#### 8.3.4.2. *DNA mutation mechanisms*

Inheritable alterations in gene expression occur via four mechanisms (Figure 8.4). The first results from DNA mutation. The mutation of one base into another (in this case, G into C) may suffice to include another amino acid in the protein, with potentially significant consequences for its secondary and tertiary structures, and hence its functionality. The second is a deletion of part of one or more genes. The result is inactivation of the gene(s) thus amputated. The third mechanism is a rearrangement of gene order, which can have consequences for gene expression. The final mechanism is the amplification of a gene into several copies (from two to several dozen). This results in a considerable increase in the production of the protein (often enzymes) induced by the gene. This case is frequently encountered in the acquisition of tolerance to various toxins such as insecticides in mosquitoes. Taking the last two mechanisms together, three types of mutation can be classically identified: point mutations, chromosomal mutations and ploidy mutations.

#### 8.3.4.3. *Chromosome mutation types*

There are three main types of mutation: point mutations, chromosomal mutations and ploidy mutations.

Point mutations concern only one or a limited number of nucleotides. They occur in isolated genes when, under the action of external factors, they mutate spontaneously (without the trigger being known). Since a gene is made up of an average of 2,000 nucleotides, a single nucleotide change can result in a mutation by addition, subtraction or replacement of one or more nucleotides. Two cases are possible: either the normal base is replaced by an analogous base in enolic or ketonic form, eventually causing a change in the complementary base; or by chemical modification of the normal bases, again causing a change in pairing. The spontaneous mutation rate for a given gene is approximately $10^{-8}$ per generation. At the individual level, the mutation rate is considerable, estimated at 10–40% of human germ cells per generation.

Chromosomal mutations correspond to a change in the order of genes on the chromosome, following a break in the DNA strand and abnormal restructuring. Most often, the mutation corresponds to the loss of a chromosomal fragment (deletion). It can also be a translocation, with exchange of fragments between chromosomes, or an inversion. Chromosomal mutations can affect a large part of a chromosome.

Ploidy mutations are cytogenetic manifestations resulting from mechanical accidents to the chromosome during mitosis. Two mechanisms are possible. The first results from chromosome maldistribution, leading to abnormalities in chromosome number. Cells are normally diploid. Following abnormal mitosis, cells may be polyploid (several sets of chromosomes) or aneuploid (absence or additional presence of one or more chromosomes, leading to an odd number of chromosomes). The second mechanism corresponds to chromosome fragmentation leading to structural anomalies (translocations, inversions, acentric fragments). In addition, the integrity of cellular genetic material can be affected by the formation of micronuclei. These are nuclear entities independent of the nucleus, present in the cytoplasm, originating from whole chromosomes or fragments separated from the rest of the chromosomes, during anaphase.

Chromosome rearrangements, when they occur in somatic cells, can induce disturbances at the individual level (altered translation or transcription of a gene, formation of abnormal proteins, etc.) but have no consequences for the individual's descendants. In the germ line, however, the situation is quite different: even a balanced chromosomal rearrangement can lead to an erroneous distribution of genetic material at meiosis. Only stable aberrations (translocations, inversions) are passed on to offspring, and can be detected after several post-irradiation cell generations, while more complex aberrations (dicentric and acentric rings) are difficult to pass on.

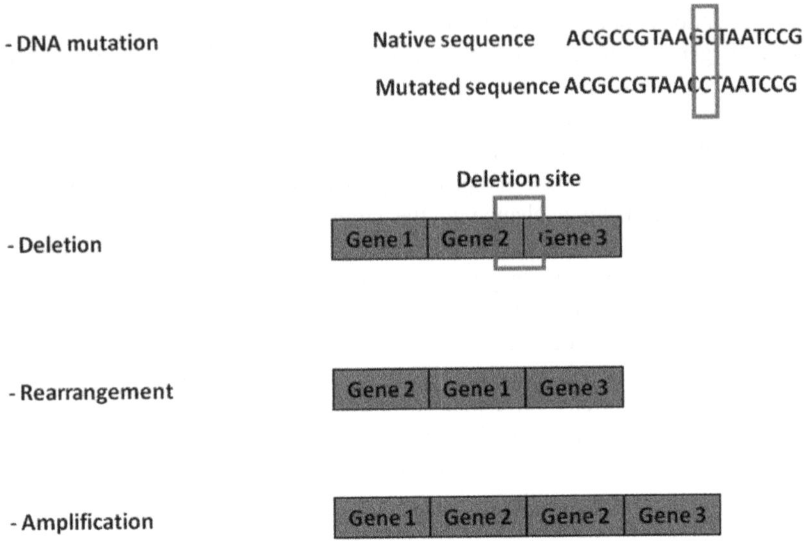


**Figure 8.4.** *The various mechanisms of heritable alterations in gene expression*

### 8.3.4.4. *Consequences of mutations*

Three types of mutations are differentiated according to their biological effects. A mutation is non-deleterious if it gives rise to a new allele that has no effect on the offspring's health or fertility. A mutation is recessive deleterious if individuals carrying two mutated alleles do not reproduce. A mutation is dominant deleterious if only one mutated allele in the genotype is required for the individual to be unable to reproduce [MOR 08].

### 8.3.4.5. *Mutations caused by ionizing radiation*

Radiation can damage human chromosomes, both in vivo and in vitro [HOL 68].

IR induces chromosomal aberrations in directly exposed cells, as well as in non-irradiated neighboring cells. These clastogenic effects of irradiation could be due in part to radiation-induced plasma factors. Thus, non-irradiated cells exposed in vitro to irradiated serum (15 minutes after irradiation) showed a 36- to 48-fold increase in hyperdiploidy ($p = 0.0001$) and a 2- to 2.2-fold increase in chromatid gaps and breaks ($p < 0.01$), but none of the chromosome-type aberrations observed in radiation-exposed cells [FAG 84].

Exposure to γ-rays at low-dose and low-dose-rate radiation accounts for most occupational, medical and environmental irradiation. This type of exposure triggers specific molecular responses in human keratinocytes. More than 140 specific genes have been identified, most of which were modulated within 48 hours. Seventeen known genes were analyzed in more detail [FRA 05].

IR can induce sister chromatid exchanges (SCE). Cells irradiated with alpha particles from an external plutonium-238 source show a significant increase in SCE frequency with doses as low as 0.31 mGy. In contrast, a dose of approximately 2.0 Gy was required to produce a similar increase in SCE by X-rays. These results indicate that genetic damage can be induced by low doses of alpha radiation in cell nuclei that are not actually crossed by an $\alpha$ particle [NAG 92].

## 8.4. The consequences of irradiation at cellular level

### 8.4.1. *Cell structure*

The structure of a human cell includes a lipid membrane that perfectly delimits its environment. Similarly, the nucleus includes a membrane and contains the chromosomes. Outside the nucleus, in the plasma, are various cellular organelles with different functions. These include in particular the mitochondria and the ER.

### 8.4.2. *The cell cycle*

The cell cycle is the set of stages that constitute and delimit the life of a cell. This cycle is made up of two periods, a growth period in which the cell grows and duplicates its genetic material (interphase) and an M period in which it divides (mitosis) to give rise to two identical daughter cells (in the case of mitosis). The period of growth or interphase is made up of three phases: the G1 phase of cell growth and normal metabolic activity, the S phase of DNA replication (the amount of DNA is doubled in preparation for mitosis) and the G2 phase preparing for mitosis (synthesis of enzymes, organelles, etc.).

In the case of sex cells, mitosis is replaced by meiosis. Meiosis has a dual purpose: on the one hand, to mix the paternal and maternal genomes, thus ensuring maximum genetic variation, and on the other hand, to produce single-chromatid haploid cells for sexual reproduction.

### 8.4.3. *Cell division*

Cell division in somatic cells is called mitosis and for germ cells meiosis.

#### 8.4.3.1. *Somatic cell division or mitosis*

In human cells, cell division or mitosis takes place in five phases. Division often lasts several hours. The first phase, or interphase, is the period between two divisions, during which the nucleus and cell are metabolically active, renewing their constituents. The second phase, or prophase, is characterized by RNA accumulation and tighter spiraling of chromatin filaments.

The third phase, or metaphase, sees the chromosomes spiralize and take on their divided form. Longitudinal splitting into two chromatids is complete. The chromatids group together at the spindle equator to form the equatorial plate.

The fourth phase, or anaphase, is relatively rapid. The chromatids separate from each other and rise towards the poles, pulled by the centromere.

The fifth and final phase, or telophase, sees the two chromatid groups return from split to interphase. The nuclear membrane is reconstituted with elements of the ER. The cell body undergoes equatorial constriction, creating two independent cellular entities [VOG 67].

Mitosis therefore comprises three different but coordinated processes. These are the identical duplication of DNA, where chromatids cleave and then split, the division of half-chromatids into two identical batches in the two daughter nuclei (karyokinesis) and the division of cytoplasm between the two daughter cells (cytokinesis).

At all stages of mitosis, abnormalities can occur, such as daughter chromosomes remaining in the original nucleus, which then has twice as many chromosomes. Normal diploid cells (2n) thus become successively tetraploid (4n), octoploid (8n), etc.

#### 8.4.3.2. *Germ cell division or meiosis*

If fertilization took place between diploid cells, the number of chromosomes in the zygote doubles with each generation. There must be chromatic reduction. This takes place before fertilization, and at the latest when the gametes are formed, so that they contain a single haploid chromosome set. This reduction takes place during meiosis, by means of two successive maturation divisions, during which homologous chromosomes come together and then separate, to be distributed randomly. The result is possible genetic recombination at this point.

### 8.4.4. *Targeted effects of ionizing radiation*

For decades, the classic paradigm was that only cells affected by IR would eventually suffer an adverse effect as a result of DNA breakage. Once the break had been repaired, there were three possible outcomes. In the first case, the repair was correct, and the cell continued its destiny, subdivided and suffered no consequences. Only the energy balance was negative, since any new molecule manufacture or DNA repair had an energy cost. In the second case, the repair was very faulty, and the cell was no longer viable. It then entered programmed death or apoptosis. If a large number of cells were affected by a high dose of irradiation, the result was tissue necrosis, leading to the complete loss of activity in the offending organ, or even the death of the individual. In the third case, repair is imperfect, a mutation remains, and after a more or less long latency period and a certain number of cell divisions, a cancerous cell appears, initiating a cancerous tumor.

The traditional model is therefore based on the theory of "stochastic breakage" of one or both strands of the DNA double helix. According to this model, high doses can cause breaks, potentially fatal to the cell by damaging both DNA strands, whereas low doses of IR would essentially cause SSB, easily repaired, resulting in no permanent damage. Available evidence makes this classical model less and less acceptable, as exposure to low doses of IR appears to have carcinogenic effects, even after years or decades, both in exposed individuals and in subsequent generations. Moreover, cells that have survived low-dose exposure, although apparently normal, accumulate damage that becomes evident in their progeny, such as non-clonal chromosomal aberrations, which can be found even in cells not directly irradiated due to the exchange of molecular signals, and complex tissue reactions involving neighboring or distant cells. For all these reasons, a paradigm shift is needed, based on evidence and epigenetics [BUR 18].

### 8.4.5. *Non-targeted effects of ionizing radiation*

A multitude of biological effects observed from the 1990s onwards in various in vivo and in vitro cell culture experiments indicated that low-dose/low-fluence IR has significantly different biological responses to high-dose radiation. This challenged the classical paradigm.

Cell growth, differentiation and death are largely driven by extracellular signaling through cell interactions with other cells and with the extracellular matrix. These interactions are in turn modulated by cytokines and growth factors, i.e. the microenvironment. Proximity effects induced by IR and genomic instability are, respectively, positive and negative manifestations of this homeostatic process. The side effects that mainly manifest themselves after low-dose or inhomogeneous radiation exposure are extracellular signaling pathways that modulate cell repair and death programs [BAR 01].

#### 8.4.5.1. *The proximity or "bystander" effect*

The proximity or BE (or neighborhood effect) is observed in cells not directly affected or not penetrated by radiation, and involves the emission of signals by irradiated cells. As a result, the volume at risk is greater than the volume irradiated. RIGI can occur in "bystander" cells, i.e. in cells not directly impacted by alpha particles [KAD 92]. Experiments with protective grids [LOR 98] demonstrate the secretion of intercellular communication factors by junctions between irradiated cells (dying cells) and non-irradiated cells.

Conventionally exposed fibroblasts showed dose-dependent production of micronucleated and apoptotic cells three days after irradiation for both X-rays and $\alpha$ particles, with a high relative biological effect (RBE) value for low-dose $\alpha$ particles. Microbeam targeting of individual $\alpha$ particles to four cells in a population produced more micronucleated and apoptotic cells than expected on the basis of direct effect alone. The study by Prise et al. [PRI 98] provides direct evidence of the production of cell-to-cell transmissible effects between affected and unaffected cells individually exposed to charged particles.

With the availability of a precision single-particle microbeam, it is possible to demonstrate, unequivocally, the presence of a BE with numerous biological endpoints. These studies provide clear evidence that irradiated cells can induce a bystander mutagenic response in neighboring cells not directly impacted by the $\alpha$ particles, and that cell-to-cell communication processes play an essential role in mediating the bystander phenomenon [HAL 03].

The BE therefore amplifies the effect of small doses of radiation. However, there is an adaptive response, whereby exposure to a low level of DNA stress renders cells resistant to further exposure. This reduces the effect of low doses of radiation. Thus, when the single-cell microbeam targeted the nuclei of 10% of mouse fibroblast cells with 1–12 alpha particles, more cells were inactivated than were actually impacted by the $\alpha$ particles. The magnitude of this proximity effect increased with the number of particles per cell. An adaptive dose of 0.02 Gy of $\gamma$ radiation, delivered six hours in advance, canceled out about half the proximity effect produced by $\alpha$ particles [SAW 01].

In a human epithelial cell line, doses of 0.01–0.5 Gy show cell death solely due to the BE. The magnitude of the effect is relatively constant and appears to saturate at doses between 0.03 and 0.05 Gy [SEY 00].

The first demonstration of a link between the BE of IR and the induction of genomic instability in vivo was made by Watson et al. [WAT 00].

In healthy human tissues, unirradiated cells up to 1 mm away from irradiated cells showed a significant increase in adverse effect compared to controls, with a mean increase in effect of 1.7-fold for micronuclei and 2.8-fold for apoptosis [BEL 05].

The BE was also observed in vivo in mouse tumor tissue where the DNA was labeled with iodine-125. This radionuclide had only a sub-cellular impact, yet neighboring unlabeled cells reacted. This phenomenon has a significant impact on the current dogma for assessing the therapeutic potential of internally administered radionuclides [XUE 02].

Observations in the literature demonstrating the induction of radiation effects in optically coupled but chemically separated organisms raise the hypothesis that alternative communication mechanisms may exist that have not yet been explored. [LE 17] demonstrates that intercellular communication via electromagnetic signaling stimulated by low-dose IR exists. This takes place via biophotons which are stimulated in particular by β-radiation. This has been shown to reduce cell survival. Ultraviolet electromagnetic radiation is therefore an additional mechanism for intercellular communication between biological systems disturbed by low doses of IR.

The transfer of radio-induced microRNA (nuclear miR-34c) via extracellular vesicles into non-irradiated cells can induce bystander oxidative stress [RAS 18].

The BE therefore has an impact on both the estimation of the risks of exposure to IR and radiotherapy. The amplification of radiation-induced cytotoxic and genotoxic effects by the BE contrasts with observations of adaptive responses, which are generally induced following exposure to low doses of radiation with low LET, and which tend to attenuate radiation-induced damage [AZZ 04].

Over the past 30 years or so, evidence has been generated in vitro, indicating that non-target cells in irradiated cell cultures also undergo significant biochemical and phenotypic changes, which are often similar to those observed in target cells. These neighboring cells undergo stressful effects, including oxidative and oncogenic effects. The bystander phenomenon may be responsible for the induction of second cancers that appear after radiotherapy [AZZ 13].

Several mechanisms involving secreted soluble factors, oxidative metabolism, gap junction intercellular communication and DNA repair have been proposed to regulate radiation-induced proximity effects [AZZ 13].

In addition to chromosomes, other sub-cellular organs are likely to be affected by IR. It is well known that mitochondria and the ER play an important role in the radiation response, but their functions in the radiation-induced BE are largely uncertain. Using a staining technique (Mito/ER-tracker), it was demonstrated that mitochondria were clearly activated after nuclear irradiation and that ER mass increased after cytoplasmic irradiation [DON 20].

#### 8.4.5.2. *Abscopal effects and clastogenic factors*

*Abscopal* effects are the effects of irradiation on tissues far from the irradiated site. They include the induction of genomic instability, cell death and malignant transformation of normal tissues. They can also have a beneficial effect, with an antitumor effect induced by radiotherapy outside the irradiation field, leading to regression of metastases after irradiation of the primary tumor. This beneficial effect was first observed in a 76-year-old Japanese man suffering from hepatocellular carcinoma. This cancer regressed after radiotherapy for thoracic vertebral bone metastasis (dose of 36 Gy to the bone). Spontaneous regression of hepatocellular carcinoma is a rare phenomenon. Abscopal regression of tumors resulting from the effect of irradiation of one tissue on a distant non-irradiated tissue is also rare [OHB 98].

Abscopal effects are thought to be due to the presence of clastogenic factors in extracellular fluids. A clastogen is a genotoxic substance capable of causing breaks in a DNA molecule, i.e. chromosomal aberration.

The appearance of chromosomal aberrations in lymphocytes from a non-irradiated mouse cultured with plasma from an irradiated mouse has been demonstrated and attributed to clastogenic factors. Since then, clastogenic factors have been found in humans in numerous circumstances. Thus, Goh and Summer [GOH 68] found that the plasma of patients subjected to irradiation is capable of inducing chromosomal disruption of normal leukocytes in other, non-irradiated individuals. Scott [SCO 69] reported that a substance either produced or activated by total body irradiation remains capable of affecting chromosomes for long periods. This has also been observed in individuals irradiated in radiotherapy [HOL 68], in survivors of Hiroshima and Nagasaki [PAN 77] and in liquidators at Chernobyl [EME 94]. The appearance of clastogenic factors has also been detected in the plasma of children exposed at Chernobyl [EME 97].

Clastogenic factors were detected in the plasma of 32 civilian workers from Armenia, who had been hired as "liquidators" around the Chernobyl atomic power plant in 1986, and also in 15 liquidators who had emigrated from the former Soviet Union to Israel. The first Armenian group, with the highest average irradiation dose (0.6 ± 0.6 Gy), were more clastogenic than those in the second group exposed to 0.2 ± 0.2 Gy. The number of aberrations in the test cultures was 17.9 ± 2.9% and

10.5 ± 3.8%, respectively, compared with 5.7 ± 3.2% in cultures exposed to reference ultrafiltrates from Armenian blood donors. Clastogenic activity was consistently inhibited by superoxide dismutase [EME 94].

The abscopal effect is not tumor-specific and is dose-dependent. The key mediator of the radiation-induced abscopal effect is the p53 gene, involved in the control of the cell cycle and apoptosis (programmed cell death). Observations suggest that pathways downstream of p53 are important in triggering this specific response [CAM 03].

## 8.5. Quantifying the effects of ionizing radiation using biomarkers

A distinction is traditionally made between general and specific biomarkers. General biomarkers provide information on the health status of individuals in a given population [AMI 13, AMI 17]. Examples include condition indices, gonado-somatic indices, growth indices and energy indices.

Biomarkers can be classified according to their action into defense biomarkers and damage biomarkers. Defense biomarkers (adaptive effect) correspond to biological phenomena whose appearance is induced by the presence of a xenobiotic in the environment. Examples include metallothionein induced by certain metals including their radionuclides, cytochromes P450 induced by PAHs and PCBs, and AChE induced by pesticides. Damage biomarkers testify to disturbances at various levels (physiological, histological, mutagenesis, etc.) induced by the presence of a xenobiotic in the environment.

Biomarkers are used in many scientific disciplines for a variety of applications [AMI 15]. In particular, they are used to monitor human health, as well as that of flora and fauna subjected to chemical pollution. The literature on biomarkers for monitoring human health is abundant [GUP 19, PAT 23].

### 8.5.1. *Biomarkers specific to the effects of ionizing radiation*

There are biomarkers that respond to exposure to low doses of external and internal irradiation in situations of radiological contamination of the environment. The chemical properties of radionuclides play a decisive role in their metabolism by living organisms. Nevertheless, and contrary to what generally happens, in the event of pollution, the changes observed in organisms exposed to IR are not, with a few exceptions such as uranium, due to the chemical properties of the pollutants, but to their nuclear properties. In general, therefore, it is not the concentration of pollutants that needs to be considered, but their type and intensity of radiation.

Radionuclide exposure biomarkers do not necessarily reflect an adaptive or deleterious effect, but their presence does reflect radioactive contamination of the individuals producing the biomarker. The quantity of the exposure biomarker in the organism is proportional to its level of radioactive contamination [AMI 14].

#### 8.5.1.1. *Metallothionein*

Metallothionein is a ubiquitous protein in the living world. Its main characteristic is its high content of a particular amino acid, cysteine, which contains a thiol group. In humans, cysteine accounts for over 33% of all amino acids. This composition imposes a specific three-dimensional structure in two domains $\alpha$ and $\beta$ and enables this molecule to trap several metal atoms per molecule. The atoms are attached by several thiols, which are brought together by the molecule's spatial folding [AMI 06].

The primary role of this molecule is to ensure homeostasis of metals essential to living beings, such as zinc and copper, which are catalysts for numerous metalloenzymes. This role has been hijacked by organisms to combat the toxicity of other metals and metalloids such as mercury, cadmium and various radionuclides [CAI 99].

#### 8.5.1.2. *The Comet assay and genotoxicity*

Radiation-induced DNA breaks can also be detected using the Comet assay performed on peripheral lymphocytes. This simple test involves isolating the cell nuclei, which are then subjected to an electric field. DNA fragments migrate more rapidly and the nucleus elongates, taking on the shape of a comet under the microscope. Images of the deformed nuclei are analyzed using image processing software. A number of parameters quantify this deformation (percentage of DNA in the head and tail of the comet, distance between the barycenter of the head and tail, etc.). This test was carried out on humans exposed to atomic testing at the Semipalatinsk site[2] [CHE 06].

Medical personnel represent the largest group of workers professionally exposed to IR[3]. Chronic exposure to low doses of radiation can result in DNA damage and genotoxicity associated with an increased risk of cancer. To monitor the health of this workforce, relevant studies with various biomarkers of genotoxicity have been carried out.

---

2 This military site in Kazakhstan was used by the Soviets on August 29, 1949 for their first nuclear test. It was subsequently the site of 456 nuclear explosions between 1949 and 1989.

3 In France, in 2021, medical and veterinary staff accounted for 59.7% of workers exposed to IR (234,284 out of 392,180 workers).

The frequencies of chromosomal aberrations and micronuclei were significantly different between all radiation-exposed workers (3.99…4.93) and all unexposed workers (1.41…1.86) for total aberrant cells and micronuclei frequencies. This was not the case for ring chromosomes and nucleoplasmic bridges. Although less frequently used, stable translocations, SCE and comet analysis parameters were also statistically different between radiation-exposed and non-exposed workers [BAU 21].

Recent epidemiological studies have highlighted the adverse health effects of exposure to low-dose, low-dose-rate IR. For example, studies on pediatric patients have reported increased risks of leukemia and brain tumors after doses of 30–60 mGy from CT scans. The European consortium DoReMi Low (Low Dose Research to Multidisciplinary Integration) has previously examined biomarkers for potential use in epidemiological studies of the effects of IR. In the final stages of development, one biomarker was identified as a priority for further research. This biomarker is the profile of mRNA transcripts specific to IR [HAL 17].

Dubessy et al. [DUB 98] isolated subpopulations of two human cell lines (KB and FaDu) from the upper aerodigestive tract carcinomas surviving high doses of irradiation. However, the overexpression of the GST-u and p53 genes observed by the authors does not appear to be a marker of radiosensitivity.

#### 8.5.1.3. *Biodosimetry*

Early knowledge of the absorbed dose is of paramount importance in radiological emergencies, in order to focus available resources solely on those subjects requiring urgent medical treatment. Several biomarkers have been established to measure dose: dicentric or micronucleus counting are the main techniques used, but they only give results after two to three days.

To obtain faster results, other biomarkers were selected, such as blood counts, radio-inducible proteins, the Comet assay, H2AX foci, immune response analysis, characterization of circulating exosomes and also, for validation, the number of micronuclei. In addition, new tools have been added, such as oxidative stress or protein biomarkers (amylase, Flt3-L) [GIO 22].

Assessment based on biodosimetry in certain high-risk occupational groups can be carried out using peripheral blood cells as samples to test and validate the specificity and sensitivity of biomarkers. This would enable non-invasive monitoring of the long-term health effects of low-dose IR [PAN 21].

#### 8.5.1.4. *Bromodeoxyuridine*

5-Bromodeoxyuridine (BrdU) is a halogenated thymidine analog, recognized since the 1960s as an excellent radiosensitizer. The most widely held hypothesis about BrdU's radiosensitizing effect is that it increases the number of single and DSB when incorporated into the cell's DNA and exposed to IR. However, new research seems to challenge previous observations. These latest studies have confirmed that BrdU is a good radiosensitizer, as it increases radiation-induced damage in DNA. But it is when incorporated into a single-stranded region that BrdU is radiosensitized [LAM 09].

#### 8.5.1.5. *Bile acids*

It is vital to find biomarkers that can measure the extent and severity of damage to key physiological systems. The digestive system is highly radiosensitive. An interesting way of finding biomarkers would be to measure bile acid levels in blood or feces [SCA 01].

#### 8.5.1.6. *Serum biomarkers*

The aim is to discover clinical biomarkers of exposure to IR. Using sophisticated techniques, computerized analyses of protein spectra could distinguish samples from unexposed patients from samples from radiation-exposed patients with a sensitivity of 91–100% and a specificity of 97–100% using various classifier models. Ménard et al. [MEN 06] detected 23 protein/peptide fragments uniquely present in the serum of the group exposed to IR.

### 8.5.2. *Biomarkers specific to thyroid cancer*

Thyroid hormone biosynthesis is under the tonic control of thyrotropin (TSH), while iodide recycling capacity is very important for normal thyroid function. During TH biosynthesis and release, new molecules are involved in the process [CAR 17].

The only proven etiological factor for thyroid cancer is childhood exposure to IR. Thyroid carcinogenesis is species-specific. There are both spontaneous and radiation-induced thyroid cancers. In mice, thyroid cancers are rare, and the only radiation-induced cancers are mainly follicular, whereas in humans, both spontaneous and radiation-induced cancers are predominantly papillary.

To identify molecular markers that might represent a radiation induction signature, Ory et al. [ORY 11a] compared microarray 25K transcriptome profiles of a set of 28 thyroid tumors, which included 14 follicular thyroid adenomas (FTAs) and 14 papillary thyroid carcinomas (PTCs), either sporadic or following external radiotherapy in childhood. They identified a signature composed of 322 genes that discriminate radiation-induced tumors (FTAs and PTCs) from their sporadic counterparts. The identification of specific molecular markers deregulated in radiation-induced thyroid tumors is important for etiological diagnosis, as neither histological features nor genetic alterations can distinguish between sporadic and radiation-induced tumors. Ory et al. [ORY 11b] demonstrated molecular similarities and specificities in tumors arising after external exposure to high doses of radiation, such as radiotherapy, and in post-Chernobyl tumors arising after internal iodine-131 contamination (106 specific gene signatures). The team has since found that five genes are common to both signatures (post-radiotherapy and post-Chernobyl thyroid tumors). Overall, our data suggest that thyroid tumors that developed following external exposure or internal contamination to $^{131}$I shared common molecular features related to DNA repair, oxidative stress and ER, allowing them to be classified as radiation-induced tumors in relation to their sporadic counterparts, irrespective of dose and dose rate, suggesting that there may be a "general" radiation-induced signature of thyroid tumors [ORY 13]. This team uses predictive classification methods based on learning [UGO 11].

These results strongly suggest the existence of a long-term imprint of IR exposure in normal tissue, in line with the identification of molecular signatures discriminating between exposed and unexposed normal tissues. This imprint could be due to the development of chronic oxidative stress in the thyroid following exposure to IR [MEJ 20].

## 8.6. Conclusions

The field of low-dose radiobiology has made considerable progress in recent years, moving from relatively simple hypotheses in the 1980s to greater complexity in the 1990s, with both targeted and non-targeted effects (NTE)

IR not only activates the DNA damage response (DDR) in irradiated cells, but also induces BEs in cells not directly targeted by radiation [RAS 18].

It is now well accepted that the effects of exposure to IR can be seen far beyond the boundaries of the tissue directly irradiated. Irradiation can affect neighboring cells, causing a BE. The effects of IR are mainly due to the phenomenon of genome instability, manifested by chromosomal aberrations, genetic mutations, delayed cell death and aneuploidy [KOV 08].

The communication of information between irradiated and non-irradiated control cell populations and the subsequent expression of radiation-like responses in the non-irradiated population, formally termed the radiation-induced control effect, is a well-established phenomenon in the study of radiobiology. Intercellular communication of proximity signals is known to occur via the exchange of soluble factors across biological fluids and by means of molecule transfer between adjacent cells via gap junctions [LE 17].

However, this paradigm shift due to the development of science has not been accompanied by the application of this developing science to environmental health protection. The precautionary approach to radiation protection is still based on a radiation protection model that assumes a linear relationship between dose (i.e. energy deposition) and effect (in this case, the probability of a harmful DNA interaction leading to mutation). This model does not take into account NTE such as BEs or delayed effects, which occur in offspring cells whose offspring did not directly receive the energy deposition of the dose. This means that if ambient radiation doses are measured, for example, 25 years after the Chernobyl accident, they represent only part of the dose effect, as the contribution of NTE is not taken into account [MOT 17].

The challenges for future research are to find an early warning (or biomarker) of radiosensitivity and individual exposure to cellular changes. Such a biomarker could be used to detect and better understand mechanisms of toxic action and inter/intra-species susceptibility to radiation in a context of environmental risk assessment and management [HOR 19].

The classical model of the harmful effects of IR, based on the theory of "stochastic breakage" of one or both strands of the DNA double helix, has been strongly challenged. According to this model, high doses can cause breaks, potentially fatal to the cell by damaging both DNA strands, while low doses of irradiation essentially cause SSB, easily repaired, resulting in no permanent damage. The evidence available makes this classical model less and less acceptable, as exposure to low doses of IR appears to have carcinogenic effects, even after years or decades, both in exposed individuals and in those who are not.

Moreover, cells that have survived low-dose exposure, although apparently normal, accumulate damage that becomes evident in their offspring, such as non-clonal chromosomal aberrations, which can be found even in cells not directly irradiated, due to the exchange of molecular signals and complex tissue reactions, involving neighboring or distant cells. For all these reasons, a paradigm shift is needed, based on evidence and epigenetics [BUR 18].

In conclusion, tissue responses in terms of cell signaling, subsequent cytokine/chemokine cascades [LOR 01, LAI 07] and their potential effects on inflammatory reactions [LOR 03] will also have an impact on the long-term outcome of exposure. In addition, such responses may well have an impact on the non-cancer effects, including cardiovascular effects [BHA 08] and fatigue [JAC 03] that are associated with radiation exposure. Also, Morgan and Sowa [MOR 09] consider that any non-cancer effects would shift the radiation risk profile towards supra-linearity and impose greater importance on the adverse effects of irradiation than previously thought.

It seems possible to tell whether or not the occurrence of thyroid cancer is related to exposure to IR. The social impact of the implementation of this data will be very significant in terms of the treatment of occupational injuries suffered by workers exposed to IR, or of public victims involved in accidents such as those at Chernobyl or Fukushima. This would put an end to all suspicions, such as those that have arisen in France concerning the inference of thyroid damage due to Chernobyl fallout.

Although there is a broad consensus regarding the harmful effects of high doses of radiation, the biological effects of low doses at low LET are not well understood. Low doses are known to promote oxidative stress, yet there is emerging evidence that this stress causes heritable modifications via epigenetic mechanisms (DNA methylation, histone modification, non-coding RNA regulation). These epigenetic modifications lead to permanent cellular transformations without altering the underlying DNA nucleotide sequence. This evidence suggests that low-dose-induced oxidative stress provides a mechanistic link between low doses and epigenetic gene regulation [THA 17].

## 8.7. References

[AIN 09] AINSBURY E.A., BOUFFLER S.D., DÖRR W. et al., "Radiation cataractogenesis: A review of recent studies", *Radiation Research*, vol. 172, pp. 1–9, 2009.

[AMI 06] AMIARD J.C., AMIARD-TRIQUET C., BARKA S. et al., "Metallothioneins in aquatic invertebrates: Their role in metal detoxification and their use as biomarkers", *Aquatic Toxicology*, vol. 76, no. 2, pp. 160–202, 2006.

[AMI 13] AMIARD-TRIQUET C., AMIARD J.C., RAINBOW P.S., *Ecological Biomarkers: Indicators of Ecotoxicological Effects*, CRC Press, Boca Raton, 2013.

[AMI 14] AMIARD, J.C., CHENAL C., GAZAL S. et al., *La qualité radiologique de l'environnement. Stratégies et méthodes de surveillance*, EDP Sciences, Les Ulis, 2014.

[AMI 15] AMIARD-TRIQUET C., AMIARD J.-C., MOUNEYRAC C., *Aquatic Ecotoxicology. Advancing Tools for Dealing with Emerging Risks*, Elsevier, London, 2015.

[AMI 17] AMIARD J.C., AMIARD-TRIQUET C., *Biomarqueurs en écotoxicologie aquatique*, Lavoisier, Tec&Doc, Paris, 2017.

[AVE 00] AVERBECK D., "Mechanisms of repair and radiation-induced mutagenesis in higher Eukaryotes", *Cancer Radiotherapy*, vol. 4, pp. 335–354, 2000.

[AYP 11a] AYPAR U., MORGAN W.F., BAULCH J.E., "Radiation-induced genomic instability: Are epigenetic mechanisms the missing link?", *International Journal of Radiation Biology*, vol. 87, no. 2, pp. 179–191, 2011a.

[AYP 11b] AYPAR U., MORGAN W.F., BAULCH J.F., "Radiation-induced epigenetic alterations after low and high LET irradiations", *Mutation Research*, vol. 707, pp. 24–33, 2011b.

[AZZ 04] AZZAM E.I., LITTLE J.B., "The radiation-induced bystander effect: Evidence and significance", *Human & Experimental Toxicology*, vol. 23, pp. 61–65, 2004.

[AZZ 13] AZZAM E.I., DE TOLEDO S.M., HARRIS A.L. et al., "The ionizing radiation-induced bystander effect: Evidence, mechanism, and significance", in SONIS S.T., KEEFE D.M. (eds), *Pathobiology of Cancer Regimen-Related Toxicities*, pp. 35–61, Springer Science, New York, 2013.

[BAR 01] BARCELLOS-HOFF M.H., BROOKS A.L., "Extracellular signaling through the microenvironment: A hypothesis relating carcinogenesis, bystander effects, and genomic instability", *Radiation Research*, vol. 156, no. 5, Part 2, pp. 618–627, 2001.

[BAR 07] BARILLET S., Toxicocinétique, toxicité chimique et radiologique de l'uranium chez le poisson zèbre (*Danio rerio*), Thesis, Université Paul Verlaine de Metz, 2007.

[BAU 21] BAUDIN C., BERNIER M.O., KLOKOV D. et al., "Biomarkers of genotoxicity in medical workers exposed to low-dose ionizing radiation: Systematic review and meta-analyses", *International Journal of Molecular Sciences*, vol. 22, p. 7504, 2021.

[BEL 05] BELYAKOV O.V., MITCHELL S.A., PARIKH D. et al., "Biological effects in unirradiated human tissue induced by radiation damage up to 1 mm away", *PNAS*, vol. 102, no. 40, pp. 4203–14208, 2005.

[BEL 20] BELLI M., TABOCCHINI M.A., "Ionizing radiation-induced epigenetic modifications and their relevance to radiation protection", *International Journal of Molecular Sciences*, vol. 21, p. 5993, doi:10.3390/ijms21175993, 2020.

[BER 13] BERNAL A.J., DOLINOY D.C., HUANG D. et al., "Adaptive radiation-induced epigenetic alterations mitigated by antioxidants", *FASEB Journal*, vol. 27, pp. 665–671, 2013.

[BHA 08] BHATTI P., SIGURDSON A.J., MABUCHI K., "Can low-dose radiation increase risk of cardiovascular disease?", *Lancet*, vol. 372, pp. 697–699, 2008.

[BUR 18] BURGIO E., PISCITELLI P., MIGLIORE L., "Ionizing radiation and human health: Reviewing models of exposure and mechanisms of cellular damage. An epigenetic perspective", *International Journal of Environmental Research and Public Health*, vol. 15, p. 1971, 2018.

[CAI 99] CAI L., SATOH M., TOHYAMA C. et al., "Metallothionein in radiation exposure: Its induction and protective role", *Toxicology*, vol. 132, pp. 85–98, 1999.

[CAM 03] CAMPHAUSEN K., MOSES M.A., MÉNARD C. et al., "Radiation abscopal antitumor effect is mediated through p53", *Cancer Research*, vol. 63, pp. 1990–1993, 2003.

[CAR 17] CARVALHO D.P., DUPUY C., "Thyroid hormone biosynthesis and release", *Molecular and Cellular Endocrinology*, vol. 458, pp. 6–15, 2017.

[CHE 01] CHEN W.L., HWANG J.S., HU T.H. et al., "Lenticular opacities in populations exposed to chronic low-dose-rate gamma radiation from radiocontaminated buildings in Taiwan", *Radiation Research*, vol. 156, 1, pp. 71–77, 2001.

[CHE 06] CHENAL C., LEGUE F., NOURGALIEVA K., "Delayed effects of low level acute irradiation and chronic environmental radioactive contamination on DNA lymphocytes of people living in Dolon, a settlement located in the vicinity of the Semipalatinsk nuclear test site (Kazakhstan)", *Science of the Total Environment*, vol. 369, nos. 1–3, pp. 91–98, 2006.

[CHO 08] CHODICK G., BEKIROGLU N., HAUPTMANN M. et al., "Risk of cataract after exposure to low doses of ionizing radiation: A 20-year prospective cohort study among US radiologic technologists", *American Journal of Epidemiology*, vol. 168, no. 8, pp. 620–631, 2008.

[CHY 09] CHYLACK L. T. Jr, PETERSON L.E., FEIVESON A.H. et al., "NASA study of cataract in astronauts (NASCA). Report 1: Cross-sectional study of the relationship of exposure to space radiation and risk of lens opacity", *Radiation Research*, vol. 172, pp. 10–20, 2009.

[COR 20] CORPET A., "Épigénétique : nous sommes plus que la séquence de nos gènes", *Médecine et philosophie*, hal-02490179, vol. 2, 2020.

[DON 20] DONG C., TU W., HE M. et al., "Role of endoplasmic reticulum and mitochondrion in proton microbeam radiation-induced bystander effect", *Radiation Research*, vol. 193, pp. 63–72, 2020

[DUB 98] DUBESSY C., MERLIN J.L., MARCHAI C., "Modifications de l'expression des gènes GST-u et p53 dans des lignées tumorales cellulaires humaines O.R.L. après irradiation gamma : induction, études cellulaires et moléculaires", *Journal de chimie physique*, vol. 95, pp. 718–723, 1998.

[EME 94] EMERIT I., LEVY A., CERNJAVSKI L. et al., "Transferable clastogenic activity in plasma from persons exposed as salvage personnel of the Chernobyl reactor", *Journal of Cancer Research and Clinical Oncology*, vol. 120, pp. 558–561, 1994.

[EME 97] EMERIT I., QUASTEL M., GOLDSMITH J. et al., "Clastogenic factors in the plasma of children exposed at Chernobyl", *Mutation Research*, vol. 373, pp. 47–54, 1997.

[FAG 84] FAGUET G.B., REICHARD S.M., WELTER D.A., "Radiation-induced clastogenic plasma factors", *Cancer Genetics and Cytogenetics*, vol. 12, pp. 73–83, 1984.

[FRA 05] FRANCO N., LAMARTINE J., FROUIN V. et al., "Low-dose exposure to g rays induces specific gene regulations in normal human keratinocytes", *Radiation Research*, vol. 163, pp. 623–635, 2005.

[FRI 06] FRIEDBERG E.C., WALKER G.C., SIEDE W. et al., *DNA Repair and Mutagenesis*, ASM Press, Washington, DC, 2006.

[FUJ 13] Fujii Y., Genet M.D., Roybal E.J. et al. "Comparison of the bromodeoxyuridine-mediated sensitization effects between low-LET and high-LET ionizing radiation on DNA double-strand breaks", *Oncology Reports*, vol. 29, pp. 2133–2139, 2013.

[GIO 22] Giovanetti A., Marconi R., Bartoleschi C. et al., "Learning from NATO biomarker research for humans", in Wood M.D., Mothersill C.E., Tsakanova G. et al. (eds), *Biomarkers of Radiation in the Environment. Robust Tools for Risk Assessment*, Springer, Dordrecht, pp. 107–122, 2022.

[GOH 68] Goh K.O., Sumner H., "Breaks in normal human chromosomes: Are they induced by a transferable substance in the plasma of persons exposed to total-body irradiation?", *Radiation Research*, vol. 35, no. 1, pp. 171–181, 1968.

[GUP 19] Gupta R.C, *Biomarkers in Toxicology*, 2nd edition, Elsevier, Amsterdam, 2019.

[HAL 03] Hall E.J., Hei T.K., "Genomic instability and bystander effects induced by high-LET radiation", *Oncogene*, vol. 22, pp. 7034–7042, 2003.

[HAL 17] Hall J., Jeggo P.A., West C. et al., "Ionizing radiation biomarkers in epidemiological studies – An update", *Mutation Research/Reviews in Mutation Research*, vol. 771, pp. 59–84, 2017.

[HOE 01] Hoeijmakers J.H., "Genome maintenance mechanisms for preventing cancer", *Nature*, vol. 411, pp. 366–374, 2001.

[HOL 68] Hollowell J.G., Gayle Littlefield L., "Chromosome damage induced by plasma of X-rayed patients: An indirect effect of X-Ray* (33295)", *P.S.E.B.M.*, vol. 129, pp. 240–244, 1968.

[HOR 19] Horemans N., Spurgeon D.J., Lecomte-Pradines C. et al., "Current evidence for a role of epigenetic mechanisms in response to ionizing radiation in an ecotoxicological context", *Environmental Pollution*, vol. 251, pp. 469–483, 2019.

[ILN 09] Ilnytskyy Y., Koturbash I., Kovalchuk O., "Radiation-induced bystander effects in vivo are epigenetically regulated in a tissue-specific manner", *Environmental and Molecular Mutagenesis*, vol. 50, pp. 105–113, 2009.

[JAC 02] Jackson S.P., "Sensing and repairing DNA double-strand beaks", *Carcinogenesis*, vol. 23, pp. 687–696, 2002.

[JAC 03] Jacobsen P.B., Thors C.L., "Fatigue in the radiation therapy patient: Current management and investigations", *Seminars in Radiation Oncology*, vol. 13, pp. 372–380, 2003.

[JAC 09] Jackson S.P., Bartek J., "The DNA-damage response in human biology and disease", *Nature*, vol 461, pp. 1071–1078, 22 October 2009.

[JAC 11] Jacob S., Nouvelle donne pour le cristallin. Aspects scientifiques, GT-CIPR, IRSN, 8 December 2011.

[KAD 92] Kadrim M.A., Macdonald D.A., Goodhead D.T. et al., "Transmission of chromosomal instability after plutonium a-particle irradiation", *Letters to Nature*, vol. 355, pp. 738–740, 1992.

[KOV 08] KOVALCHUK O., BAULCH J.E., "Epigenetic changes and nontargeted radiation effects. Is there a link?", *Environmental and Molecular Mutagenesis*, vol. 49, pp. 16–25, 2008.

[LAI 07] LAIAKIS E.C., BAULCH J.E., Morgan W.F., "Cytokine and chemokine responses after exposure to ionizing radiation: implications for the astronauts", *Advances in Space Research*, vol. 39, pp. 1019–1025, 2007.

[LAM 09] LA MADELEINE C., Utilisation de l'essai comète et du biomarqueur y-H2AX pour détecter les dommages induits à l'ADN cellulaire par le 5-bromodeoxyuridine post-irradiation, Thesis, Université de Sherbrooke, 2009.

[LAU 05] LAURENT C., Rôle du stress oxydatif dans le développement des effets cellulaires radio-induits au niveau cutané : application aux irradiations localisées accidentelles, Thesis, Université de Versailles Saint-Quentin-en-Yvelines, 2005.

[LE 17] LE M., Investigating the generation of biophotons induced by low-dose beta-irradiation and their role in the radiation-induced bystander effect, Thesis, McMaster University, 2017.

[LIN 00] LINDAHL T., BARNES D.E., "Repair of endogenous DNA damage", *Cold Spring Harbor Symposia on Quantitative Biology*, vol. 65, pp. 127–133, 2000.

[LOB 05] LÖBRICH M., RIEF N., KÜHNE M. et al., "*In vivo* formation and repair of DNA double-strand breaks after computed tomography examinations", *PNAS*, vol. 102, no. 25, pp. 8984–8989, 2005.

[LOR 98] LORIMORE S.A., KADHIM M.A., POCOCK D.A. et al., "Chromosomal instability in the descendants of unirradiated surviving cells after a-particle irradiation", *Proceedings of the National Academy of Sciences of the USA*, vol. 95, pp. 5730–5733, 1998.

[LOR 01] LORIMORE S.A., COATES P.J., SCOBIE G.E. et al., "Inflammatory-type responses after exposure to ionizing radiation in vivo: A mechanism for radiation-induced bystander effects?", *Oncogene*, vol. 20, pp. 7085–7095, 2001.

[LOR 03] LORIMORE S.A., WRIGHT E.G., "Radiation-induced genomic instability and bystander effects: Related inflammatory-type responses to radiation-induced stress and injury? A review", *International Journal of Radiation Biology*, vol. 79, pp. 15–25, 2003.

[MA 10] MA S., HU X., JIAO B. et al., "Low-dose radiation-induced responses: Focusing on epigenetic regulation", *International Journal of Radiation Biology*, vol. 86, no. 7, pp. 517–528, 2010.

[MEJ 20] MEJLACHOWICZ D., Développement de modèles *in vitro* et *in vivo* pour analyser la réponse aux radiations ionisantes du tissu thyroïdien normal humain, Thesis, Université Paris-Saclay, 2020.

[MÉN 09] MÉNAGER M.T., GARNIER-LAPLACE J., GOYFFON M., *Toxicologie nucléaire environnementale et humaine*, Lavoisier, Tec&Doc, Paris, 2009.

[MÉN 06] MÉNARD C., JOHANN D., LOWENTHAL M. et al., "Discovering clinical biomarkers of ionizing radiation exposure with serum proteomic analysis", *Cancer Research*, vol. 66, no. 3, pp. 1844–1850, 2006.

[MIN 04] MINAMOTO A., TANIGUCHI H., YOSHITANI N. et al., "Cataract in atomic bomb survivors", *International Journal of Radiation Biology*, vol. 80, no. 5, pp. 339–345, 2004.

[MON 05] MONLEAU M., Biocinétique et toxicologie de l'uranium chez le rat après inhalations aiguës et répétées, Thesis, Université de la Méditerranée, Aix Marseille II, 2005.

[MOR 08] MORGENTHALER S., *Génétique statistique*, Springer-Verlag, Berlin, 2008.

[MOR 09] MORGAN W.F., SOWA M.B., "Non-targeted effects of ionizing radiation: Implications for risk assessment and the radiation dose response profile", *Health Physics*, vol. 97, no. 5, pp. 426–432, 2009.

[MOT 12] MOTHERSILL C., SEYMOUR C., "Reflections in mutation research. Changing paradigms in radiobiology", *Mutation Research, Reviews in Mutation Research*, vol. 750, pp. 85–95, 2012.

[MOT 17] MOTHERSILL C., RUSIN A., SEYMOUR C., "Low doses and non-targeted effects in environmental radiation protection; where are we now and where should we go?", *Environmental Research*, vol. 159, pp. 484–490, 2017.

[NAG 92] NAGASAWA H., LITTLE J.B., "Induction of sister chromatid exchanges by extremely low doses of a-particles", *Cancer Research*, vol. 52, pp. 6394–6396, 1992.

[NAK 13] NAKASHIMA E., NERIISHI K., MINAMOTO A. et al., "Radiation dose responses, thresholds, and false negative rates in a series of cataract surgery prevalence studies among atomic bomb survivors", *Health Physics*, vol. 105, no. 3, pp. 253–260, 2013.

[NER 07] NERIISHI K., NAKASHIMA E., MINAMOTO A. et al., "Postoperative cataract cases among atomic bomb survivors: radiation dose response and threshold", *Radiation Research*, vol. 168, pp. 404–408, 2007.

[OHB 98] OHBA K., OMAGARI K., NAKAMURA T. et al., "Abscopal regression of hepatocellular carcinoma after radiotherapy for bone metastasis", *Gut*, vol. 43, pp. 575–577, 1998.

[ORY 13] ORY C., UGOLIN N., HOFMAN P. et al., "Comparison of transcriptomic signature of post-Chernobyl and postradiotherapy thyroid tumors", *Thyroid*, vol. 23, no. 11, pp. 1390–1400, 2013.

[ORY 11a] ORY C., UGOLIN N., LEVALOIS C. et al., "Gene expression signature discriminates sporadic from post-radiotherapy-induced thyroid tumors", *Endocrine-Related Cancer*, vol. 18, no. 1, pp. 193–206, 2011.

[ORY 11b] ORY C., UGOLIN N., SCHLUMBERGER M. et al., "Discriminating gene expression signature of radiation-induced thyroid tumors after either external exposure or internal contamination", *Genes (Basel)*, vol. 3, no. 1, pp. 19–34, 2011.

[PAN 77] PANT G.S., KAMADA N., "Chromosome aberrations in normal leukocytes induced by the plasma of exposed individuals", *Hiroshima Journal of Medical Sciences*, vol. 26, nos. 2–3, pp. 149–154, 1977.

[PAN 21] PANERA N., CAMISA V., BRUGALETTA R. et al., "Blood cell gene expression profiles: A narrative review of biomarkers and effects of low-dose ionizing radiation exposure", *Journal of Health and Social Sciences*, vol. 6, no. 3, pp. 349–366, 2021.

[PAT 23] PATEL V.B., PREEDY V.R., RAJENDRAM R., "Biomarkers in toxicology", in PATEL V.B., PREEDY V.R. (eds), *Biomarkers in Disease: Methods, Discoveries and Applications*, vol. 9, Springer, Cham, 2023.

[PRA 02] PRADEL J., BEUTIER D., DELLERO N., "La radioactivité naturelle, source de repères", *Revue des ingénieurs*, 15–19, May–June 2002.

[PRI 98] PRISE K.M., BELYAKOV O.V., FOLKARD M. et al., "Studies of bystander effects in human fibroblasts using a charged particle microbeam", *International Journal of Radiation Biology*, vol. 74, no. 6, pp. 793–798, 1998.

[RAF 05] RAFNSSON V., OLAFSDOTTIR E., HRAFNKELSSON J. et al., "Cosmic radiation increases the risk of nuclear cataract in airline pilots a population-based case-control study", *Archives of Ophthalmology*, vol. 123, pp. 1102–1105, 2005.

[RAS 18] RASTOGI S., HWANG A., CHAN J. et al., "Extracellular vesicles transfer nuclear Abl-dependent and radiation-induced miR-34c into unirradiated cells to cause bystander effects", *Molecular Biology of the Cell*, vol. 29, pp. 2228–2242, 1 September 2018.

[ROT 03] ROTHKAMM K., LOBRICH M., "Evidence for a lack of DNA double-strand break repair in human cells exposed to very low X-ray doses", *Proceedings of the National Academy of Sciences of the USA*, vol. 100, pp. 5057–5062, 2003.

[SAW 01] SAWANT S.G., RANDERS-PEHRSON G., METTING N.F. et al., "Adaptive response and the bystander effect induced by radiation in $C_3H$ 10T½ cells in culture", *Radiation Research*, vol. 156, pp. 177–180, 2001.

[SCA 01] SCANFF P., GRISON S., MARAIS T. et al., "Le biomarqueur d'atteinte par les rayonnements ionisants, un complément indispensable à la dosimétrie en situation accidentelle. Démarche et exemple appliqué au système digestif", *Congrès national de la radioprotection*, SFRP, Tours, FR0108368, 2001.

[SCO 69] SCOTT D., "The effect of irradiated plasma on normal human chromosomes and its relevance to the long-lived lymphocyte hypothesis", *Cell and Tissue Kinetics*, vol. 2, pp. 295–305, 1969.

[SEY 00] SEYMOUR C.B., MOTHERSILL C., "Relative contribution of bystander and targeted cell killing to the low-dose region of the radiation dose-response curve", *Radiation Research*, vol. 153, no. 5, pp. 508–511, 2000.

[THA 17] THARMALINGAM S., SREETHARAN S., KULESZA A.V. et al., "Low-dose ionizing radiation exposure, oxidative stress and epigenetic programing of health and disease", *Radiation Research*, vol. 188, pp. 525–538, 2017.

[UGO 11] UGOLIN N., ORY C., LEFEVRE E. et al., "Strategy to find molecular signatures in a small series of rare cancers: validation for radiation-induced breast and thyroid tumors", *PLoS One*, vol. 6, no. 8, p. e23581, 2011.

[VAU 16] VAURIJOUX A., Étude des conséquences génétiques et épigénétiques consécutives à la signalisation persistante des dommages radio-induits de l'ADN, Université Paris-Saclay, 2016.

[VOG 67] VOGEL G., ANGENMANN, H., *Atlas de biologie*, Stock, Paris, 1967.

[WAT 00] WATSON G.E., LORIMORE S.A., MACDONALD D.A. et al., "Chromosomal instability in unirradiated cells induced in vivo by a bystander effect of ionizing radiation", *Cancer Research*, vol. 60, pp. 5608–5611, 2000.

[WOR 07] WORGUL B.V., KUNDIYEV Y.I., SERGIYENKO N.M. et al., "Cataracts among chernobyl clean-up workers: Implications regarding permissible eye exposures", *Radiation Research*, vol. 167, pp. 233–243, 2007.

[XUE 02] XUE L.Y., BUTLER N., MAKRIGIORGOS G.M. et al., "Bystander effect produced by radiolabeled tumor cells in vivo", *Proceedings of the National Academy of Sciences of the USA*, vol. 99, no. 21, pp. 13765–13770, 2002.

# 9

# Health Effects of Ionizing Radiation at the Individual Level

## 9.1. Introduction

Ethically, it is impossible to carry out toxicology tests on humans to quantify the harmful effects of ionizing radiation. It is also difficult to extrapolate the results of toxicological tests on other mammals to humans. As a result, the only solutions are to quantify, as best we can, the results of accidental exposures of nuclear workers, occurring during gamma or criticality irradiation, or those of workers and the public subjected to a major nuclear accident such as those at Chernobyl or Fukushima. In all these cases, the challenge is to assess these exposures as accurately as possible. For the effect of low or very low doses (< 5 milligrays), it is essential to use epidemiological studies with large cohorts made up of an exposed population and a control population (see Chapter 11). This is the case, for example, of the study conducted over 75 years ago on the life expectancy of Japanese survivors of the nuclear blasts on the cities of Hiroshima and Nagasaki on August 6 and 9, 1945.

### 9.1.1. *Historical study of Hiroshima and Nagasaki survivors*

A great deal of data on the health effects of radiation on human populations comes from the epidemiological follow-up of the cohort of survivors of the two atomic bombs dropped on Hiroshima and Nagasaki over Japan in August 1945. Analysis of the effects of ionizing radiation on this cohort, known as the Life Span Study (LSS)[1], was the foundation on which radiation protection was organized worldwide.

1 This was the cohort of survivors, as of October 1, 1950, led by the USA and Japan.

Following several meetings in the USA and Hiroshima, a plan was drawn up by a group chaired by Thomas Francis (Professor of Epidemiology, University of Michigan), involving the creation of well-defined groups of exposed and unexposed people, making it possible to determine disease occurrence rates and monitor their evolution over time. These recommendations were approved by a conference on November 27, 1955.

The epidemiological studies that followed described the development of excess cancers caused by ionizing radiation and quantified the intensity of the risk as a function of the radiation dose received. Monitoring of the development of effects observed in the cohorts studied, which is still ongoing [SAM 20], provides fairly comprehensive data on the expression of radiation-induced risks over time, as a function of dose received, sex, age at exposure and age reached by survivors.

### 9.1.2. *Assessment of doses received by survivors of bombs dropped on Japan*

To establish a "dose–effect" relationship between the doses delivered during the explosions and their impact on health, it was necessary to calculate the individual doses received by each of the survivors alive on October 1, 1950, the date on which the LSS cohort began to be monitored. These highly complex individual dosimetric assessments were carried out four times: in 1957, 1965, 1986 and 2002. A full-scale test carried out in 1955 in the Nevada desert, during which gamma and neutron dosimetric measurements were taken, led to the creation of the first provisional dosimetric system, called DT57D, which was implemented in 1957.

However, the first complete assessments of these individual doses date back to 1965. Called TD65, they were based on a number of experimental data: a second and then a third trial (1957 and 1958) in the Nevada desert, where measurements were taken inside two, then seven Japanese-style houses, built with traditional Japanese materials and arranged in different orientations, in order to experimentally evaluate their shielding effects [JON 75]. In 1962, a small nuclear reactor placed at the top of a tower [ZER 88] was also used for complementary dosimetric measurements.

From 1986, the third dose assessment was based on the calculation of the spectral energy distributions of the gamma rays and neutrons produced when the two bombs (different in nature) were fired at the point of explosion. The changes in neutron energy distribution brought about by the high humidity of the air in Japan in August were also taken into account. These doses, referred to as DS86, were found to be higher in Hiroshima (high neutron component) than in Nagasaki – data in line with the number of cancers observed in each of the two bombed cities [SHI 90].

It became clear, however, that the activation measurements made on various materials at Hiroshima were not consistent with the neutron data from the DS86 system. In December 2000, a joint US–Japanese group of experts was formed to reassess the doses and resolve these inconsistencies. On March 15, 2003, a fourth dosimetric system was created, the DS02[2], to replace the DS86.

This new calculation system involved a complete re-examination of all the parameters used in particle emission calculations, which are then used to calculate individual doses delivered to survivors. The calculated spectra of emitted particles for DS02 include 200 neutron energy classes and 43 gamma ray classes (instead of 38 and 27 for DS86).

These qualitative and quantitative data related to the two particle spectra (gammas and neutrons) evolve as a function of distance from the hypocenter[3] (or "zero point") and the attenuation of radiation by the various protective shields (air and its water load, concrete, wood, metals)[4]. From the initial spectral shapes of gammas and neutrons, the spectra of incident particles are calculated for each individual, in order to assess the doses delivered to organs. This dose calculation also depends on the orientation of each survivor in relation to the point of explosion of the bomb (irradiation from the front, side or back) and their body position (standing, sitting, lying down)[5]. DS02 also takes into account exposure caused by neutron activation of the soil, contamination of the soil in both cities by radioactive aerosol fallout and, in the case of Hiroshima, the black rain[6] that fell shortly after the explosion.

The changes in dose estimates brought about by DS02 are smaller than expected: an increase of approximately 10% in gamma radiation estimates for both cities. The anticipated sharp increase in the neutron component at Hiroshima, for low-dose survivors, did not materialize. However, DS02 improves on the DS86 assessments in many details, including the specifics of the radiation emitted by the bombs, the shielding effects by structures in the city and the ground components made

2 TD65: Tentative Dosimetry 1965, DS86: Dosimetry System 1986, DS02: Dosimetry System 2002.

3 The hypocenter is the projection of the bomb's center of gravity on the ground during the explosion.

4 Aerial photos of the cities, taken in preparation for the bombings, were used to assess the geometry and characteristics of the screens formed by the various buildings.

5 Several written testimonies testify to the fact that survivors remember exactly where they were when the flash of the bomb then the noise and blast of the explosion surprised them, sometimes injuring them [HAC 15, TÔH 12].

6 A large number of houses and wooden bridges burned down, and the rain that fell after the blast leached the atmosphere at altitude, bringing radioactive ash and aerosols to the ground. These deposits, combined with the activation of the soil, resulted in external irradiation for the survivors.

radioactive by neutron activation. Due to the increase in calculated “gamma” doses, the estimates of solid cancer risk[7] per sievert and the curvilinear dose response for leukemia are both decreased by approximately 8% [PRE 04].

## 9.2. The adverse effects of ionizing radiation depend on the age at exposure

Epidemiological studies conducted on atomic bomb survivors show that the nature of the effects, their probability of occurrence and the time between exposure to ionizing radiation and the manifestation of the damage vary greatly with the dose received, the victim’s sex, age at exposure and age reached, from those irradiated in utero to young children, adolescents, young adults, adults and the elderly.

### 9.2.1. *Adverse effects of ionizing radiation on the embryo and fetus*

Prenatal development comprises three periods of fetal maturation: the pre-embryonic period from fertilization to three weeks gestation, the embryonic period from four to eight weeks gestation and the fetal period from nine weeks to birth. As early as the embryonic period, a neural plate develops, which is the origin of the nervous system, which, after maturation, produces the neural tube, which becomes longer with each passing week. The upper part of this neural tube forms a bulge and closes at approximately six or seven weeks, giving rise to the brain, while the lower part becomes the spinal cord and nervous network [EFE 23]. It is during this important phase of development that serious damage occurs when the fetus is irradiated.

By the end of the 8th week, the first version of the brain and spinal cord are in place. The brain is half the size and weight of the fetus. During the first trimester, no fewer than 5,000 neurons are formed every second, forming the baby’s gray matter. These nerve cells enable the brain to mature over the months and years ahead. These periods of high reproductive activity, when tissue morphology and function are not yet definitively fixed, are highly radiosensitive (Bergonié and Tribondeau law, 1906).

In 1956, medical epidemiologists Alice Stewart and Georges Kneale first warned of the potential risks associated with the use of ionizing radiation for diagnostic purposes, such as X-rays in pregnant women and children. They revealed a link between cases of leukemia occurring before the age of 10 and the mother’s exposure

---

7 By convention, radiation-induced cancers in all organs except the red blood-forming marrow are called “solid” cancers.

to X-rays in early pregnancy. Their work, based on data from the Oxford Survey of Childhood Cancers, showed that the risk of cancer was highest when in utero exposure occurred during the first trimester. The excess cancer risk arising from this radiological examination was directly related to the fetal dose [STE 70]. This work has led to the gradual discontinuation of prenatal X-ray screening in several countries.

Other studies have confirmed this risk and its importance, such as that of Doll and Wakeford, for whom radiation doses of the order of 10 mGy, received by the fetus in utero, cause a consequent increase in the risk of childhood cancer [DOL 97]. It has subsequently been widely reported that prenatal exposure to ionizing radiation can interfere with embryonic and fetal development, depending on the dose and gestational age at which exposure occurs [DES 07].

#### 9.2.1.1. *Mortality due to in utero exposure*

The initial study conducted on the female population of Nagasaki aimed to measure the impact of in utero exposure. However, in the absence of individual dosimetric data, distance from the hypocenter was used as the exposure parameter. The study involved a group of 1,774 women of childbearing age, exposed at less than 2,000 meters from the hypocenter, and a second group of 1,774 women, also of childbearing age, but at a distance of 4,000–5,000 meters from the hypocenter (exposure close to zero). Women in the latter group were matched by age to those in the group exposed at less than 2,000 m.

Of these two groups (3,458 women), 211 were pregnant. Of these, 98 mothers were within 2,000 m of the bomb's hypocenter, and 30 of them developed signs or damage caused by exposure to ionizing radiation (alopecia, desquamation, oropharyngeal lesions, purpura and petechiae[8]). Of these, 68 mothers did not develop any major signs of exposure. The remaining 113 mothers, who were located between 4,000 and 5,000 m from the hypocenter, constituted the control group [YAM 90].

Four radiation-induced disorders were selected to assess the impact of radiation doses on survivors: the frequency of clinically recognized mental disorders, the decrease in intelligence measured by conventional intelligence tests, school performance and the occurrence of unprovoked seizures [YAM 90].

Of the 30 pregnant women within 2,000 m of the hypocenter who developed signs of irradiation, seven pregnancies (23%) ended in fetal death and six (20%) in neonatal or infant death. Of this group, five surviving children (17%) were

8 Small, purplish-red hemorrhagic spots that appear under the skin. Purpura is also a subcutaneous hemorrhage, a symptom of significant irradiation.

diagnosed with one or more of the following: mental disorders, eye disorders and urinary incontinence. Only 12 surviving children (40%) did not suffer from any of these disorders. The overall morbidity–mortality rate for these 30 women was a high 60%.

For the group whose mothers were less than 2,000 m from the hypocenter but did not develop major signs of irradiation, fetal and infant mortality was 8.8%, with a morbidity of 1.5% for surviving children. Fetal and infant mortality, for the group whose mothers were between 4,000 and 5,000 m from the hypocenter and received little or no irradiation, was 6.2%, with 0.9% morbidity for surviving children.

The mean height and cranial circumference of children born to mothers showing signs of radiation damage were significantly smaller than in children born to mothers in the control group. For mothers exposed during the second and third trimesters of gestation, the frequency of fetal and neonatal deaths was significantly higher in the offspring of mothers with radiation damage. Table 9.1 shows all the results [YAM 90].

| Distance from hypocenter | Pregnant women | Miscarriages | Miscellaneous deaths* (in thousands) | Mortality rate | Morbidity rate | Unallocated |
|---|---|---|---|---|---|---|
| < 2 km with signs** | 30 | 3 | 10 | 43.3% | 5 (16.7%) | 12 (40%) |
| < 2 km without signs | 68 | 1 | 5 | 8.8% | 1 (1.5%) | 61 (89.7%) |
| 4 km to 5 km | 113 | 2 | 5 | 6.2% | 1 (0.9%) | 105 (92.9%) |
| Total | 211 | 6 | 20 | 12.3% | 7 (3.3%) | 178 (84.3%) |

*Various deaths = fetal, neonatal or infant death.

**Signs[9] = alopecia, oropharyngeal lesions, purpura and petechiae.

**Table 9.1.** *Impact (mortality–morbidity) of exposure to ionizing radiation in utero in Nagasaki (according to [YAM 90])*

9 In the days or weeks following the atomic bombings, highly exposed survivors experienced fever, nausea, vomiting, inappetence, bloody diarrhea, epilation and purpura, sores in the throat or mouth, and caries and ulceration of the gums around the teeth [YAM 90]. The latency of onset of these symptoms varied, but they generally occurred earlier in highly exposed individuals. This information was collected between 1956 and 1961, and hair loss was the effect best remembered by the victims.

The 1990 report by Yamazaki and Schull showed a high mortality rate among fetuses whose mothers, exposed less than 2,000 m from the hypocenter, showed symptoms of exposure, such as hair loss [YAM 90].

However, these data are difficult to assess, partly because of the possible effects of blast and thermal radiation, but also because of the indirect effects of hemorrhage, infection and malnutrition and the absence of individual dosimetry.

#### 9.2.1.2. *Mortality of survivors irradiated in utero*

The mortality risks of people exposed in utero (2,463) during the Hiroshima and Nagasaki bombings were monitored from October 1950–2012 by Sugiyama et al. [SUG 21]. Individual estimates of maternal uterine dose (DS02) were used. Head size, birth weight and parental survival status were assessed as potential indicators of radiation effect. There were 339 deaths (216 males and 123 females), including solid cancers (137), lympho-hematopoietic cancers (8), non-cancerous diseases (134), external causes (56) or unknown (4).

In men, the excess relative risk[10] per gray [ERR/Gy (95% confidence interval (CI))] was increased for non-cancer mortality (1.22), but not for solid cancer mortality. In women, ERR/Gy were strongly increased for solid cancer (2.24), non-cancer mortality (2.86) and mortality from external causes (2.57).

Prenatal radiation exposure is a risk factor that strongly increases solid cancer mortality in women, but not in men. However, mortality from non-cancerous disease is increased for both sexes.

The effect of radiation exposure from atomic explosions on non-cancerous disease and mortality from external causes in those exposed in utero was associated with small head size, low birth weight and parental loss [SUG 21].

#### 9.2.1.3. *Radiation-induced cancers due to in utero exposure*

In utero exposure to radiation is known to increase the risk of childhood cancers, and childhood exposure is associated with an increased risk of cancer in adulthood. However, epidemiologists wanted to know whether in utero exposure induced cancer risks in adulthood.

An initial cancer mortality study by Delongchamp et al., carried out between October 1950 and May 1992, included 807 atomic bomb survivors exposed in utero and 5,545 people exposed during childhood to more than 10 millisieverts (10 mSv) [DEL 97]. The comparison group comprised 10,453 people exposed to doses of less

10 Excess relative risk = (observed risk - expected risk) x (expected risk)$^{-1}$.

than 10 mSv or none at all. Analyses were limited mainly to cancer deaths between the ages of 17 and 46.

No deaths from solid cancers were recorded in men exposed in utero. However, significant risks were found for cancers of the digestive system and for cancers specific to women, which did not differ significantly from those observed in those exposed as children. The significant difference in solid cancer mortality between the sexes is unexplained. No deaths from leukemia were observed at high doses, a result that differs markedly from that of people exposed as children [DEL 97].

In a second study, with a larger cohort, Preston et al. [PRE 08] examined the incidence rates of solid cancers in survivors exposed in utero (2,452 cases) or under 6 years of age (15,388 cases) during the bombings. The age of the survivors in the larger cohort studied ranged from 12 to 55 years. In the group exposed in utero, 94 cancers were diagnosed and 649 in the group exposed in early childhood.

The ERR in adulthood increased with dose, both for those exposed in utero (at age 50, ERR = 1.0 per Sv) and in early childhood (at age 50, ERR = 1.7 per Sv).

Both in utero and early childhood groups showed statistically significant increases in the incidence rates of solid cancers in adulthood with the doses received. This increase in radiation-induced cancer risk was significantly higher for irradiation at less than 6 years of age than in utero. However, further follow-up of both populations is required to reach any conclusions [PRE 08]. The accumulation of data, since Alice Stewart's [STE 70] and Doll's [DOL 97] work on these disorders, shows that there may be no threshold for cancer induction in the case of in utero irradiation.

#### 9.2.1.4. *Morbidity due to in utero exposure*

Several published studies on survivors exposed in utero to the atomic bombings of Hiroshima and Nagasaki showed, as early as 1952, that significant exposure to ionizing radiation during gestation has serious and irreversible harmful effects on brain development: severe mental disorders, later epileptic seizures, reduced intelligence quotient (IQ) and school performance.

These effects occur in survivors exposed from 8 to 15 weeks after ovulation. This period corresponds to a time in embryo development when the production of neuronal elements is most rapid and immature neurons enter a migration phase, reaching their final destination [OTA 84, YAM 90]. The probability of mental disorders resulting from these exposures is linear with no threshold (44% per gray). If this effect had a threshold, it would be between 0.1 and 0.2 gray [YAM 90].

A second risk period appears from 16 to 25 weeks after ovulation. The probability of the effect also follows a linear distribution, but with a threshold close to 0.7 gray and a reduced rate of damage compared with the first period (14.5% per Gy) [YAM 90].

Radiation doses received in utero can also cause microcephaly associated with mental disorders, as well as a reduction in other anthropometric measurements (height, weight, chest circumference, etc.). This effect suggests that ionizing radiation-induced microcephaly is linked to generalized growth delay [OTA 98, SCH 99]. The risk of observing this reduction in head circumference between 0 and 7 weeks post-ovulation increases significantly with dose, but no increase in the risk of severe mental disorders is noted during this period [OTA 84, OTA 93]. Similarly, no risk of microcephaly was observed for exposures during the third trimester [OTA 93].

Of the cohort of 1,566 people exposed in utero during the Hiroshima and Nagasaki bombings, studied by Otake and Schull [OTA 93], 1,473 children had their head circumference measured at least once between the ages of 9 and 19. Of these, 62 had a head circumference below that expected for their age (4.2%). Of the 26 cases with severe mental disorders[11], 15 children had reduced head circumference (58%). The mean IQ values of the cases of those with mental disorders, with and without microcephaly, were 63.8 and 68.9, respectively. These values were significantly lower than 96.4, which was the IQ value for microcephaly cases without severe mental disorders, and 107.8, which was the mean value for the whole sample [OTA 93]. Magnetic resonance imaging of the brains of some survivors with mental disorders revealed a large region of abnormally located gray matter, suggesting an abnormality in neuronal migration [OTA 98].

### 9.2.2. *The harmful effects of ionizing radiation on children*

The effects of ionizing radiation on children are likely to be very different from those on adults, for a number of physiological reasons. We have already seen that the doses received by the fetus are highly damaging in terms of morbidity and mortality and that they may differ according to the age and sex of the victims.

Children run a greater risk than adults of developing cancer after exposure to ionizing radiation, due to their developing bodies and long life expectancy after exposure. Children therefore require special attention following a nuclear accident or when radiation is used for medical diagnosis or treatment [KUT 16].

11 In 1988, Otake et al. reported 30 cases of severe mental disorders diagnosed before the age of 17, out of 1,598 cases studied, of which 1,068 (67%) received a dose of less than 10 mGy [OTA 88].

#### 9.2.2.1. *Children's anatomical and physiological development*

Several specific aspects of anatomical development and physiology can influence the response to ionizing radiation. These include, for example, the long period of human growth, which accounts for almost 25% of the total lifespan, and the different comparative anatomy of children and adults. During development, the growth of each system or organ is unique. This is the case, for example, with the anatomical development and physiology of the child's brain, characterized by a long period of development and a succession of specific radiosensitive events. This organ accounts for 10% of birth weight and only 2% of adult weight. The development of cortical grey matter takes place in several production peaks around the age of 1–2 years, then the rate remains high until the age of 12, and then declines until the age of 16. However, recent research suggests that the ventral temporal cortex does not thin during childhood, but rather becomes more myelinated [NAT 19].

Similarly, the development of the various sexual organs takes place at very precise, hormone-sensitive periods, which are also highly radiosensitive. Ova, for example, are fully developed in the first trimester after conception. Irradiation of the mother during this period can render the daughter permanently sterile. The number of follicles is approximately 295,000 at birth, 180,000 at age 13, 65,000 at age 25, 16,000 at age 35 and 1,000 at menopause [WAL 11].

In humans, as in all species, individual susceptibility varies greatly with age, with young life stages (embryo, fetus, infancy) being more sensitive to the majority of physical and chemical stresses [AMI 17]. Indeed, young children often have a higher metabolism than adults, with more intense ventilation, a relatively more abundant diet and still imperfect detoxification mechanisms. On the contrary, older individuals may also be more susceptible than younger adults, due to wear and tear on all physiological functions, particularly protective ones (skin, immune system, etc.). The critical stages for a pollutant may vary according to gender. Early stages are also the most sensitive, particularly at specific times of the year. An individual's general state of health (nutrition, illness, hormonal status, etc.) will also have a major impact on their susceptibility to a given stress.

#### 9.2.2.2. *Radionuclide dosimetry in children*

In terms of external exposure, children have smaller body diameters. Therefore, for a given flux of incident particles, the internal doses received at organ level will be slightly higher than those delivered to the same organs in adults[12]. Also, they are smaller in size than adults and are therefore closer to the ground. As a result, the

12 The total effective dose calculated for the year 2015 and related to the number of children in the exposed population varied from 0.97 mSv for children under one year old to 0.49 mSv for children aged 1–5 and 0.30 mSv from 6 to 10 [IRS 18].

dose received from radionuclides in the soil will be higher than in adults. The correction factor is approximately 1.2 for children.

For internal exposure, age-dependent conversion coefficients for effective dose per unit of physical dose ($Sv.Gy^{-1}$) for the two primordial radionuclide chains (uranium and thorium) and $^{40}K$ vary with age, from approximately 0.9 before age 4, 0.75 between ages 4 and 14 and 0.7 in adulthood [NCR 09].

The doses received when radionuclides are ingested can be higher in children. For a given radionuclide and the same ingested activity, organ doses are higher in children due to the size and mass of their organs, the fact that these organs are closer together, their diet (milk[13]) and metabolic differences (2–3% decrease in basal metabolism per decade after adulthood). However, the dose delivered is not inversely proportional to weight. For example, for 1 Bq of iodine-131 ingested by a child under 2 years of age, the dose delivered is $1.8.10^{-7}$ sievert per becquerel and $2.2.10^{-8}$ (Sv/Bq) for an adult, i.e. a ratio equal to 8.2, whereas thyroid weights are 1 gram for a 1-year-old child[14] and 20 g for an adult man (16 g for a woman), i.e. a ratio equal to 20. These dose reductions can be explained by the higher transit speeds of the elements through the organs and the whole body in young children. However, the intake of inhaled air[15] and the quantity of food ingested are generally lower in children than in adults.

### 9.2.2.3. *Main differences in the effects of ionizing radiation on children*

During brain development, high in utero doses of ionizing radiation can reduce the number of neurons and lower IQ (as seen in child survivors of Hiroshima and Nagasaki), disrupt the development of myelin around axons or even destroy these mantles.

The respiratory system develops throughout life. The lungs comprise 20 million alveoli at birth, 150 million by age 2 and 300–400 million by age 7. Before the age of 5, the diameter of the peripheral airways is smaller than that of the median airways, which increases resistance to air entry. Mucociliary transport velocity is also lower in children, resulting in greater retention of inhaled aerosols. Children have fewer underlying pulmonary pathologies or lesions associated with senescence.

---

13 The potassium content of milk is 1.7 g/100 g, and with 31.3 Bq of $^{40}K$ per gram of potassium, the radioactive load of milk is equal to 53 Bq/100 g.

14 1 ± 0.1 g at one year, 1.8 g ± 0.3 g from one to two years, 2.6 g ± 0.9 g from three to four years, 4.6 ± 1.5 from five to nine years and 10.2 g ± 2.5 g from 10 to 14 years [ICR 75].

15 0.2 liters in babies, 0.55 at one year, one to four years, 2.2 to eight years and 5.5 liters in adults [ICR 75].

Bone growth varies according to gender. Therefore, cartilage is present in greater quantity in children[16] until the age of 18 for girls or 20 for boys. Cartilage does not concentrate the long-lived alpha-emitting radionuclides that are the main cause of radiation-induced bone cancers[17]. Bone marrow, a blood-forming organ, is distributed throughout the bones in young children and is more limited in adults [BEY 15].

The digestive system also differs significantly from that of adults. For example, the diet of newborns and infants is essentially milk-based, containing iodine and strontium[18]. As young children put everything within reach into their mouths, soil ingestion is not unusual, and in the case of contaminated soils, children can incorporate radionuclides, and their dose coefficients ($Sv.Bq^{-1}$) are higher than those of adults. However, digestive transit varies considerably with age. It is significantly faster in children, especially infants. Therefore, transit time through the colon in adults varies on average from 39 h (20–30 years) to 66 h (75–85 years) [MAD 04]. The average total transit time for adults is between 24 and 40 h and for infants aged 0–6 months between 10 and 12 h. Transit speed also varies according to the type of food [FAB 64].

A study of ophthalmological damage caused by ionizing radiation in survivors of the atomic bombing of Hiroshima and Nagasaki was carried out in 1978–1980. It showed that a radiosensitive aging effect was stronger in children aged under 15 at the time of the bombing. It was observed for both axial opacities and subcapsular changes posterior to Hiroshima [CHO 83].

According to Shuryak, carcinogenesis processes differ with the age of the exposed individual. The radiation-induced cancer risks of individuals exposed at a young age are dominated by initiation processes, whereas those of people exposed at an older age are more influenced by the promotion of pre-existing precancerous cells [SHU 10]. This difference may lead to differences in dose response depending on age.

Children are more sensitive to radiation than adults, in particular with a higher relative risk of cancers, including leukemia, brain cancer, breast cancer, skin cancer

16 Cartilage in newborn babies varies from 30 to 36% of body weight, compared with around 1.5% in the adult. The average cartilage weight is 1,100 g for the adult male and 900 g for the adult female [ICR 75].

17 The first radiation-induced bone sarcomas were mainly caused by radium-226 (Radium Girls in the USA), and then by plutonium. Bone sarcoma is caused by internal irradiation of the "skin" of the bone (the endosteum) by alpha radiation from radium-226.

18 0.25 mg of stable iodine per liter of milk and 0.1–0.2 mg of stable strontium per gram of calcium, measured in powdered milk [MAD 65].

and thyroid cancer. This is partly due to the radiosensitivity of their developing organs and tissues [KUT 16].

Radiosensitivity in children varies according to the type of tissue considered. For approximately 25% of cancer types (leukemia, thyroid, skin, breast or brain cancers), children are indeed more radiosensitive [LEB 14].

#### 9.2.2.4. *Children exposed to natural sources of ionizing radiation*

A study carried out in Switzerland estimated domestic radon levels for each individual home, using a model developed and validated on the basis of some 45,000 measurements taken throughout the country. A total of 997 cases of childhood cancer were included in the study. In comparison to children exposed to a radon concentration below the median (< 77.7 $Bq.m^{-3}$), the adjusted hazard ratios for children with exposure at the 90th percentile (≥ 139.9 $Bq.m^{-3}$) were 0.93 (95% CI 0.74, 1.16) for all cancers, 0.95 (95% CI 0.63 to 1.43) for all leukemias, 0.90 (95% CI 0.56 to 1.43) for acute lymphoblastic leukemia and 1.05 (95% CI 0.68 to 1.61) for central nervous system (CNS) tumors. The authors concluded that there is no evidence that domestic radon exposure is associated with childhood cancer, despite relatively high radon levels in Switzerland [HAU 13].

A subsequent Swiss study, involving 2,093,660 Swiss children under the age of 16, included in the census (1990–2000), with follow-up to 2008, identified 1,782 incident cases of cancer, including 530 with leukemia, 328 with lymphoma and 423 with a tumor of the CNS. A radiation model was used to predict terrestrial and cosmic radiation dose rates at residential locations. Hazard ratios for each one-millisievert increase in cumulative external radiation dose were 1.03 (95% CI 1.01, 1.05) for all cancers, with 1.04 (95% CI 1.00, 1.08) for leukemia, 1.04 (95% CI 1.00, 1.08) for CNS tumors and 1.01 (95% CI 0.96, 1.05) for lymphoma. The authors concluded that background radiation may contribute to childhood cancer risk, including for leukemia, lymphoma and CNS tumors [SPY 15].

A large-scale, record-based case–control study testing associations between childhood cancer and natural background radiation was conducted by Kendal et al. [KEN 12]. Cases (27,447) born and diagnosed in Great Britain between 1980 and 2006 and matched cancer-free controls (36,793) were drawn from the National Childhood Tumour Register. Radiation exposures were estimated for the mother's place of residence at the child's birth. There was a 12% ERR of childhood leukemia per millisievert of cumulative dose to red bone marrow, due to gamma radiation. The analogous association for radon was not significant. Associations with other childhood cancers were not significant for either exposure. The statistically significant risk of leukemia reported in this study is consistent with predictions for a high dose rate. The study supports the extrapolation of high-dose-rate risk models to prolonged exposures to natural background levels [KEN 12].

#### 9.2.2.5. *Children exposed to anthropogenic activities in the nuclear field*

The studies we do have are few in number and mainly European. Yet the populations that have suffered significant doses are substantial. First and foremost are the child survivors of the Hiroshima and Nagasaki atomic bomb blasts (LSS), where 22,708 children under 10 (21.5% of the cohort) and 23,079 aged 10–19 (21.9%) received a dose during the explosions [GRA 17].

Then, there are the children living near nuclear test sites, such as the Marshall Islands, where 39% of the population who received a high dose were under 10 years of age at the time of the "Bravo" test, and those in Semipalatinsk[19] with 4,500 people aged under 20. This is the case for the child population near the Nevada Test Site, which was the subject of a case–control study on leukemia developed in children living in Utah. Also, near Mayak[20], whose "Children's registry" lists 90,000 individuals born in Ozyorsk or arriving before their 15th birthday from 1934 to 2008. Then, there are the children who lived close to areas that have also suffered a major nuclear accident, such as Chernobyl (classified as level 7 on the International Nuclear Event Scale [INES]), where 27% of those evacuated during the Chernobyl accident were aged between 0 and 17, or Fukushima, where 13% of those contaminated were aged between 0 and 14 [UNS 13].

#### 9.2.2.6. *Children exposed to scanners*

Finally, there are diagnostic examinations, such as those performed with CT scanners, which are more penalizing in terms of dose than conventional imaging, and which particularly concern children. Some 5–9 million CT scans are performed on children in the United States every year. The use of CT scans in adults and children has increased eightfold since 1980, with annual growth estimated at approximately 10% per year [NCI 18]. This annual number of scans performed on children is very significant, as in the UK only 283,919 were performed in 18 years (1985–2002) and in France 121,437 in 16 years (2000–2015) [GRA 17]. When these figures are weighted by population size, the annual number of scans performed on children in the USA is more than 100 times higher than in France and the UK.

---

19 From 1949 to 1955, the Semipalatinsk nuclear test range was the main Soviet airborne nuclear test site. It is located in Kazakhstan. The effects of these tests are said to have caused health problems for more than 1.5 million inhabitants of the region.

20 Built between 1945 and 1948, the Mayak nuclear complex is a major Russian nuclear site where eight plutonium-bearing reactors were constructed (1948–66) and where spent fuel was reprocessed. Radioactive effluents were discharged directly into the nearby Tetcha River. On September 29, 1957, a huge explosion and fire in a tank containing radioactive solutions (hydrogen from radiolysis) contaminated large areas of the country (accident classified at level 6 on the INES, with around 200 deaths and thousands displaced).

A retrospective cohort study by M. Pearce et al. [PEA 12], which included patients who had not been diagnosed with cancer at the time of their first CT scan, investigated the effects of this type of exposure before the age of 22, between 1985 and 2002, in 81 health departments in Great Britain. Reconstructions of doses received in the bone marrow and brain were carried out using phantoms[21] representing five age categories: 0, 5, 10, 15 and 20 years. The incidence of intracranial tumors (benign and malignant), leukemia (acute and chronic) and mortality were observed up to 2008. Over 176,000 patients were followed up, for approximately 10 years for leukemia and seven years for brain tumors. With one to more than five scans per child, 284,000 scan procedures were recorded. Doses, depending on age, were estimated at between 28 and 44 mGy for brain scans and between 2 and 9 mGy for bone marrow scans. For thoracic and pelvic CT scans, bone marrow doses were estimated at between 2 and 4 mGy. Over the period under review, 74 cases of leukemia and 135 cases of intracranial tumors were recorded.

The analyses show a significant increase in cancer incidence with cumulative dose. For leukemia, an ERR of 0.036 per mGy is observed, and 0.023 per mGy for brain tumors. The use of scanners in children, assuming cumulative doses of approximately 50 mGy, could lead to a near tripling of the risk of leukemia, and doses of approximately 60 mGy could triple the risk of brain cancer. While the authors indicate that the results obtained for leukemia appear to be consistent with those expected by transposition of results from the cohort of Hiroshima–Nagasaki survivors, those obtained for brain tumors are more than three times higher than those expected [IRS 12].

Another epidemiological study, conducted in Australia by Mathews et al. [MAT 13], involved a cohort of 10.9 million people, identified from Australian Medicare records, aged 0–19 years as at January 1, 1985 or born between the beginning of 1985 and the end of 2005. All exposures to Medicare-funded scans between 1985 and 2005 were identified for this cohort. Cancers diagnosed in cohort members up to the end of 2007 were obtained through linkage with national cancer records.

A total of 60,674 cancers were recorded, including 3,150 in 680,211 people exposed to CT scanners at least one year before any cancer diagnosis. The average follow-up time after exposure was 9.5 years. The overall incidence of cancer was

21 Anthropometric device representing a human organism with a skeleton, a lung and all other organs, made of materials equivalent to human tissue (same hydrogen, carbon and oxygen content) and of the same density. The phantom is made up of a stack of "slices", with small cavities to accommodate small dosimeters (e.g. FLi), to measure the variation in dose as a function of depth, and then to calculate the dose to the organs, in order to assess the effective dose.

24% higher in exposed people[22] than in those not exposed, after controlling for age, sex and year of birth.

The authors observed a dose–response relationship, and the incidence rate ratio (IRR) increased by 0.16 (0.13 to 0.19) for each additional scan. The IRR was higher for exposure at younger ages (P <0.001 for trend). For the groups 1–4 years, 5–9 years, 10–14 years and 15 years or more since first exposure, the IRRs were 1.35 (1.25–1.45), 1.25 (1.17–1.34), 1.14 (1.06–1.22) and 1.24 (1.14–1.22), respectively.

The IRR increased significantly for many types of solid cancer (digestive organs, melanoma, soft tissue, female genitalia, urinary tract, brain and thyroid), for leukemia, myelodysplasia and certain other lymphoid cancers. There was an excess of 608 cancers in people exposed to CT scans (147 brain cancers, 356 other solid cancers, 48 leukemia or myelodysplasia and 57 other lymphoid cancers). The average effective dose per examination was estimated at 4.5 mSv [MAT 13].

These three studies, which cover a total of several hundred thousand children, adolescents and young adults exposed to CT scanners, present converging results. They show a significant increase in the incidence of several types of cancer as the cumulative dose increases, for relatively short follow-up periods (7–10 years) during which not all the effects have yet been expressed. Although the clinical benefits should outweigh the low absolute risks, all the authors of these studies conclude that radiation doses from scanners should be kept as low as possible, and that alternative procedures, not involving ionizing radiation, should be considered where possible [PEA 12, IRS 12, MAT 13].

#### 9.2.2.7. *Morbidity and mortality in children who survived Hiroshima and Nagasaki*

The cancer mortality of a group of children exposed in childhood[23] (0–14 years), survivors of the Hiroshima and Nagasaki bombs, was compared to that of an unexposed control group (population of Japan[24] in 1945) with a 40-year follow-up (1950 to 1990) by H. Goto et al. [GOT 12]. The standardized mortality ratios (SMRs) calculated for these survivors varied according to dose level. They were significantly elevated for both sexes, for all deaths, while showing differences for some organs. SMRs were significantly higher in exposed boys, for all deaths, all cancers, leukemia and liver cancer, and in exposed girls for all cancers, solid cancers, liver cancer and breast cancer. For certain organs, these high SMRs were observed at low or even very low doses.

22 The incidence rate ratio (IRR) was 1.24 (95% confidence interval 1.20–1.29; P < 0.001).

23 There were 13,601 boys and 13,218 girls aged 0 to 14 in 1945.

24 This represents 13.54 million boys and 13.25 million girls [GOT 12].

In boys, SMRs were significantly higher for all deaths and for liver cancers in those exposed to very low doses (<5 mSv), and for all cancers, solid cancers and liver cancers[25] in those exposed to low doses (5–100 mSv). In girls, SMRs were significantly higher for liver cancer and uterine cancer in those exposed to low doses, and for leukemia[26], solid cancers, stomach cancer and breast cancer in those exposed to high doses (>100 mSv) [GOT 12].

The authors point out that the low SMRs observed for radiation-induced leukemia at low doses may be due to the fact that many children died of leukemia between August 1945 and October 1950 (when the LSS cohort was set up), since the incubation period for pediatric cancers is shorter than that for adults.

The analysis of the LSS of Japanese atomic bomb survivors found 310 deaths from leukemia in the period 1950–2000 among 86,611 people in the LSS cohort. In this study, the ERR for leukemia in children under 10 at the time of exposure reached a huge maximum of approximately 70 for a dose of 1 Gy, while the ERR for survivors exposed at age 30 and over was close to 2 [KUT 16].

Thyroid cancer was the most common type of solid tumor among atomic bomb survivors, aged under 20 at the time of the explosions. In their study, Thompson and colleagues catalogued all tumor cases for the period 1958–1987 and found that thyroid cancer was particularly common in children under 10 years of age in 1945, with the effect decreasing with age [THO 94].

Among survivors of the atomic bombings of Hiroshima and Nagasaki, the risk of cancer incidence was very high in young children, close to 5,000 cases per 100,000 in girls and 2,500 in boys, then gradually decreasing when exposure occurred at an older age. In adulthood, the difference between males and females was less pronounced, and the risk was less than 1,000 cases per 100,000 [HRI 11]. The ERR/Gy among Hiroshima and Nagasaki survivors (1950–2003) of all types of cancer is always higher when exposure occurred at a young age [OZA 12]. For leukemia in particular, the differences are also very marked and point in the same direction [RIC 09].

Data from 1958 to 1998, the LSS cohort shows that childhood exposure could increase the risk of cancer of the corpus uteri [PRE 07].

---

25 The SMRs for liver cancer mortality at very low doses were 2.809 (1.386–5.241) for girls and 2.400 (1.759–3.210) for boys.

26 The SMRs for leukemia at high doses were very high: 4.619 (2.604–7.723) for girls and 7.097 (4.660–10.439) for boys, aged 0–14 in 1945.

#### 9.2.2.8. *Morbidity and mortality in children with high exposure levels*

A number of epidemiological studies have also demonstrated an increase in thyroid cancers in children and young adults exposed to high thyroid doses (> 1 Gy) during childhood due to the iodine-131 released during the Chernobyl accident [YAM 14]. The first cases of thyroid cancer were observed as early as five years after the accident. The highest cancer rates were observed in children aged 5 or under at the time of exposure, suggesting a vulnerability of the thyroid gland in this age group. Risks decrease with age, and it is generally accepted that the risk of developing radiation-induced thyroid cancer decreases when exposure occurs after the age of 15 [KUT 16].

Ron and colleagues carried out a meta-analysis of five cohort studies (atomic bomb survivors, children treated for ringworm, two studies of children irradiated for tonsillar hypertrophy and infants irradiated for thymus hypertrophy) and two case–control studies (patients with cervical cancer and childhood cancer). The combined studies included nearly 120,000 people (about 58,000 exposed to a wide range of doses and 61,000 unexposed subjects) and nearly 700 thyroid cancers [RON 95].

For people exposed to radiation before the age of 15, linearity best describes dose response, even up to 0.10 Gy. For childhood exposures, the ERR/Gy was 7.7 (95% CI 2.1, 28.7). The ERR began to decrease approximately 30 years after exposure, but was still high at 40 years. Risk also decreased significantly with age at exposure, with little apparent risk after age 20. The thyroid gland in children has one of the highest risk coefficients of all organs and is the only tissue with convincing evidence of a risk of about 1.10 Gy [RON 95].

### 9.3. Radiation-induced cancers in adults

Most of the "dose-effect" data available, relating to people of both sexes and all ages (from in utero exposure to old age), comes from epidemiological studies carried out on survivors of the atomic bombs dropped on Japan on August 6 and 9, 1945. They concern various somatic effects of a deterministic (non-stochastic) nature, such as radiation burns that occur when exposed individuals have received a radiation dose exceeding a threshold value (or range of doses) or stochastic (random) effects such as radiation-induced cancers. The occurrence of these cancers is stochastic in nature and, generally, follows a linear law with no threshold. In this case, the probability of occurrence increases linearly with the increase in dose received by the exposed organism.

The International Commission on Radiological Protection (ICRP) has since divided these effects into tissue reactions (formerly non-stochastic and deterministic) and stochastic effects. Effective dose limits aim to reduce the risk of stochastic effects (cancerous/healing effects) and are based on risk coefficients assuming linear dose response with no threshold and dose and dose rate effectiveness. The equivalent dose limits set by the ICRP aim to avoid tissue reactions (cataracts and non-cancerous skin changes) and are based on a threshold dose. However, the boundary between these two categories is becoming blurred, as in the case of radiation-induced cataracts. A new approach is suggested by Hamada and Fujimichi [HAM 14], who propose classifying radiation effects according to whether or not they are life-threatening.

### 9.3.1. *The effects of high doses, acute radiation syndrome*

Acute radiation syndrome is a violent attack on the organism due to external irradiation of most, or even all, of the body by a high dose of penetrating radiation (gamma, neutrons), delivered in a very short time (measured in seconds or minutes). Survivors of the Hiroshima and Nagasaki atomic bombs, firefighters responding to the Chernobyl No. 4 reactor core explosion in May 1986 and a number of technicians accidentally exposed in irradiators (cobalt-60, electric radiation generators) or during criticality accidents have all suffered serious or even fatal injuries.

The variations in individual radiosensitivity that have been demonstrated in patients treated with radiotherapy mean that, rather than the notion of a threshold dose for the appearance of an acute radiological symptom, we should consider a dose range above which the effect may occur. The main radiological syndromes concern the red bone marrow (hematopoietic), the gastrointestinal system, the cardiovascular system and the CNS. In the case of the latter two systems, which only manifest themselves at very high doses (several tens of grays), the onset of symptoms heralds death, which occurs in the days following exposure. In accident situations, and in the absence of dosimetric data, emergency medical measures are based on the symptoms observed, their rate of onset and the development of the complete blood count (CBC).

The classic symptoms of irradiation are nausea, vomiting, anorexia, alopecia and, for higher doses, diarrhea. Depending on the size of the dose, these symptoms may appear from a few minutes to a few days after exposure. The shorter the symptom onset time, the higher the dose received. Hematopoietic syndrome is characterized by a drop in the number of blood cells, which can lead to infections due to a fall in the number of white blood cells, bleeding due to a reduction in the number of platelets and anemia due to a fall in the number of red blood cells. These symptoms may appear after an acute dose equal to or higher than the dose range

0.2–0.7 gray received by the whole body. Gastrointestinal damage occurs at higher doses, close to the lethal dose (6–8 grays).

The sorting of liquidators irradiated at Chernobyl (May 1986) was based mainly on CBC data. A sharp drop in blood components led to referral of the exposed person to a sterile room and blood transfusions, either total or selective (packed red blood cells or platelets), or even a bone marrow transplant.

Thirteen people exposed to radiation at Chernobyl, with total body doses estimated at between 5.6 and 13.4 Gy, received bone marrow transplants. Only two recipients, who received radiation doses estimated at 5.6 and 8.7 Gy, were alive more than three years after the accident. The remaining 11 died of various causes, including radiation burns (five) [BAR 89].

Acute syndrome due to internal exposure is very rare, but we can cite the case of Alexander Litvinenko's criminal poisoning with polonium-210 in London, UK, in 2006. The 43-year-old presented to hospital with acute abdominal pain, diarrhea and vomiting. He was hospitalized due to dehydration and persistent gastrointestinal symptoms. Within six days, the patient developed thrombocytopenia and neutropenia. After two weeks, in addition to bone marrow failure, he showed signs of alopecia and inflammation of the mucous membranes of the mouth and digestive system. His condition continued to deteriorate, and within three weeks, he had reached multivisceral failure requiring ventilation, hemo-filtration and cardiac assistance, associated with a drop in consciousness. He died on day 23 [NAT 16].

According to the ICRP [ICR 12], the threshold values for mortality are approximately 1 Gy of acute dose without medical care and 2–3 Gy with good medical care. Prolonged doses of 4–8 Gy over one week, or 10–14 Gy accumulated over one to three months, are unlikely to result in death.

### 9.3.2. *The survivors of Hiroshima and Nagasaki*

Some 284,000 survivors of the atomic bomb were counted in the 1950 national census. Of these, approximately 190,000 who were still living in Hiroshima or Nagasaki at the time of the census were used as the basis for the selection of the 94,000 survivors in this cohort [GRA 17].

Table 9.2 describes the two populations with known or unknown doses (93,669) present in the bombed cities and still alive in October 1950 [OZA 12]. The first studies concerned this basic cohort, called the Life Span Study (LSS), within which other cohorts will be defined.

| | Known doses | Unknown doses | NIC | Cohort monitored | Total cohort |
|---|---|---|---|---|---|
| Hiroshima | 58,494 | 3,442 | 20,179 | 78,673 | 82,115 |
| Nagasaki | 28,117 | 3,616 | 6,350 | 34,467 | 38,083 |
| Total | **86,611** | **7,058** | **26,529** | **113,140** | **120,198** |
| Men | 35,687 | 3,287 | 11,143 | 46,830 | 50,117 |
| Women | 50,924 | 3,771 | 15,386 | 66,310 | 70,081 |

**Table 9.2.** *Members of the initial LSS cohort alive in October 1950, according to their dosimetric status [OZA 12]*

NOTE ON TABLE 9.2.– *NIC (Not in the City): residents of Hiroshima and Nagasaki absent from the city, or more than 10 km from the hypocenter, at the time of the explosions. People exposed, but with unknown doses, are those for whom it was not possible to calculate the dose received, for various reasons (uncertain location and/or physical position during the explosion).*

Over the period 1950–2003, 58% of the 86,611 members of the LSS cohort (with DS02 dose estimates) died. The risk of all causes of death was associated with radiation dose. Cancer mortality risk was significantly increased for most major sites, including breast, bladder and ovary, lung, colon, stomach, liver, gallbladder, esophagus, while rectum, pancreas, uterus, prostate and renal parenchyma did not show significantly increased risks. The mean ERR/Gy by gender was 0.42 [95% CI 0.32, 0.53] for all solid cancers at 70 years after exposure at 30 years, based on a linear model. The risk increased by approximately 29% per decade of decreasing age at exposure (95% CI 17%, 41%) [OZA 12].

Verification of the vital status of LSS members is based on the national family register of the Japanese system (Koseki). As confirmed incidence of solid cancers was not possible until the creation in 1958 of tumor registries related to the population of Hiroshima and Nagasaki, analyses of incidence data were limited to a subset of LSS cohort members who were alive and had not developed cancer before January 1, 1958 [PRE 07].

After excluding those who had died or developed cancer before January 1 1958, as well as those who could not be traced using Koseki (86), the LSS solid cancer incidence cohort consisted of 105,444 living subjects with no known history of cancer at the start of follow-up (93.1% of the LSS cohort) (Table 9.3). For

80,205 members of the cohort (out of 86,611), individual dose estimates were available, and a further 25,239 (out of 26,529) were not in either city at the time of the bombings (zero doses). The cohort of exposed and unexposed individuals is 105,444 (out of 113,140).

| | **NIC D = 0** | **Doses D > 0** | **Known doses** | **Unknown doses** | **Total cohort** |
|---|---|---|---|---|---|
| Hiroshima | 19,249 | 54,152 | 73,401 | 3,148 | 76,549 |
| Nagasaki | 5,990 | 26,053 | 32,043 | 3,325 | 35,368 |
| Total | **25,239** | **80,205** | **105,444** | **6,473** | **111,917** |
| Men | 10,488 | 32,422 | 42,910 | 2,954 | 45,864 |
| Women | 14,751 | 47,783 | 62,534 | 3,519 | 66,053 |

**Table 9.3.** *Members of the LSS cohort on January 1, 1958, by dosimetric status (according to [GRA 17])*

Epidemiologists have analyzed trends in the incidence and mortality of survivors from leukemia and radiation-induced solid cancers[27]. The earliest incidence was that of leukemia, with the first excesses appearing in the second and third years following the atomic bombings.

Among the first analyses of the cohort initiated in 1958, that of Thomson et al. [THO 94] concerned 8,613 primary solid cancers diagnosed between 1958 and 1987 among the 79,972 individuals in the cohort. A statistically significant excess risk for all solid cancers was demonstrated, with an excess relative risk at 1 Sv (ERR1Sv) equal to 0.63.

Breast cancer showed the highest ERR, at a dose of one sievert (ERR1Sv = 1.59), followed by thyroid (ERR1Sv = 1,15), bladder (ERR1Sv = 1.02), ovary (ERR1Sv = 0.99), lung (ERR1Sv = 0.95), colon (ERR1Sv = 0.72) and stomach (ERR1Sv = 0.32). There was evidence of an increase in neural tissue tumors (excluding the brain) in people exposed to the bombs before the age of 20. For the first time, radiation was associated with liver cancer (ERR1Sv = 0.49) and non-melanoma skin (ERR1SV = 1.0) for cancer incidence in the LSS cohort [THO 94].

27 The so-called *solid* organs are characterized by volume and mass, while the red blood-forming marrow is, in adults, a tissue distributed in the vertebrae, ribs, clavicles, sternum, pelvic bones and skull, and in the trunk-like extremities of the femurs and humeri.

The work of Ron et al. [RON 94] confirms these data for 1958–1987 and points out that cancers of the salivary gland and thyroid were also in excess. They note that no significant effect of radiation, either in terms of cancer incidence or mortality, was observed for cancers of the pharynx, rectum, gallbladder, pancreas, nose, larynx, uterus, prostate or kidney [RON 94]. However, with longer latency periods, cancers appear at significant rates. From 1950 to 2003, the risk of cancer mortality increased significantly for most major sites, with, in addition to breast, bladder, ovary, lung, colon and stomach, those of liver, gallbladder and esophagus, while rectum, pancreas, uterus, prostate and renal parenchyma still showed no significantly increased risk [OZA 12].

At the end of 2009, 42,138 members of the LSS cohort (33.6% men and 40.5% women) were alive. The majority of those alive (83.4%) were those who had been exposed in childhood (less than 10 years of age).

From January 1958 to the end of 2009, a total of 24,448 first primary cancers were diagnosed among the 105,444 cohort members (exposed and unexposed). After excluding hematopoietic cancers (1,290) and cancers diagnosed only at autopsy (620), the study by Grant et al. [GRA 17] included 22,538 solid cancers, 992 of which were associated with radiation exposure.

For women, dose response was consistent with linearity, with an estimated ERR of 0.64 per Gy (95% CI 0.52 to 0.77). For men, a significant upward curvature over the whole dose range as well as over restricted dose ranges was observed and, consequently, a linear-quadratic model was used, resulting in an ERR of 0.20 (95% CI 0.12 to 0.28) at 1 Gy and an ERR of 0.010 (95% CI -0.0003 to 0.021) at 0.1 Gy [GRA 17].

Average age at diagnosis ranged from 60.1 years for cervical cancer to 74.5 years for non-melanoma skin cancers.

### 9.3.3. *Radiation-induced cancers as a function of gender*

Studies of cancer incidence and mortality in the cohort of survivors of the Hiroshima and Nagasaki bombings show significant differences between the sexes and in the intensity of the disease in various age groups.

#### 9.3.3.1. *Radiation-induced breast cancer in women*

From the earliest studies, it was breast cancer that appeared to have the highest ERR: equal to 1.59 for a dose of one sievert (ERR.$Sv^{-1}$) with a significant excess for exposures occurring in the first two decades of life, whereas for all solid cancers, the ERR.$Sv^{-1}$ was equal to 0.63 [THO 94].

During the period 1950–1985, an incidence survey among atomic bomb survivors, the LSS, identified 807 cases of breast cancer and 20 second breast cancers. A strongly linear response[28] with radiation dose was found, with the highest ERR among survivors aged under 20 at the time of the bombings. Sixty-eight of the cases were under 10 at the time of exposure, confirming the excess risk due to exposure during infancy and childhood. A much lower dose response was observed in women exposed at the age of 40 and over [TOK 94].

Another LSS, which looked at 1,059 cases of breast cancer diagnosed between 1950 and 1990, confirmed the linear and statistically highly significant response to the dose received. In the analysis, particular attention was paid to the modification of the dose–response relationship to radiation as a function of age at exposure (E) and attained age (A). The ERR for a dose of 1 Sv decreased with increasing values of E and A [LAN 03].

Therefore, for a given dose, the younger the age of exposure, the higher the risk, and the older the age of the survivor, the lower the risk of breast cancer. Exposure before the age of 20 is associated with a higher ERR at 1 Sv than exposure at an older age[29]. It has also been observed that the ERR per 1 sievert decreases with attained age, with by far the greatest drop around the age of 35 [LAN 03]. Estimating the risk of breast cancer incidence must necessarily take into account both the age at exposure and the age reached by the individual.

The most recent studies cover the 62,534 women in the LSS cohort (exposed to known and non-exposed doses) and the 1,470 breast cancers identified between 1958[30] and 2009 [GRA 17, BRE 18]. Table 9.4 shows the distribution, by dose range, of women in the known-dose cohort (including zero-dose NICs) who were alive on January 1, 1958.

28 The incidence of breast cancer increased in direct proportion to the increase in dose received.

29 Of 397 new cases of breast cancer diagnosed between 1998 and 2009, 75% were exposed before the age of 20.

30 January 1, 1958 was chosen as the date on which the Hiroshima and Nagasaki cancer registry was set up.

| Doses in gray | Members of the cohort (%) |
|---|---|
| < 0.005 | 36,155 (57.8) |
| 0.005–0.1 | 16,336 (26.1) |
| >0.1–0.2 | 3,462 (5.5) |
| >0.2–0.5 | 3,625 (5.8) |
| >0.5–1.0 | 1,854 (3.0) |
| >1.0–2.0 | 849 (1.4) |
| >2.0 | 253 (0.4) |
| Total | 62,534 (100) |

**Table 9.4.** *Doses received by women in the LSS cohort alive on January 1, 1958 [GRA 17]*

A strong linear dose response was found, with an ERR of 1.12 per Gy for 70-year-old women after 30 years' exposure [BRE 18]. Despite the small number of excess male breast cancers (10), the data also suggest a dose–response relationship (ERR = 5.7 not statistically significant) [BRE 18].

Breast cancer incidence rates after radiation exposure in eight large cohorts[31] are described and compared by Preston et al. [PRE 02]. They collected 1,502 breast cancers in 77,527 women (of whom approximately 35,000 were exposed) with 1.8 million woman-years of follow-up. Excess rates were similar in these cohorts. The results support dose–response linearity for breast cancer. They underline the importance of age at exposure and attained age. They suggest similar risks for acute and fractional high-dose-rate exposures, with much lower effects due to low doses [PRE 02].

### 9.3.3.2. *Radiation-induced ovarian cancer*

The study by Tokuoka et al. [TOK 87] of radiation-induced benign and malignant tumors recorded from 1950 to 1980 showed that the estimated minimum latency period for radiation-induced ovarian cancer appears to be 15–20 years. Age-adjusted ovarian cancer incidence rates showed a statistically significant increase with increasing dose. This dose–response relationship was significant ($P < 0.01$) only in the second half of the study period, 1965–1980. The effect of radiation was greatest in the youngest age group at the time of the explosion. For the youngest age group, those exposed under the age of 20, the relative risk tended to

31 Atomic bomb survivors, American patients with acute post-partum mastitis and Swedish patients with benign breast disease treated with radiotherapy, and American infants with thymus enlargement (thymus irradiated to reduce its size).

decrease with attained age, although cases were few [TOK 87]. However, while the study of radiation-induced ovarian cancers shows that their dose-dependent increase was statistically significant from 1950 to 1980, it ceased to be so when the study was extended to 2009. Utada M. et al. [UTA 21] assessed the radiological risk of ovarian cancer, between 1958 and 2009, among 62,534 female atomic bomb survivors in the LSS cohort, by estimating the ERR/Gy for total ovarian cancer and according to tumor type. The authors identified 288 primary ovarian cancers, including 77 type 1[32] epithelial cancers, 75 type 2 epithelial cancers, 66 indeterminate epithelial cancers and 70 other cancers. Although not significant, dose was positively associated with the risk of total ovarian cancer [ERR/Gy = 0.30, 95% CI -0.22 to 1.11]. No significant trend in ERR was observed as a function of time since exposure or age at exposure [UTA 21]. However, for Nagasaki survivors, the incidence rate of ovarian cancer (1973–1987) between the ages of 40 and 49 was higher than that of unexposed individuals during all periods studied [NAK 94].

#### 9.3.3.3. *Radiation-induced cancers of the prostate*

A study of the incidence of prostate cancer among men in the LSS cohort, 851 incident cases of prostate cancer were observed among 41,544 male subjects, during follow-up from 1958 to 2009 [MAB 21]. More than 50% of cases were diagnosed in people who were under 20 years of age at the time of the bombings, and who had reached or were close to the age of increased risk of prostate cancer during the last decade of follow-up. In the dose–response analyses, the period corresponding to the emergence of prostate-specific antigen (PSA) screening in the general population was taken into account, as were the effects of attained age and birth cohort. Between 2005 and 2009, a PSA test was included in the biennial health examinations of the Adult Health Study.

The authors of the study found a significant linear dose response, with an estimated ERR/Gy of 0.57 (95% CI 0.21, 1.00, P = 0.001). It was suggested that the ERR decreased with increasing age at exposure. The observed dose response reinforces the evidence for an effect of radiation on the risk of prostate cancer incidence in atomic bomb survivors [MAB 21].

### 9.3.4. ***Effects on red bone marrow***

Excess leukemia was the first radiation-induced cancer observed in A-bomb survivors. It was a Japanese doctor from Hiroshima, Takuso Yamawaki, who first noted an increase in leukemia cases in his clinical practice in the late 1940s. This led

32 The tumor is said to be epithelial if it affects cells on the outer surface of the organ (90% of cases). Other non-epithelial tumors of the ovary are rare ovarian malignancies.

to the creation of a register of leukemia and related disorders, and the first reports of high leukemia risk published in the early 1950s.

The risks of radiation-induced leukemia differ in two ways from those of most solid cancers: firstly, radiation causes a very significant increase in leukemia rates (albeit with a smaller number of cases, as leukemia is relatively rare in the Japanese population, even among highly exposed survivors); and secondly, the increased risk appears earlier after exposure, particularly in children. The excess of leukemias began to appear two years after exposure and peaked approximately six to eight years after exposure.

The analyses carried out by Richardson et al. [RIC 09] focused on 310 leukemia deaths observed during the period 1950–2000 among 86,611 people (exposed and unexposed) in the LSS cohort. Analyses focused on the variation in leukemia mortality as a function of time since exposure, age at exposure, city and sex. The ERR/Gy of acute myeloid leukemia is well described by a quadratic dose–response function that peaks at approximately 10 years after exposure. Mortality from acute lymphatic leukemia and chronic myeloid leukemia is best described by a linear dose–response function that does not vary with time since exposure.

Adult T-cell leukemia[33] was not associated with the estimated bone marrow dose. The excess of leukemia deaths due to exposure to ionizing radiation is estimated at 103 deaths out of the 310 observed.

In the most recent decade of observation (1991–2000), the estimated attributable fraction of leukemia deaths among survivors exposed to more than 5 milligrays was 0.34, suggesting that the effect of atomic bombing on leukemia mortality persisted in this cohort for five decades.

In an unexposed Japanese population, the lifetime risk of leukemia is about seven cases per 1,000 people. For typical LSS survivors, who received a dose of 5 milligray or more (an average dose of approximately 0.2 Gy), the lifetime risk of leukemia increases to approximately 10 cases per 1,000 (a relative risk close to 1.5)[34].

## 9.4. Adverse effects of ionizing radiation on different tissues and organs

For a very long time, radiation protection was based solely on the occurrence of cancers caused by ionizing radiation. However, such radiation is capable of altering

33 Adult T-cell leukemia is a rare, highly aggressive cancer of the T cells of the immune system.

34 See: https://www.rerf.or.jp/en/programs/roadmap_e/health_effects-en/late-en/leukemia/.

human health by triggering non-cancerous diseases. Our knowledge on this subject is advancing steadily. The ICRP published revised recommendations for a radiation protection system in Publication 103 [ICR 12]. This included consideration of the harm resulting from the non-carcinogenic health effects of radiation. These effects, previously referred to as "deterministic effects", are now called "tissue reactions". Depending on the organ or physiological system, their response is more or less sensitive, as shown by the ICRP recommendations for acute dose thresholds (Table 9.5).

However, for certain organs, these 2012 recommendations are highly overestimated in the light of scientific advances. This is particularly true for damage to the lens of the eye and the cardiovascular system (see Table 9.5 and Chapter 11).

| Organ | Dose (Gy) | Organ | Dose (Gy) |
|---|---|---|---|
| Hematopoiesis | 0.5 | Cardiovascular | 0.5 |
| Intestine | 6 | Cerebrovascular | 0.5 |
| Cataract of the lens | 0.5 | Femoral heads and ribs | 50* |
| Pneumonitis | 6.5 | Kidney | 7–8 |
| Spinal cord | 50* | Skeletal muscle | 55* |

* If the fractionated dose is 2 Gy.

**Table 9.5.** *Acute dose thresholds recommended in 2012 by the ICRP [ICR 12]*

In multicellular organisms, there is usually specialization of various cell types with specific functions. They are grouped into tissues, and several tissues form an organ. Several organs work together to fulfill a specific physiological function, such as respiration, digestion, thought or sexuality. Various physiological systems (metabolic, endocrine, reproductive, immune and behavioral) can be disrupted by the presence of radionuclides in cells or by the ionizing radiation they emit.

Ionizing radiation can have an impact on metabolism by disrupting certain metabolic pathways, such as those for steroids (vitamin D and cholesterol), iron or xenobiotics. Irradiation also has a major influence on energy metabolism by increasing metabolic cost and/or decreasing energy acquisition [AMI 08].

Various metals disrupt the endocrine system. This is also the case for certain radionuclides such as uranium and cesium-137 (Gueguen et al., pp. 309–322, in: [MEN 09]). Metals, including radionuclides, can alter reproductive success by impacting reproductive organs, gametes, sexual behavior or embryonic development.

Cadmium and uranium are the most studied elements (Henner et al., pp. 323–335, in: [MEN 09]).

Human behavior can be disturbed by uranium, cesium, cadmium and cobalt, resulting in neurodegenerative pathologies (Lestaevel et al., pp. 351–365, in: [MEN 09]).

The suppressive effects of high doses of ionizing radiation on the immune system are well documented. At low doses, however, uncertainties remain, and both stimulatory and suppressive effects have been observed.

The ICRP has divided effects at tissue and organ level into stochastic effects and tissue reactions (formerly known as non-stochastic and deterministic effects). However, the boundary between these two categories is becoming blurred. For this reason, Hamada and Fujimichi [HAM 14] propose a new approach that distinguishes between radiation effects that are life-threatening and those that cause non-life-threatening diseases. In this case, the criteria for assessing risk management are mortality and incidence, respectively. Therefore, for cataracts and circulatory diseases, epidemiological evidence now suggests a linear non-threshold (LNT) dose response for doses well below those recommended by the ICRP (Table 9.5).

In the first decade after the explosions in Japan, a high incidence of leukemia was the first sign of post-irradiation cancers. This incidence peaked seven years after the atomic bombings, before gradually declining without abating 70 years later. The following decades saw the emergence of solid cancers.

### 9.4.1. *Effects on the circulatory system*

The risk of cardiovascular mortality in Hiroshima bomb survivors shows a statistically significant dose–response relationship with initial radiation dose, according to the RERF[35]. In this study, dosimetry was based solely on the initial radiation emitted during the explosion. The dose–response relationship for cardiovascular mortality was calculated using distance from the hypocenter as a surrogate indicator for the radiation dose received.

This cohort study was carried out at Hiroshima University, with Hiroshima atomic bomb survivors, starting in 1970. It involves 29,605 atomic bomb survivor subjects, who were exposed to Hiroshima within 3.5 km of the hypocenter and alive on January 1, 1970. These survivors were followed up until December 31, 2010. To analyze cardiovascular mortality, the cohort data were stratified by sex and age at the time of the bombing and by 10-year age groups. The risk of mortality from

35 Radiation Effects Research Foundation.

cardiovascular disease in atomic bomb survivors was slightly positive, with a statistically significant dose–response relationship. The results of the study by Hara et al. also indicate a high excess risk among the oldest men and women in August 1945 exposed near the hypocenter. According to Hara et al. [HAR 16], the age-related mortality difference in women could be explained by a protective effect of female sex hormones on the circulatory system in young women at the time of the atomic explosion (at the time of bombing) without further clarification.

However, the reasons for the high risk of cardiovascular mortality observed in male Hiroshima survivors remain unknown [HAR 16]. The results on the incidence of heart disease among survivors exposed in childhood are less clear-cut. For Shimizu et al. [SHI 10], there was no significant age-related change in heart disease at the time of exposure. In the study by Shimizu et al. [SHI 10], which looked at 19,054 deaths from circulatory disease among the 86,611 lifetime study members (with one dose of DS02 available), approximately 9,600 participants died of stroke and 8,400 of heart disease between 1950 and 2003. For stroke, the estimated ERR/Gy was 9% (95% CI 1% to 17%, P = 0.02) based on a linear dose–response model, but an indication of possible upward curvature suggested a relatively low risk at low doses.

For heart disease, the estimated ERR/Gy was 14% (6%–23%, P < 0.001). A linear model provided the best fit, suggesting an excess risk even at lower doses. However, the dose–response effect over the restricted dose range of 0–0.5 Gy was not significant. Doses above 0.5 Gy are associated with a high risk of stroke and heart disease. Stroke and heart disease together account for about a third of the excess radiation-related mortality other than cancer among atomic bomb survivors [SHI 10].

A retrospective analysis (Childhood Cancer Survivor Study) involving 14,358 young people under the age of 21, survivors for more than five years of a cancer that occurred before the age of 21, analyzed the cumulative incidence of adverse cardiac outcomes. This incidence continued to increase up to 30 years after diagnosis. The comparison group included 3,899 siblings of cancer survivors. Adult survivors of childhood and adolescent cancer are therefore significantly more likely than siblings to be at high risk of cardiovascular disease [MUL 09].

### 9.4.2. *Effects on the digestive system*

In the Sakata et al. study, which covers a follow-up of the LSS cohort from 1958 to 2009, the authors used dose estimates from the latest dosimetric system (DS02) and took into account the effects of smoking and alcohol consumption. They also performed dose–response analyses for different organs of the digestive system.

Examining 52 years of data, they found 394 oral cavity/pharynx cancers, 486 esophageal cancers and 5,661 stomach cancers in the entire LSS cohort (105,444 subjects) [SAK 19].

The radiological risk of oral cavity and pharyngeal cancer, other than salivary gland cancer, was elevated, but not significantly so. On the contrary, salivary gland cancer showed a very high linear dose response, with an ERR of 2.54 per Gy. Radiation risk decreased significantly with age at exposure (−66% per decade).

The dose response for esophageal cancer was statistically significant according to a simple linear, linear-quadratic and quadratic model. The linear-quadratic and quadratic models described the data better than a simple linear model, and of the two, the quadratic model showed a slightly better fit. The gender difference in linear ERR was not statistically significant.

The risk of stomach cancer increased significantly with dose. The sex-specific mean ERR at age 70 was 0.33 per Gy (95% CI 0.20 to 0.47). The ERR decreased significantly with increasing attained age, but not with age at exposure, and was higher in women than in men [SAK 19].

Cancers of the salivary glands, esophagus and stomach showed a significant increase in risk with radiation dose. Adjustment for lifestyle factors had virtually no impact on estimates of the effect of radiation [SAK 19].

### 9.4.3. *Effects on the reproductive system*

In addition to radiation-induced cancers (section 9.3.3), gonadal irradiation can lead to reduced fertility. This may be temporary or permanent, depending on the dose to the organ and the patient's age. The testicle is one of the body's most sensitive organs to radiation, due to the rapid division of the germinal epithelium. Radiation-induced testicular dysfunction occurs in a dose-dependent manner.

In humans, the deterministic late effects of ionizing radiation on the testis are mainly due to the persistence of disorders of spermatogenesis and/or endocrine disorders at a distance from irradiation. The morphological, physiological and radiobiological specificities of the human testis, as well as the numerous factors related to irradiation conditions, make it difficult to assess late effects and tolerance doses [BAR 97].

Doses of 0.10 Gy to the testicles can affect the number of germ cells (spermatogonia) and are likely to cause temporary sterility. Immature spermatogonia are the most sensitive to radiological damage. Radiation doses of less than 0.8 Gy lead to an abnormally low concentration of spermatozoa in male semen, while doses of 0.8–2 Gy can lead to total and irreversible loss of spermatozoa (azoospermia). On average, return to pre-irradiation seminal parameters can be observed within 10–24 months [CHA 14].

In general, malignancy and its treatment can lead to a temporary or permanent reduction in infertility. A recent review of the literature shows that with fractional treatment, a total dose in excess of 2.5 Gy results in a total loss of testicular function, whereas with single-fraction treatment, radiation doses in excess of 6 Gy are required to cause a total cessation of sperm production (azoospermia) [AHM 22].

For the ovaries, a radiation dose ≤2 Gy can reduce ovarian reserve by 50%. Sterilization by irradiation of the ovaries is immediate. There is no latency period, as in men [AHM 22]. The number of oocytes decreases progressively throughout human maturation, and the effect of ionizing radiation in causing sterility appears to be correlated with the number of oocytes [PAU 10]. Ovarian doses for permanent sterility range from 2.5 to 6 Gy. There has been some recovery of fertility even during major radiological accidents, as observed in Chernobyl reconstruction workers [LIN 16].

Exposure to ionizing radiation during childhood can lead to altered uterine vascularization, reduced uterine volume and elasticity, fibrosis and necrosis of the uterine wall (myometrium), atrophy and insufficiency of the endometrium. These exposures have a gonadotoxic action with long-term effects that include ovarian failure, pubertal arrest and resulting infertility [MAR 18].

Irradiation can have a profound effect on reproductive function. Irradiation modalities (total dose, number of fractions and duration) play an important role in the radiobiological effect on the tissues concerned and vary according to the different tissues and organs. Irradiation of the CNS may affect the timing of the onset of puberty, lead to hyperprolactinemia or cause gonadotropin deficiency if the hypothalamo–hypophyseal axis is involved in the radiation field [OGI 93]. The uterus may also be damaged by radiotherapy. As radiation has a significant impact on reproductive potential, fertility preservation procedures must be carried out before and/or during cancer treatment. Fertility preservation strategies have been in use for some years and have recently diversified thanks to advances in reproductive biology [MAR 18].

In the course of his career, J-C Zerbib has had occasion to observe, during accidental exposure to low doses, temporary sterility in women for doses well below (a few milligrays) those which lead to partial destruction of the oocyte stock. This is probably due to an effect on the maturation of oocytes leading to the egg. This is the case of a PhD student who was doing crystallography research with an X-ray generator that was leaking and irradiating her pelvis. This accidental irradiation of a few milligrays resulted in temporary sterility lasting around five months. The second example concerns flight attendants, who were subject to annual doses of approximately 2 $mSv.year^{-1}$. When a pregnancy did not materialize, they asked to work on the ground for a year, despite the loss of salary. As the majority of the planned pregnancies were resolved within this timeframe, this shows that there has been a temporary sterilization, which has diminished with the cessation of cosmic exposure.

### 9.4.4. *Effects on the lens*

Cataracts are an opacification of the crystalline lens which leads to a reduction in visual acuity. It is known that the action of thermal radiation from incandescent glass or metal and that of ionizing radiation cause occupational cataracts[36]. "Natural" cataracts occur between the ages of 65 and 75, apart from other conditions such as diabetes. The three main types of lens opacity are nuclear cataract (NC), cortical cataract (CC) and posterior subcapsular cataract (PSC).

Radiation-induced cataracts, which occur earlier than natural cataracts, have been known for a century. Marie Curie was one of the first known victims, with a bilateral cataract (caused by radiation from radium-226 and its descendants) which left her practically blind in 1921, at the age of 54 [LOR 92].

Historically, the thresholds proposed by the ICRP, which still considers this effect to be deterministic with a threshold, have been steadily reduced. The European Directives (which take up the ICRP's recommendations), which had set a limit of 150 $mSv.year^{-1}$ on May 13, 1996, adopted a limit of 20 $mSv.year^{-1}$ (i.e. an occupational limit of 0.8 Sv for life) on December 5, 2013. These provisions have been transposed into the laws and regulations of the Member States of the European Community.

#### 9.4.4.1. *Cataracts of the survivors of the bombs dropped on Japan*

Numerous studies of the effect of ionizing radiation on cataract development have been carried out in survivors of the Hiroshima and Nagasaki atomic bomb blasts [MIN 04, NAK 06, NER 07].

An ophthalmological study of age and radiation-related ophthalmological damage in survivors of the atomic bomb in Hiroshima and Nagasaki was carried out in 1978–1980. A stronger radiation-sensitive aging effect was observed in those

36 Occupational cataracts are listed in Table no. 6 for radiation-induced cases and in Table no. 71 for cases caused by thermal irradiation.

under 15 years of age at the time of the bombing for both axial opacities and posterior subcapsular changes in Hiroshima [CHO 83].

In Hiroshima and Nagasaki survivors, subcapsular opacities are 1.5–2 times more sensitive in the young [NAK 06]. In children (aged 5–17) living in a contaminated area of Chernobyl, the exposed group (29–86 mSv) showed a small (3.6%) but statistically significant ($p = 0.0005$) excess of subclinical posterior subcapsular lens changes similar in shape to those identified in atomic bomb survivors [DAY 95]. The odds ratios (ORs) at 1 Sv were 1.12 (95% CI 0.94, 1.30) in NCs, 1.29 (95% CI 1.12, 1.49) in CCs and 1.41 (95% CI 1.21, 1.64) in PSCs [MIN 04]. The subcapsular form can be considered the dominant form of radiation-induced cataracts.

#### 9.4.4.2. *Cataracts in Chernobyl liquidators*

At Chernobyl, opacities in young men aged 12–17 increased by a factor of 2–3, but in these cases, dosimetry is unreliable and confounding factors are not quantified. Occupational exposure of 8,607 workers involved in the removal of radioactive debris from the roof of Chernobyl's damaged No. 4 reactor and No. 3 reactor (liquidators) resulted in subcapsular or posterior cortical cataracts, characteristic of radiation exposure, which were present in 25% of subjects, with an OR/1 Gy of 1.70 [1.22; 2.38] (all types combined) [WOR 07].

Several studies have demonstrated radiation-induced cataracts at doses measured in mGy or tens of mGy, which may call into question the existence of a threshold, or even the deterministic nature attributed a priori to this effect.

Since 2007, epidemiological data have argued against a high threshold dose for protection against radiation-induced cataracts. An analysis was conducted to determine this "threshold" among 3,761 survivors of the Hiroshima and Nagasaki bombs, who underwent medical examinations between 2000 and 2002 and for whom radiation dose estimates were available, including 479 cases of operated cataracts. A non-significant dose threshold of 0.1 Gy was found. The prevalence of operated cataracts among survivors increased significantly with dose.

The threshold estimate (0.1 Gy) and its upper limit (0.8 Gy) for cataract prevalence are much lower than the 2–5 Gy threshold usually assumed by the radiation protection community (ICRP) and are statistically compatible with a total absence of a threshold [NER 07]. This view is shared by Wassilieff, for whom the very existence of a threshold is no longer a certainty, since the pathogenesis of radiation-induced cataracts is less deterministic than previously thought and more stochastic [WAS 09].

Ainsbury et al. [AIN 09] also suggested that the threshold for cataract development is certainly lower than previously estimated (approximately 0.5 Gy) or that radiation-induced cataracts could in fact be described by a linear model with no threshold. In a literature review, Shore [SHO 16] considers that the few studies available provide only limited support for the ICRP hypothesis, and that there is no convincing evidence of a decrease in cataractogenic effects at low dose rates.

### 9.4.4.3. *Occupational cataracts*

Cataracts are observed in people subjected to cosmic radiation (aviators, astronauts), scattered X-rays (cardiology and interventional radiology) and the various β and γ radiations encountered in several nuclear facilities by workers classified as being exposed to ionizing radiation.

The incidence of cataracts in 295 cosmonauts was observed for cosmic radiation doses in excess of 8 mSv [WAS 09]. In 171 astronauts, lens doses ranging from 15 to 129 mSv caused a significant increase in opacities ($p = 0.015$). Within the astronaut group, cataract size (PSC) was greater in subjects who received higher doses of cosmic radiation ($p = 0.016$) [CHY 09].

Among 445 airline pilots, according to the simplified WHO classification system of the four types of cataracts identified, 71 pilots had NCs, 102 had cortical lens opacification, 69 had central optic zone opacification and 32 had posterior subcapsular lens opacification. The OR for the risk of NC was 3.02 (95% CI 1.44–6.35) for pilots versus non-pilots, The association between pilots' exposure to cosmic radiation and the risk of NC, adjusted for age, smoking and tanning habits indicates that cosmic radiation may be a causal factor for radiation-induced cataract in commercial airline pilots [RAF 05].

In 114 people exposed in contaminated buildings in Taiwan, a significant dose-dependent increase in the number of lens defects in the under-20s was demonstrated [CHE 01].

Among US medical personnel, all 35,705 cataract-free US X-ray technicians aged 24–44 were followed. For workers in the highest (mean, 60 mGy) versus lowest (mean, 5 mGy) category of occupational lens dose, the adjusted hazard ratio for cataract was 1.18 (95% CI 0.99, 1.40) [CHO 08].

Interventional cardiologists have a very high risk of cataracts. The O'CLOC (Occupational Cataracts and Lens Opacities in interventional Cardiology) study was launched in France by the IRSN [JAC 10, JAC 12] to test the existence of an increased risk of cataracts in interventional cardiologists compared with a reference unexposed group. This was done using a cross-sectional exposed/unexposed epidemiological study. The first conclusion is that there is a significant risk of crystalline opacities, at the posterior subcapsular level, in the population of French interventional cardiologists compared with unexposed workers. The second conclusion of the O'CLOC study is based on a retrospective assessment of doses to the eye, ranging from 25 mSv to 1,600 mSv in two categories: hemo-dynamists and rhythmologists. After an average working life of 20 years, over 28% of hemodynamicists and 19% of rhythmologists had already exceeded the new ICRP threshold of 0.5 Sv for radiation-induced crystalline opacities. Approximately 60% of cardiologists had been exposed to eye radiation in excess of 20 mSv.year$^{-1}$ (ICRP recommended annual dose limit) at least once since 2000. With the former limit of 150 mSv.year$^{-1}$, only one cardiologist was above this level. In this study, the risk of posterior subcapsular opacities was calculated to be almost four times higher in interventional cardiologists than in the unexposed control group [JAC 12]. These results are consistent with the risk of posterior subcapsular lens opacities observed in the O'CLOC study. These results are consistent with those obtained during the European epidemiological study on radiation-induced lens opacities among interventional cardiologists (EURALOC) project (2015–2017) [DOM 18, COV 18, STR 18].

A review of these various studies indicates that posterior subcapsular opacities, and to a lesser extent CCs, are the signatures of radiation-induced cataracts.

With the lowering of the dose limit, in the absence of protection, the lens of the eye will probably be the critical organ in cardiology and interventional radiology. Passive dosimetry systems do exist, but will require further development before they are fully operational. Standardized calibration procedures have yet to be defined. Active dosimetry systems are very limited. Protective devices (80–90% effective) do exist, such as leaded goggles and visors, overhead screens or tabletop screens. Practices can be optimized (tube position, practitioner position, beam quality, field size). Implementing these means/practices will require effort [JAC 12].

The work of Neriishi et al. [NER 07] shows that recent evidence argues against a high threshold dose for radiation-induced vision-impairing cataractogenesis. They carried out an analysis to determine the most appropriate threshold model for 3,761 A-bomb survivors who underwent medical examinations between 2000 and 2002 and for whom radiation dose estimates were available, including 479 cases of post-operative cataract. Analyses showed a statistically significant increase in the prevalence of post-operative cataracts. A non-significant dose threshold of 0.1 Gy (95% CI <0–0.8) was found. The prevalence of post-operative cataracts in A-bomb survivors increased significantly with A-bomb radiation dose. Both the estimate (0.1 Gy) and the upper limit (0.8 Gy) of the dose threshold for the prevalence of operative cataracts were much lower than the 2–5 Gy threshold usually assumed by the radiation protection community and were statistically compatible with the absence of a threshold [NER 07].

Current ICRP thresholds for damage to vision range from 2 to 10 grays at low LETs and two to three times less at high LETs. The dose limit for the crystalline lens of the eye is 0.12 Sv. Presumably, much lower radiation doses than those adopted by the ICRP have an effect on the eye and the Commission should review the radiation protection standards for the crystalline lens of the eye. Furthermore, the ICRP has always considered cataracts to be the result of a deterministic effect of ionizing radiation and has defined a high threshold, deemed protective, for this effect. The results of several studies on the dose–effect relationship of cataracts raise the question of whether this health impact is not the result of a non-threshold dose–effect relationship.

### 9.4.5. *Effects on the integumentary system*

Irradiation of the skin, generally acute and localized, results in effects formerly known as "deterministic", since they occur in all victims as soon as the dose received reaches a threshold dose or dose range.

After irradiation of the skin with a single dose of over 2 grays, the first effect is the appearance of redness (erythema or dry epidermitis). The effect is early and fleeting. It appears in the hours following irradiation and may later manifest itself as dry desquamation and depilation. If the dose is higher (6–12 Gy), the erythema will be followed by phlyctene, a lifting of the skin under which a clear liquid is found, which appears around three weeks after acute irradiation. A more rapid onset indicates the severity of exposure. This manifestation, comparable to that of a thermal burn, is also referred to as a "radiation burn". For an even higher dose range, necrosis of the basal layer of the epidermis can occur. This serious damage is not reversible and, as in the case of skin cancer, requires surgical treatment (excision-grafting).

These situations arise in industrial or medical accident situations. The appearance of redness, especially if extensive, is an important warning sign in any suspected radiological accident. It should immediately trigger further biological dosimetry tests[37]. All published radiological and nuclear accidents worldwide since 1944 are recorded in a special register set up in 1975, under the responsibility of the AEC, by the Radiation Emergency Assistance Center in Oak Ridge (USA). Other published reports concern the work of authors such as Dr. Jammet and Dr. Gongora (199 accidents, 1956–1981, worldwide) [JAM 81] and Dr. Rodrigues de Oliviera (199 accidents, 1956–1981, worldwide). Rodrigues de Oliviera (180 accidents,

37 Measurement and monitoring of blood components. In the event of neutron irradiation, measurements of induced radioactivity on the victim must be carried out (sodium-24 in the whole body and phosphorus-32 in the skin products [nails, hair]).

1945–1985) [DEO 87], by Zerbib [ZER 94] (1945–1993), by Dr. Nenot (1950–2000) [NEN 01, NEN 07] and by Amiard (1945–2006) [AMI 19a, AMI 19b].

### 9.4.6. *Effects on the central nervous system*

Radiation exposure is one of the few toxic substances known to be associated with the risk of CNS tumors. Alina Brenner et al. [BRE 20] assessed the radiation risks of primary glioma, meningioma, schwannoma and other or unspecified tumors in the LSS cohort. Cases diagnosed between 1958 and 2009 were verified using the Hiroshima and Nagasaki population cancer registries. There were 285 CNS tumors (67 gliomas, 107 meningiomas, 49 schwannomas and 64 other tumors) among 105,444 individuals with brain radiation dose estimates contributing to 3.1 million person-years of observation. Based on a simple linear model with no effect modification, the ERR/Gy was 1.82 (95% CI 0.51 to 4.30) for meningioma, 1.67 (95% CI 0.12 to 5.26) for glioma, 1.45 (95% CI -0.01 to 4.97) for schwannoma and 1.40 (95% CI 0.61 to 2.57) for all CNS tumors as a group. For each tumor type, the dose–response relationship was consistent with linearity and appeared to be stronger in men than in women, particularly for meningioma. More than 60 years after the bombings, radiation-related risks for CNS tumors continue to be high [BRE 20].

### 9.4.7. *Effects on the respiratory system*

As part of the LSS of Japanese atomic bomb survivors, the risk of lung, laryngeal and other respiratory cancers was assessed in 105,444 subjects followed from 1958 to 2009. During this period, Cahoon Elisabeth et al. [CAH 17] identified 2,446 cancers of the lung, 180 of the larynx and 115 other respiratory sites (trachea, mediastinum and other ill-defined sites) as primary cancers. Smoking data were used to study the joint effects of radiation and smoking. For non-smokers, the average ERR/Gy of lung cancer (at 70 years after exposure at 30 years) was estimated at 0.81 (95% CI 0.51 to 1.18) with a female/male ratio of 2.83. There was no evidence of curvature in the radiation dose–response relationship, overall or by gender. The risk of lung cancer increased with the number of years of smoking and decreased with the time since smoking cessation, irrespective of the level of radiation exposure.

The ERR/Gy for lung cancer was significantly higher for light to moderate smokers than for heavy smokers. Of 2,446 cases of lung cancer, 113 (5%) could be attributed to radiation exposure. Of the 1,165 cases of lung cancer in smokers, 886 (76%) could be attributed to smoking. Although there is little evidence of an effect of radiation on laryngeal cancer, an increased risk is observed for other cancers of the respiratory system, but this is not statistically significant [CAH 17].

Data on radiation-induced lung cancer, resulting from internal irradiation following inhalation of soluble and insoluble alpha- and beta-emitting radionuclides, are reviewed by the authors. Emphasis is placed on the radiotoxicity of alpha emitters, which is an order of magnitude more dangerous than that of beta/gamma radiation, and on recent data showing that the more homogeneous the irradiation of the lungs, the greater the carcinogenic risk, which contradicts what has been called the hot particle theory [COG 86].

## 9.5. The transgenerational effects of ionizing radiation

The hereditary transmission of the genetic information carried by DNA can be disrupted in two main ways. The first is the result of poor copying of parental DNA (denial, repetition, etc.). The filial DNA then presents a physical defect with a single mutation or numerous mutations in the nucleic acid sequence. In the second case, the order of the nucleic acids is not changed, but some nucleic acids undergo a chemical modification such as methylation. This is called epigenetic modification. Physical mutations and epigenetic modifications can be caused by chemical and/or physical stresses. Among these stresses, ionizing radiation plays a significant role.

### 9.5.1. *Hereditary transmission of genetic information*

The question of potential long-term or hereditary effects in humans exposed to low doses (or dose rates) of ionizing radiation is a major concern. Lorimore et al. [LOR 98] observed chromosomal instability in the clonal descendants of hematopoietic stem cells after irradiation of murine bone marrow with $\alpha$ particles. Similarly, modifications of so-called epigenetic marks such as histone modifications, DNA methylation and non-coding RNAs result in altered transcriptomes and proteomes, without directly changing the DNA sequence. Moreover, some of these environmentally induced epigenetic changes tend to persist across generations and, consequently, epigenetic modifications are considered to be the vectors of environmental influence on the genome [HOR 19]. Furthermore, cells appearing in surviving colonies may carry heritable lethal defects that appear, perhaps only after many successful divisions. This would explain the well-known observation that previously irradiated but apparently healed tissue subsequently proves extremely sensitive to further exposure to irradiation or cytotoxic drugs [SEY 86].

Alterations at the molecular level of DNA (sequential for genetic mutations or epigenetic marks) are transmitted over several generations, but do they have consequences for the health of descendants?

### 9.5.2. *Findings on hereditary effects*

Radiotherapy and chemotherapy are treatments that damage DNA. They are capable of causing cancer in treated patients through their mutagenic actions on somatic cells. The question is whether such treatments can also lead to the transmission of genetic damage to offspring, as observed in animal experiments.

Our knowledge comes from studies of several thousand survivors of childhood cancer, treated decades ago. So far, the results are reassuring, and no hereditary damage has been detected. However, reproductive effects, including infertility, are common consequences of cancer treatment and may represent damage to germ cells. Further studies need to be undertaken [BYR 99].

Data collected in Hiroshima and Nagasaki over the past 40 years on the children of atomic bomb survivors and on children from an appropriate control population are analyzed on the basis of recently revised estimates of radiation doses. No statistically significant effects are found for eight different indicators. These estimates suggest that humans are less sensitive to the genetic effects of radiation than had been assumed on the basis of past extrapolations from experiments on mice [NEE 91].

In Finland, it was found that women who have been treated for cancer tend to give birth prematurely, but there is no increase in the number of cancers in their offspring [MAD 10].

A large multicenter cohort study was conducted on the 4,699 children of people treated for cancer (1,128 men and 1,627 women). Treated patients were under 21 years of age at the time of their initial cancer diagnosis (between early 1970 and late 1986) and had survived at least five years after diagnosis. After quantifying the doses delivered to the testicles and ovaries during their radiotherapy, the authors of this retrospective study showed that these doses did not entail a risk of congenital anomalies. Similarly, after quantifying chemotherapy (doses of alkylating agents), they showed that chemotherapy treatment was also not significantly associated with abnormalities in the children of patients treated [SIG 12].

A group of 40,487 people born between May 1946 and December 1984 (who were cancer-free in January 1958) with one or both of whose parents were Hiroshima or Nagasaki survivors, was studied by Izumi et al. [IZU 03]. Using information from the tumor registry, they analyzed data on cancer incidence in the group from 1958 to 1997, and examined the effects of paternal and/or maternal irradiation on their offspring. During the 39 years of follow-up, 575 cases of solid tumors and 68 cases of hematopoietic tumors were diagnosed. The median age at diagnosis was 39.7 years. Median doses were 143 millisieverts for 15,992 exposed fathers and 133 millisieverts for 10,066 exposed mothers. Cancer incidence was no higher in subjects with exposed parents than in reference subjects (dose range 0–4 millisieverts), nor did incidence rates increase with increasing dose. For 3,568 subjects with two exposed parents, the adjusted relative risk for all cancers was 0.97 (95% CI 0.70–1.36) [IZU 03].

The risk of major congenital malformations and perinatal death in children of atomic bomb survivors (71,603) was also analyzed, using gonadal dose estimates from the DS02 dosimetry system and analytical methods to characterize dose–response relationships. Analyses showed that parental exposure to radiation was associated with an increased risk of major congenital malformations and perinatal death, but estimates were imprecise for direct radiation effects and most were not statistically significant [YAM 21].

Grant et al., with a follow-up of 62 years (median follow-up 54.3 years, IQR (interquartile range) 45–59.3), studied the risk of death in children of atomic bomb survivors who had received non-zero gonadal doses. A total of 5,183 deaths from the disease were observed. The mean age of the 68,689 surviving children at the end of follow-up was 53.1 years (SD 7.9), of whom 23% (15,623) were over 60. The mean dose to parents with non-zero gonadal doses was 264 mGy. No increase in the risk of death from cancer or non-cancerous diseases was observed in relation to parental gonadal exposure. No change in radiological risk was observed according to attained age or time elapsed between parental exposure and birth [GRA 15].

In conclusion, the descendants of cancer patients who were treated with radiotherapy, and those of survivors of the Hiroshima and Nagasaki bombs, do not appear to be at risk of childhood cancer or congenital malformations. Exposure of one or both parents does not show that cancers, malformations or early deaths affect their offspring in a statistically significant way. However, a number of authors suggest that these analyses of the hereditary impact of parents exposed to ionizing radiation should be followed up [FEN 01, GRA 15, IZU 03, MAD 10, SIG 12, YAM 21].

## 9.6. Individual variability in the adverse effects of ionizing radiation

There are three aspects of individual response to ionizing radiation: radiosensitivity for early or late tissue adverse events after radiotherapy on normal tissues (non-cancer effects attributable to cell death), radiosusceptibility for radiation-induced cancers and radiodegeneration for non-cancer effects that are often attributable to mechanisms other than cell death (e.g. cataracts and circulatory diseases) [FOR 16].

## 9.7. Conclusions

The deleterious effects of irradiation vary with the age of exposure (in utero, infancy, puberty, adulthood, old age), the age reached and the sex of the person exposed to ionizing radiation. This last parameter is reflected in significant differences in the organs affected, the intensity of the risk and its persistence with increasing age.

Table 9.6 shows the excess incidence or mortality of radiation-induced cancers due to exposure to ionizing radiation, as a function of age at exposure (in utero to adulthood).

<table>
<tr><th>Age at exposure</th><th>Male cancers</th><th>Female cancers</th><th>Type of exposure [References]</th></tr>
<tr><td>In utero</td><td>No solid cancers</td><td>Digestive system and female-specific cancers</td><td>Atomic bombs [DEL 97, PRE 08, SUG 21]</td></tr>
<tr><td>0–14 years</td><td>All cancers (with LD), liver (with VLD), leukemia, thyroid</td><td>Solid cancers, stomach, liver (with LD), leukemia, breast, thyroid, uterus (with LD)</td><td>Atomic bombs [BRE 18, GOT 12, KUT 16, RON 95]</td></tr>
<tr><td>Less than 16 years old</td><td>All cancers, CNS and thyroid cancers</td><td>Leukemia, lymphoma, thyroid cancer</td><td>Natural exposure [KEN 12, RON 95, SPY 15]</td></tr>
<tr><td>Less than 22 years old</td><td colspan="2">Lymphoid cancers, cancers of the digestive organs, cancers of the female genital organs, soft tissues, thyroid, urinary tract, leukemias, melanomas, myelodysplasias, brain tumors, neural tissue tumors, thyroid</td><td>Scanners [PEA 12, RON 95, MAT 13, THO 94]</td></tr>
</table>

VLD: very low dose (d < 5mGy), LD: low dose (5–100 mGy), CNS: central nervous system.

**Table 9.6.** *Incidence and mortality of radiation-induced cancers as a function of age at exposure to ionizing radiation*

### 9.7.1. *Effects of irradiation in utero and in early childhood*

Epidemiological studies carried out on the population of survivors of the bombs dropped on Hiroshima and Nagasaki show that radiation-induced damage is far more severe in women than in men, whether the exposure is during fetal development, infancy, puberty or adulthood.

Prenatal exposure to ionizing radiation is a risk factor that strongly increases mortality from solid cancer in women, but not in men. In men, the ERR for a dose of one gray (ERR/Gy) increased only for mortality from non-cancerous disease. In women, ERR/Gy were increased for solid cancer mortality (digestive system and female-specific), non-cancer mortality and mortality from external causes.

Although there was a significant decrease in ERR with increasing attained age, this decrease was more rapid in men than in women. A significant difference in solid cancer mortality between the sexes is unexplained.

Irradiation in utero or in early childhood has been shown to result in statistically significant increases in the incidence rates of solid cancers in adulthood. Radiation doses of the order of 10 mGy received by the fetus in utero result in a significant increase in the risk of childhood cancer. The ERR in adulthood increased with dose, both for those exposed in utero and in early childhood. However, the increase in radiation-induced cancer risk was significantly higher for irradiation at less than 6 years of age than in utero.

### 9.7.2. *Effects of radiation on children*

Analysis of cancer mortality in a group of children exposed as infants (<14 years) during the bombings in Japan, and followed for 40 years, showed that SMRs were significantly high for both sexes, for all deaths, even at low doses. The organs with the highest SMRs were different for boys (leukemia and liver cancer, even at very low doses) and girls (solid cancers, liver cancer and uterine cancer, even at low doses). For girls exposed to high doses, the highest SMRs were for leukemia, solid cancers, stomach cancer and breast cancer.

Three epidemiological studies, involving several hundred thousand children, adolescents and young adults exposed to CT scanners, have consistently shown a significant increase in the incidence of several types of cancer with cumulative dose, although the follow-up periods were relatively short (7–10 years) [IRS 12, MAT 13, PEA 12].

In Switzerland and the UK, a statistically significant risk of leukemia in children has been associated with exposure to natural sources of gamma radiation [KEN 12]. In addition to leukemia, this exposure may also contribute to the risk of other cancers, such as lymphoma and tumors of the CNS [SPY 15]. However, in the case of domestic radon exposure, there is no evidence that it is associated with childhood cancer despite relatively high radon levels [HAU 13].

Because of their developing organs and long life expectancy after exposure, children run a greater risk than adults of developing cancer after exposure to ionizing radiation. Children therefore require special attention when radiation is used for diagnosis or treatment or following a nuclear accident [KUT 16].

Several studies of the offspring of cancer patients treated with radiotherapy and those of survivors of the Hiroshima and Nagasaki bombs, with follow-up up to average adulthood (62 years), show no risk of childhood cancer or congenital malformations. Exposure of one or both parents does not show that cancers, malformations or early deaths affect their offspring in a statistically significant way.

### 9.7.3. *Effects of irradiation at puberty*

The study by Alina Brenner et al. showed that age at puberty is a powerful amplifier of the effect of radiation on breast cancer incidence. The highest risks are associated with exposure around the time of first menstruation. For a given dose, the excess risk decreases with increasing age at puberty [BRE 18].

### 9.7.4. *Effects of irradiation in adulthood*

The combined analysis of solid tumors showed a relative risk twice as high for women as for men, and a tendency for the relative risk to decrease with increasing age at exposure ($P < 0.001$). Women had a higher relative risk of cancer of the lung, the entire respiratory system and the urinary system than men [THO 94].

For all solid tumors, the excess of cancers increased with time since exposure, based on an absolute excess risk model. On average, for all ages at exposure, the relative risk decreased with time since exposure [THO 94].

As the health impact of doses varies with age at exposure and age reached in life, it has been estimated that at age 70, after exposure at 30 years, rates of radiation-induced solid cancers increase by approximately 35% per Gy for men and 58% per Gy for women (1.66 times more) [GRA 17].

In the Grant et al. study, which represents a 51-year follow-up (1958–2009), the stomach was the most common cancer site in both men and women, accounting for 29.5% of cases in men and 21.3% in women. Other common cancer sites included lung (13.8%), liver (10.7%), colon (7.5%) and rectum (4.9%) in men and breast (12.2%), colon (9.4%), lung (8.3%) and cervix (7.3%) in women [GRA 17].

Breast cancer incidence rates after radiation exposure in eight large cohorts, studied and compared by Preston et al. [PRE 02], suggest similar risks for acute and fractionated high-dose-rate exposures, with much lower effects due to low doses. The results support dose–response linearity for breast cancer. They underline the importance of age at exposure and attained age. The ERR decreases with increasing age.

Brenner and colleagues have shown that there is a strong linear dose response for breast cancer, with an ERR of 1.12 per Gy for women aged 70 after 30 years' exposure [BRE 18].

While the study of radiation-induced ovarian cancers shows that their dose-dependent increase is statistically significant from 1950 to 1980, it ceases to be so when the study is extended to 2009 [TOK 87].

Leukemia is a rare disease in Japan, and the absolute number of leukemia cases among A-bomb survivors is relatively low, although the relative risk is high. Leukemia accounts for only approximately 3% of all cancer deaths and less than 1% of all deaths, although it currently constitutes approximately 16% of all excess cancer deaths in the LSS cohort due to radiation exposure. As chronic lymphocytic leukemia (CLL) is a small fraction of leukemias in Japan, which are themselves rare diseases, it is surprising to see the ICRP deduce from the absence of a significant ERR that CLL is not a radiation-induced leukemia.

On the basis of epidemiological studies of atomic bomb survivors, with a 30-year follow-up (1958–1987), it appeared that a number of organs were not the site of radiation-induced cancer: the larynx, nose, pancreas, pharynx, prostate, rectum, kidneys, renal parenchyma, uterus, did not present significantly increased risks [RON 94, THO 94]. In studies with an increased follow-up of 22 years (1988–2009), it was found that the rectum in men and the cervix in women were statistically in excess, and even constituted a significant proportion of solid cancers (4.9% and 7.3%, respectively) [GRA 17].

### 9.7.5. *Transgenerational effects*

No increase in the risk of death from cancer or non-cancerous diseases was observed in relation to parental gonadal exposure (mean dose equal to 264 mGy) in over 68,000 children with a follow-up of 62 years. No change in radiological risk was observed according to age reached or time elapsed between parental exposure and birth [GRA 15].

Our knowledge comes from studies of several thousand survivors of childhood cancer, treated several decades ago. So far, the results have been reassuring, and no hereditary damage has been detected [BYR 99].

## 9.8. References

[AHM 22] AHMED Y., KHAN A.M.H., RAO U.J. et al., "Fertility preservation is an imperative goal in the clinical practice of radiation oncology: A narrative review", *ecancermedicalscience*, vol. 16, no. 1461, November 2022.

[AIN 09] AINSBURY E.A. et al., "Radiation cataractogenesis: A review of recent studies", *Radiation Research*, vol. 172, pp. 1–9, 2009.

[AMI 08] AMIARD J.C., AMIARD-TRIQUET C., *Les biomarqueurs dans l'évaluation de l'état écologique des milieux aquatiques*, Lavoisier, Tec&Doc, Paris, 2008.

[AMI 17] AMIARD J.C., *Les risques chimiques environnementaux. Méthodes d'évaluation et impacts sur les organismes*, 2nd edition. Lavoisier, Tec&Doc, Paris, 2017.

[AMI 19a] AMIARD J.C., *Military Nuclear Accidents: Environmental, Ecological, Health and Socio-economic Consequences*, vol. 1, ISTE London, and John Wiley & Sons, New York, 2019.

[AMI 19b] AMIARD J.C., *Industrial and Medical Nuclear Accidents: Environmental, Ecological, Health and Socio-economic Consequences*, vol. 2, ISTE, London, and John Wiley & Sons, New York, 2019.

[BAR 89] BARANOV A., GALE R.P., GUSKOVA A., et al., "Bone marrow transplantation after the Chernobyl nuclear accident", *New England Journal of Medicine*, vol. 27, no. 32, pp. 206–212, July 1989.

[BAR 97] BARDET É., "Effets tardifs des radiations ionisantes sur le testicule", *Cancer/Radiothérapie*, vol. 1, no. 6, pp. 778–780, December 1997.

[BEY 15] BEY P., "Risques de leucémies après radiothérapie (RT)", *Risques de leucémies et exposition aux rayonnements ionisants*, 9 June 2015.

[BRE 18] BRENNER A.V., PRESTON D.L., SAKATA R. et al., "Incidence of breast cancer in the life span study of atomic bomb survivors: 1958–2009", *Radiation Research*, vol. 190, no. 4, pp. 433–444, July 2018.

[BRE 20] BRENNER A.V. SUGIYAMA H., PRESTON D.L. et al., "Radiation risk of central nervous system tumors in the Life Span Study of atomic bomb survivors, 1958–2009", *European Journal of Epidemiology*, vol. 35, no. 6, pp. 591–600, June 2020.

[BYR 99] BYRNE J., "Long-term genetic and reproductive effects of ionizing radiation and chemotherapeutic agents on cancer patients and their offspring", *Teratology*, vol. 59, no. 4, pp. 210–215, 1999.

[CAH 17] CAHOON E.K., PRESTON D.L., PIERCE D.A., et al. "Lung, laryngeal and other respiratory cancer incidence among Japanese atomic bomb survivors: an updated analysis from 1958 through 2009", *Radiat. Res.*, vol. 187, no. 5, pp. 538–548, 2017.

[CHA 14] CHARLES O.E. et al., "Current practices in fertility preservation in male cancer patients", *Urology Annals*, vol. 6, no. 1, pp. 13–17, January–March 2014.

[CHE 01] CHEN W.L., HWANG J.S., HU T.H. et al., "Lenticular opacities in populations exposed to chronic low-dose-rate gamma radiation from radiocontaminated buildings in Taiwan", *Radiation Research*, vol. 156, no. 1, pp. 71–77, 2001.

[CHO 83] CHOSHI K. TAKAKU I., MISHIMA H. et al., "Ophthalmologic changes related to radiation exposure and age in adult health study sample, Hiroshima and Nagasaki", *Radiation Research*, vol. 96, no. 3, pp. 560–579, December 1983.

[CHO 08] CHODICK G., BEKIROGLU N., HAUPTMANN M. et al., "Risk of cataract after exposure to low doses of ionizing radiation: A 20-year prospective cohort study among US radiologic technologists", *American Journal of Epidemiology*, vol. 168, no. 8, pp. 620–631, 2008.

[CHY 09] CHYLACK L., PETERSON L.E., FEIVESON A.H. et al., "NASA Study of Cataract in Astronauts (NASCA). Report 1: Cross-sectional study of the relationship of exposure to space radiation and risk of lens opacity", *Radiation Research*, vol. 172, pp. 10–20, 2009.

[COG 86] COGGLE J.E., LAMBERT B.E., MOORES S.R., "Radiation effects in the lung", *Environmental Health Perspectives*, vol. 70, pp. 261–291, December 1986.

[COV 18] COVENS P., DABIN J., DE TROYER O. et al., "Track, calculate and optimise eye lens doses of interventional cardiologists using mEyeDose and mEyeDose_X", *Journal of Radiology Protection*, vol. 38, no. 2, pp. 678–687, 2018.

[DAY 95] DAY R.1., GORIN M.B., ELLER A.W., "Prevalence of lens changes in Ukrainian children residing around Chernobyl", *Health Physics*, vol. 68, no. 5, pp. 632–642, 1995.

[DEL 97] DELONGCHAMP R.R., MABUCHI K., YOSHIMOTO Y. et al., "Cancer mortality among atomic bomb survivors exposed in utero or as young children, October 1950–May 1992", *Radiation Research*, vol. 147, no. 3, pp. 385–395, March 1997.

[DEO 87] DE OLIVEIRA R.A., "Un répertoire des accidents radiologiques 1945–1985", *Radioprotection*, vol. 22, no. 2, pp. 89–135, 1987.

[DES 07] DE SANTIS M., CESARI E., NOBILI E. et al., "Radiation effects on development", *Birth Defects Research (Part C)*, vol. 81, pp. 177–182, 2007.

[DOL 97] DOLL R., WAKEFORD R. "Wakeford R, Risk of childhood cancer from fetal irradiation", *British Journal of Radiology*, vol. 70, pp. 130–139, February 1997.

[DOM 18] DOMIENIK-ANDRZEJEWSKA J., CIRAJ-BJELAC O., ASKOUNIS P. et al., "Past and present work practices of European interventional cardiologists in the context of radiation protection of the eye lens-results of the EURALOC study", *Journal of Radiology Protection*, vol. 38, no. 3, pp. 934–950, 2018.

[EFE 23] EFELYA, "Comment se développe le cerveau du fœtus ?", *Efelya.fr*, 10 January 2023.

[FAB 64] FABRY C., Schéma anatomo-physiologique du tractus gastro-intestinal à prendre en considération pour le calcul des niveaux de contamination radioactive, CEA-R-2413 report, 1964.

[FEN 01] FENIG E., MISHAELI M., KALISH Y. et al., "Pregnancy and radiation", *Cancer Treatment Review, Controversy*, vol. 27, no. 1, pp. 1–7, 2001.

[FOR 16] FORAY N., BOURGUIGNON M., HAMADA N., "Individual response to ionizing radiation", *Mutation Research Review*, vol. 770, pp. 369–386, 2016.

[GOT 12] GOTO H., WATANABE T., MIYAO M. et al., "Cancer mortality among atomic bomb survivors exposed as children", *Environmental Health and Preventive Medicine*, vol. 17, no. 3, pp. 228–234, May 2012.

[GRA 15] GRANT E.J., FURUKAWA K., SAKATA R. et al., "Risk of death among children of atomic bomb survivors after 62 years of follow-up: a cohort study", *Lancet Oncology*, vol. 16, no. 13, pp. 1316–1323, October 2015.

[GRA 17] GRANT E.J., BRENNER A., SUGIYAMA H. et al., "Solid cancer incidence among the life span study of atomic bomb survivors: 1958–2009", *Radiation Research*, vol. 187, no. 5, pp. 513–537, 20 March 2017.

[HAC 15] HACHIYA M., Journal d'Hiroshima, 6 août-30 septembre 1945, Collection TEXTO, Editions Tallandier, 2015, Paris, 1st translation into English by the University of North Carolina Press, 1955.

[HAM 14] HAMADA N., FUJIMICHI Y., "Classification of radiation effects for dose limitation purposes: history, current situation and future prospects", *Journal of Radiation Research*, vol. 55, pp. 629–640, July 2014.

[HAR 16] HARA N., SATOH K., OTANI K. et al., "High excess risk of heart disease mortality among Hiroshima atomic bomb male survivors exposed near the Hypocenter", *Hiroshima Journal of Medicine Science*, vol. 65, no. 2, pp. 45–51, September 2016.

[HAU 13] HAURI, D., SPYCHER B., HUSS A. et al., "Domestic radon exposure and risk of childhood cancer: A prospective census-based cohort study", *Environmental Health Perspectives*, vol. 121, pp. 1239–1244, 2013.

[HOR 19] HOREMANS N., SPURGEON D.J., LECOMTE-PRADINES C. et al., "Current evidence for a role of epigenetic mechanisms in response to ionizing radiation in an ecotoxicological context", *Environmental Pollution*, vol. 251, pp. 469–483, 2019.

[HRI 11] HRICAK H., BRENNER D.J., ADELSTEIN S.J., et al., "Managing radiation use in medical imaging: A multifaceted challenge", *Radiology*, vol. 258, no. 3, p. 889, 2011 March.

[ICR 75] ICRP, International Commission on Radiological Protection, ICRP No. 23, Report of the Task Group on Reference Man, 1975.

[ICR 12] ICRP, ICRP statement on tissue reactions and early and late effects of radiation in normal tissues and organs – threshold doses for tissue reactions in a radiation protection context, *Annals of the ICRP*, vol. 41, nos. 1–2, pp. 3–322, 2012.

[IRS 12] IRSN, Note de lecture par l'IRSN de la publication scientifique, Radiation exposure from CT scans in childhood and subsequent risk of leukaemia and brain tumours: a retrospective cohort study, IRSN, 13 June 2012.

[IRS 18] IRSN, Exposition des enfants aux rayonnements ionisants due aux actes d'imagerie médicale diagnostique réalisés en France en 2015, IRSN, Pôle Santé et Environnement, Service d'études et d'expertise en radioprotection 2018.

[IZU 03] IZUMI S., KOYAMA K., SODA M. et al., "Cancer incidence in children and young adults did not increase relative to parental exposure to atomic bombs", *British Journal of Cancer*, vol. 89, no. 9, pp. 1709–1713, 3 November 2003.

[JAC 10] JACOB S., MICHEL M., SPAULDING C. et al., "Occupational cataracts and lens opacities in interventional cardiology (O'CLOC study): Are X-rays involve? Radiation-induced cataracts and lens opacities", *BMC Public Health*, vol. 10, no. 537, 2010.

[JAC 12] JACOB S., MICHEL M., BRÉZIN, A.P. et al., "Ionizing radiation as a risk factor for cataract: What about low-dose effects?", *Journal of Clinical and Experimental Ophthalmology*, S1 005, pp. 1 7, 2012.

[JAM 81] JAMMET H., GONGORA R. *3ème congrès de l'Association Italienne de Radioprotection Médicale*, Saint Vincent, Italie, 13/15 September 1981.

[JON 75] JONES T.D., AUXIER J.A., CHEKA J.S. et al., "In vivo dose estimates for the A bomb survivors shielded by typical Japanese houses", *Health Physics*, vol. 28, pp. 367–381, April 1975.

[KEN 12] KENDALL G.M., LITTLE M.P., WAKEFORD R. et al., "A record-based case–control study of natural background radiation and the incidence of childhood leukaemia and other cancers in Great Britain during 1980–2006", *Leukemia*, vol. 27, pp. 3–9, 5 June 2012.

[KUT 16] KUTANZI K.R., LUMEN A., KOTURBASH I., MIOUSSE I.R., "Pediatric exposures to ionizing radiation: Carcinogenic considerations", *Environmental Research and Public Health*, vol. 13, no. 11, November 2016.

[LAN 03] LAND C.E., TOKUNAGA M., KOYAMA K. et al., "Incidence of female breast cancer among atomic bomb survivors, Hiroshima and Nagasaki, 1950–1990", *Radiation Research*, vol. 160, no. 6, pp. 707–717, 2003.

[LEB 14] LEBARON-JACOB A., UNSCEAR : Rapport sur les effets des rayonnements ionisants chez l'enfant, GT-CIPR, 15 May 2014.

[LIN 16] LINDER J., LINDER L., "Radiation accident. Isolated exposure", *Ciottone's Disaster Medicine*, 2nd edition, 2016.

[LOR 92] LORIOT N., *Irène Joliot-Curie*, Le livre de poche, 1992.

[MAB 21] MABUCHI K., PRESTON D.L., BRENNER A.V. et al., "Risk of Prostate Cancer Incidence among Atomic Bomb Survivors: 1958–2009", *Radiation Research*, vol. 195, no. 1, pp. 66–76, January 2021.

[MAD 65] MADELMONT C., MICHON G., "Quelques observations sur la composition minérale des laits en poudre", *Bulletin de l'Académie Vétérinaire*, no. 118–119, pp. 397–403, 1965.

[MAD 04] MADSEN J.L., GRAFF J., "Effects of ageing on gastrointestinal motor function", *Age Ageing,* vol. 33, pp. 154–159, 2004.

[MAD 10] MADANAT-HARJUOJA L.M., MALILA N., LÄHTEENMÄKI P.M., et al., "Preterm delivery among female survivors of childhood, adolescent and young adulthood cancer", *International Journal of Cancer*, vol. 127, pp. 1669–1679, 2010.

[MAR 18] MARCI R., MALLOZZI M., DI BENEDETTO L. et al., "Radiations and female fertility", *Reproductive Biology and Endocrinology*, vol. 16, no. 112, 16 December 2018.

[MAT 13] MATHEWS J.D., FORSYTHE A.V., BRADY Z. et al., "Cancer risk in 680 000 people exposed to computed tomography scans in childhood or adolescence: Data linkage study of 11 million Australians", *BMJ*, vol. 346, no. 7910, pp. 1–18, 2013.

[MEN 09] MÉNAGER M.T., GARNIER-LAPLACE J., GOYFFON M., *Toxicologie nucléaire environnementale et humaine*, Lavoisier, Tec&Doc, Paris, 2009.

[MIN 04] MINAMOTO A., TANIGUCHI H., YOSHITANI N. et al., "Cataract in atomic bomb survivors", *International Journal of Radiation Biology*, vol. 80, no. 5, pp. 339–345, 2004.

[MUL 09] MULROONEY D.A., YEAZEL M.W., KAWASHIMA T. et al., "Cardiac outcomes in a cohort of adult survivors of childhood and adolescent cancer: Retrospective analysis of the Childhood Cancer Survivor Study cohort", *British of Medicine Journal*, vol. 339, December 2009.

[NAK 94] NAKAMURA T., SHIMOKAWA I., HIGAMI Y., The risk of ovarian cancer in atomic bomb survivors, Nagasaki city, Japan 1973–1987, Nagasaki University, Japan 1994.

[NAK 06] NAKASHIMA E., NERIISHI K., MINAMOTO A., "A reanalysis of atomic-bomb cataract data, 2000–2002: a threshold analysis", *Health Physics*, vol. 90, no. 2, pp. 154–160, 2006.

[NAT 16] NATHWANI A.C., DOWN J.F., GOLDSTONE J. et al., "Polonium-210 poisoning: a first-hand account", *Lancet*, 2016.

[NAT 19] NATU V., GOMEZ J., BARNETT M. et al., "Apparent thinning of human visual cortex during childhood is associated with myelination", *Proceedings of the National Academy of Sciences of the USA*, vol. 116, no. 41, pp. 20750–20759, 2019.

[NCI 18] NATIONAL CANCER INSTITUTE (NCI), Radiation risks and pediatric computed tomography (CT): A guide for health care providers, 4 September 2018.

[NCR 09] NCRP, Ionizing radiation exposure of the population of the United States, Report No. 160, 2009.

[NEE 91] NEEL J.V., SCHULL W.J., AWA A.A. et al., "The children of parents exposed to atomic bombs: Estimates of the genetic doubling dose of radiation for humans", in *The Children of Atomic Bomb Survivors: A Genetic Study*, NEEL J.V., SCHULL W.J. (eds), National Academies Press, 1991.

[NEN 01] NÉNOT J.C., "Les accidents d'irradiation, 1950-2000 leçons du passé", *Radioprotection*, vol. 36, no. 4, pp. 431–450, 2001.

[NEN 07] NÉNOT J.C., GOURMELON P., Les accidents dus aux rayonnements ionisants, le bilan sur un demi-siècle, IRSN-2007/79-FR report, 15 February 2007.

[NER 07] NERIISHI K., NAKASHIMA E., MINAMOTO A. et al., "Postoperative cataract cases among atomic bomb survivors: Radiation dose response and threshold", *Radiation Research*, vol. 168, pp. 404–408, 2007.

[OGI 93] OGILVY-STUART A.L., SHALET S.M., "Effect of radiation on the human reproductive system", *Environmental Health Perspectives*, vol. 101, no. 2, pp. 109–116, July 1993.

[OTA 84] OTAKE M., SCHULL W.J. "In utero exposure to A-bomb radiation and mental retardation; a reassessment", *British Journal of Radiology*, vol. 57, no. 677, pp. 409–414, May 1984.

[OTA 88] OTAKE M., YOSHIMARU H., SCHULL W.J., Severe mental retardation among the prenatally exposed survivors of the atomic bombing of Hiroshima and Nagasaki: A comparison of the T65Dr and DS86 dosimetry systems, Technical report RERF TR 16–87, May 1988.

[OTA 93] OTAKE M., SCHULL W.J. "Radiation-related small head sizes among prenatally exposed A-bomb survivors", *International Journal of Radiation Biology*, vol. 63, no. 2, pp. 255–270, 1993.

[OTA 98] OTAKE M., SCHULL W.J. "Radiation-related brain damage and growth retardation among the prenatally exposed atomic bomb survivors", *International Journal of Radiation Biology*, vol. 74, no. 2, pp. 159–171, August 1998.

[OZA 12] OZASA K., SHIMIZU Y., SUYAMA A. et al., "Studies of the mortality of atomic bomb survivors, report 14, 1950–2003: An overview of cancer and noncancer diseases", *Radiation Research*, vol. 177, no. 3, pp. 229–243, March 2012.

[PAU 10] PAULINO A.C., CONSTINE L.S., RUBIN P. et al., "Normal tissue development, homeostasis, senescence, and the sensitivity to radiation injury across the age spectrum", *Seminar of Radiation Oncology*, vol. 20, no. 1, pp. 12–20, 2010.

[PEA 12] PEARCE M.S., SALOTTI J.A., LITTLE M.P. et al., "Radiation exposure from CT scans in childhood and subsequent risk of leukaemia and brain tumours: a retrospective cohort study", *Lancet*, vol. 380, no. 9840, pp. 499–505, 4 August 2012.

[PRE 02] PRESTON D.L., MATTSSON A., HOLMBERG E. et al., "Radiation effects on breast cancer risk: A pooled analysis of eight cohorts", *Radiation Research*, vol. 158, no. 2, pp. 220–235, August 2002.

[PRE 04] PRESTON D.L., PIERCE D.A., SHIMIZU Y. et al., "Effect of recent changes in atomic bomb survivor dosimetry on cancer mortality risk estimates", *Radiation Research*, vol. 162, no. 4, pp. 377–389, October 2004.

[PRE 07] PRESTON D.L., RON E., TOKUOKA S. et al., "Solid cancer incidence in atomic bomb survivors: 1958–1998", *Radiation Research*, vol. 168, no. 1, pp 1–64, July 2007.

[PRE 08] PRESTON D.L., CULLINGS H., SUYAMA A. et al., "Solid cancer incidence in atomic bomb survivors exposed in utero or as young children", *Journal of National Cancer Institute*, vol. 100, no. 6, pp. 428–436, 19 March 2008.

[RAF 05] RAFNSSON V., OLAFSDOTTIR E., HRAFNKELSSON J. et al., "Cosmic radiation increases the risk of nuclear cataract in airline pilots A population-based case-control study", *Arch Ophthalmology*, vol. 123, pp. 1102–1105, 2005.

[RIC 09] RICHARDSON D., SUGIYAMA H., NISHI N. et al., "Ionizing radiation and leukemia mortality among japanese atomic bomb survivors, 1950–2000", *Radiation Research,* vol. 172, no. 3, pp. 368–382, 2009.

[RON 94] RON E., PRESTON D.L., MABUCHI K. et al., "Cancer incidence in atomic bomb survivors. Part IV: Comparison of cancer incidence and mortality", *Radiation Research*, vol. 137, no. 2 Suppl., pp. 98–112, February 1994.

[RON 95] RON E. et al., "Thyroid cancer after exposure to external radiation: A pooled analysis of seven studies", *Radiation Research*, vol. 141, no 3, pp. 259–277, March 1995.

[SAK 19] SAKATA R., PRESTON D.L., BRENNER A.V. et al., "Radiation-related risk of cancers of the upper digestive tract among japanese atomic bomb survivors", *Radiation Research*, vol. 192, no. 3, pp. 331–344, September 2019.

[SAM 20] SAMET J., NIWA O., "At the 75th anniversary of the bombings of Hiroshima and Nagasaki, the Radiation Effects Research Foundation continues studies of the atomic bomb survivors and their children", *Carcinogenesis*, vol. 41, no. 11, pp. 1471–1472, 2020.

[SCH 90] SCHIMIZU Y., SCHULL W.J., KATO H., "Cancer risk among atomic bomb survivors", *JAMA*, vol. 264, no. 5, pp. 601–604, August 1990.

[SCH 99] SCHULL W.J., OTAKE M., "Cognitive function and prenatal exposure to ionizing radiation", *Teratology*, vol. 59, no. 4, pp. 222–226, April 1999.

[SEY 86] SEYMOUR C.B., MOTHERSILL C., ALPER T., "High yields of lethal mutations in somatic mammalian cells that survive ionizing radiation", *International Journal of Radiation Biology*, vol. 50, no. 1, pp. 167–179, 1986.

[SHI 90] SHIMIZU Y., SCHULL W., KATO H., "Cancer risk among atomic bomb survivors", *JAMA*, vol. 264, no. 5, August 1990.

[SHI 10] Yukiko S. et al., "Radiation exposure and circulatory disease risk: Hiroshima and Nagasaki atomic bomb survivor data, 1950–2003", *British Medical Journal*, vol. 340, no. 5349, 15 January 2010.

[SHO 16] SHORE R E., "Radiation and cataract risk: Impact of recent epidemiologic studies on ICRP judgments", *Mutation Research*, vol. 770, Pt. B, pp. 231–237, 2016.

[SHU 10] SHURYAK I., SACHS R.K., BRENNER D.J., "Cancer risks after radiation exposure in middle age", *Journal of National Cancer Institute*, vol. 102, no. 21, pp. 1628–1636, 2010.

[SIG 12] SIGNORELLO L.B., MULVIHILL J.J., GREEN D.M. et al., "Congenital anomalies in the children of cancer survivors: A report from the childhood cancer survivor study", *Journal of Clinical Oncology*, vol. 30, no. 3, pp. 239–245, 2012.

[SPY 15] SPYCHER B.D., LUPATSCH J.E., ZWAHLEN M. et al., "Background ionizing radiation and the risk of childhood cancer: A census-based nationwide cohort study", *Environmental Health Perspective*, vol. 123, no. 6, pp. 622–628, June 2015.

[STE 70] STEWART A., KNEALE G.W., "Radiation dose effects in relation to obstetric X-rays and childhood cancers", *The Lancet*, vol. 295, no. 7658, pp. 1185–1188, 1970.

[STR 18] STRUELENS L., DABIN J., CARINOU E. et al., "Radiation-induced lens opacities among interventional cardiologists: Retrospective assessment of cumulative eye lens doses", *Radiation Research*, vol. 189, no. 4, pp. 399–408, 2018.

[SUG 21] SUGIYAMA H., MISUMI M., SAKATA R. et al., "Mortality among individuals exposed to atomic bomb radiation in utero: 1950–2012", *European Journal of Epidemiology*, vol. 36, no. 4, pp. 415–428, April 2021.

[THO 94] THOMPSON D.E., MABUCHI K., RON E. et al., "Cancer incidence in atomic bomb survivors. Part II: Solid tumors, 1958–1987", *Radiation Research*, vol. 137, no. 2 Suppl., pp. S17–S67, February 1994.

[TÔH 12] TÔHARA H., *Il y a un an Hiroshima*, Editions Arléa, Paris, 2012.

[TOK 87] TOKUOKA S., KAWAI K., SHIMIZU Y. et al., "Malignant and benign ovarian neoplasms among atomic bomb survivors, Hiroshima and Nagasaki, 1950–80", *Journal of the National Cancer Institute*, vol. 79, no. 1, pp. 47–57, July 1987.

[TOK 94] TOKUNAGA M., "Incidence of female breast cancer among atomic bomb survivors, 1950–1985", *Radiation Research*, vol. 138, no.2, pp. 209–223, May 1994.

[UNS 13] UNSCEAR, Sources, effects and risks of ionizing radiation, vol. II, Annex B - Effects of radiation exposure of children, United Nations, New York, 2013.

[UTA 21] UTADA M., BRENNER A.V., PRESTON D.L. et al., "Radiation risk of ovarian cancer in atomic bomb survivors: 1958–2009", *Radiation Research*, vol. 195, no. 1, pp. 60–65, 1 January 2021.

[WAL 11] WALLACE W.H., "Oncofertility and preservation of reproductive capacity in children and young adults", *Cancer*, vol. 117, no. 10 Suppl, pp. 2301–2310, 2011.

[WAS 09] WASSILIEFF S., "Cataracte et rayonnements ionisants", *Radioprotection*, vol. 44, no. 4, pp. 505–517, 2009.

[WOR 07] WORGUL B.V., KUNDIYEV Y.I., SERGIYENKO N.M. et al., "Cataracts among Chernobyl clean-up workers: Implications regarding permissible eye exposures", *Radiation Research*, vol. 167, pp. 233–243, 2007.

[YAM 90] YAMAZAKI J.N., SCHULL W.J, "Perinatal loss and neurological abnormalities among children of atomic bomb, Nagasaki and Hiroshima revisited, 1949 to 1989", *JAMA*, vol. 264, no. 5, pp. 605–609, August 1990.

[YAM 14] YAMASHITA S., "The Fukushima Nuclear Power Plant accident and comprehensive health risk management", *Health Physics*, vol. 106, no. 2, pp. 166–180, February 2014.

[YAM 21] YAMADA M., FURUKAWA K., TATSUKAWA Y. et al., "Congenital malformations and perinatal deaths among the children of atomic bomb survivors: A reappraisal", *American Journal of Epidemiology*, vol. 190, no. 11, pp. 2323–2333, November 2021.

[ZER 88] ZERBIB J.C., "Pour un réexamen des données qui fondent la protection des travailleurs contre les rayonnements ionisants", *Colloque Nucléaire-Santé-Sécurité*, pp. 308–325, January 1988.

[ZER 94] ZERBIB J.C., "Accidents radiologiques : les retours d'expériences", *Annales de l'Association belge de radioprotection*, vol. 19, no. 4, pp. 621–643, 1994.

# 10

# Occupational Diseases Caused by Ionizing Radiation

## 10.1. History of the first radiation injuries caused by X-rays and radium

The first health damage caused by radiation was due to its use without precautions, as a result of total ignorance of the dangers presented by X-rays produced by electrical systems. Shortly afterwards, scientists in possession of the first fractions of micrograms of radium (µg or µCi) also perceived, to their cost, the risks presented by the radiation emitted by this radioactive substance. These were radioactive sources, often in powder form, obtained by chemical extraction from hundreds of kilograms of uranium ore tailings.

However, the most widespread damage was caused by X-rays, as the first radiation generators were significantly less expensive than a milligram of radium-226[1] and thus enjoyed significant commercial development. The first victims of radium were scientists, and first and foremost Marie Curie, who discovered 226Ra. In a note to the Académie des Sciences in 1901, Henri Becquerel and Pierre Curie [BEC 01] recounted the first illnesses affecting scientists working with radium, who were not yet aware of the long-term deleterious effects of extensive exposure to ionizing radiation. In 1920, at the age of 53, Marie Curie was left practically blind by a double cataract caused by external irradiation from the radium she was handling. She died on July 4, 1934, at the age of 67, of a rapidly progressing aplastic pernicious anemia, febrile according to the medical report of death [CUR 18]. This was undoubtedly due to her chronic external exposure to radiation emitted by radium-226 and its radioactive descendants.

1 "To obtain 1 to 2 mg of active radium bromide, 1 tonne of pitchblende residue, 5 tonnes of chemicals, 50 tonnes of washing water and several months of work are required" [COS 11].

In 1921, only a few dozen grams of radium had been produced since its discovery, 80% of it in the United States. However, with the discovery in 1923 of uranium ore deposits in the Belgian Congo, combined with the reduction in the cost of industrial extraction of radium from ore tailings [COS 11], anarchic development of uses followed, and diseases became apparent, two decades after Marie Curie's discovery.

### 10.1.1. *The first X-ray-induced cancers*

The first deleterious effects, caused by X-rays emitted by Crookes tubes, concerned limited parts of the body. These were mainly radiological burns, similar to thermal burns, involving the pulp of the first phalanges of the fingers or the dorsal surface of the hands, placed under X-ray beams (experimenters and technicians testing Crookes tubes on their hands).

The first sign of damage is a fleeting erythema (redness of the skin) which appears rapidly, followed by dryness of the skin: "dry" epidermitis. If the dose is higher, the erythema will be followed by phlyctene (lifting of the skin under which a translucent liquid is found). This is a radiological burn whose severity increases with the size of the dose received. Damage to blood vessels can lead to ulceration or even necrosis. In the latter case, protecting the victim's life necessitates amputations of phalanges, fingers, hands and sometimes even more.

Skin carcinogenesis, which occurs in the injured area, sometimes more than a decade after the radiological burn, will prove to be of a random nature, as it does not occur in all cases of radiodermatitis.

### 10.1.2. *A major survey of radiation-induced cancers carried out in 1914*

In her medical thesis on X-ray cancers [FEY 14], published in 1914, Dr. Sophie Feygin presents a chronology and analysis of worldwide publications since 1897, relating to the first X-ray-induced skin cancers. In France, the case of Arthur Radiguet[2], who died in 1905 as a result of generalized cancer, caused by repeated exposure to X-rays, was the first major warning to physicians practicing radiology. The number of published cases of cancer soon increased, as soon after Röntgen's discovery, Crookes tubes were used in a wide variety of applications, in total ignorance of the health effects of prolonged and multiple exposure. It was in 1911,

2 Arthur Radiguet, an engineer-constructor who was a daily fixture at demonstrations of his X-ray tube devices, ran a company in Paris that marketed a variety of devices for the popularization of science, as well as for commercial use. He won a gold medal for his devices at the Brussels Universal Exhibition in 1896.

with Hesse [HES 11, FEY 14], that the toll of damage increased to 94 radiological cancers, including those resulting from radiotherapy.

In her thesis, Feygin analyzes 104 cases of malignant neoplasia caused by X-rays, including 94 radiological tumors described in collections published in Germany, America (the most numerous) and England, to which she adds three cases described in German literature, one in Switzerland and six cases of French radiologists [FEY 14]. These 104 cancers involved some 30 doctors and their assistants, 25 professionals and six cases of patients treated by radiotherapy.

The surgical answer to radiation-induced skin cancer was found in 1897 by the American surgeon Allen Porter. The technique involved surgical excision of the affected area, followed by autografting. Although knowledge of this technique came late, it enabled Professor Lagrot to operate on a French surgeon for the first time in Europe on August 10, 1949, for skin cancer by excision-grafting [RAD 17]. In 1974, in a report on his experience (104 excision-grafts), he stated that the latency time between exposure to ionizing radiation and the onset of skin cancer was 25.5 years, with wide variations: a minimum of six years with chronic irradiation and a maximum of 44 years [LAG 74].

### 10.1.3. *The first radium-226-induced cancers*

During World War I, in which the United States was involved, hundreds of young women were employed by the US Radium Corporation, which used a luminescent paint (based on zinc sulfide to which a small quantity of radium was added[3]) to luminesce the hands and indications of various dials installed in the cockpits of military aircraft [MOO 17, MEN 12].

To paint the hands and numbers on the dials as precisely as possible, the foremen taught these young women to hone the tip of their brush between their lips. In this way, fine particles of paint deposited on the lips ended up entering the mouth. Radium, which belongs to the same family as calcium (alkaline earths), has a bone affinity like calcium. It fixed itself to the enamel of the teeth, then to the bony parts of the jaw, in the sinuses, while another fraction of the paint, ingested with the saliva, passed into the circulating blood and fixed itself to other parts of the skeleton.

The first warnings came from a dentist, Théodore Blum, who discovered necrotic gums, whose cells were destroyed by alpha particles emitted by radium fixed to tooth enamel [MEN 12]. These radium attachments lead to bone fragility (jaw fracture when a tooth was extracted) and bone sarcomas.

3 The first patent was filed in 1903 by the American Georges F. Kunz [DAV 12].

In 1922, one of the young workers, Mollie Maggia, fell ill and had to leave the workshop. Her condition rapidly worsened, leaving the doctors distraught. She died on September 12, 1922, less than a year after a succession of painful ailments.

It was a young woman, Grace Fryer, 18 years old when she was hired by the US Radium Corporation on April 10, 1917, who decided with four other colleagues, sick like herself, to lodge a complaint in 1927 against their company. The American and European press dubbed them the "radium girls" [MEN 12], [MOO 17]. Several epidemiological studies carried out on these workers, exposed before 1930, demonstrated significant increases in bone sarcomas and sinus cancers. Breast cancer has also been found to be associated with radium load [ADA 80][4].

## 10.2. The creation and difficult development of the occupational disease recognition system

The first occupational health problems were described in the work of an Italian physician, Bernardo Ramazzini (1633–1714), who analyzed those caused by 220 different trades. His work presents the symptoms caused by the various toxic substances and work postures involved [RAM 00].

The illnesses analyzed by Ramazzini were classified into two main families, still found three centuries later: illnesses caused by i) molecules which, mixed in the form of vapors or dusts with the air that workers breathed, penetrated their organs and disrupted their functions, and ii) excess or lack of exercise of certain parts of the body.

Among the molecules first cited by Ramazzini were lead and mercury, the first two toxic substances to be recognized two centuries later (1919) by the first French law on occupational diseases (ODs). As for the second family of disorders caused by over-exertion, the consequences of industrial development in the 21th century show that they are still present, three centuries later, with the various types of work involving repetitive movements (known in Europe as musculoskeletal disorders). On average, these disorders account for 86.0% of ODs recognized in France from 2011 to 2020 [MIN 21].

4 This study involved 1,180 "radium girls", 736 of whom had a known bone disease. A significant excess of breast cancers was found in those whose internal radium-226 contamination was equal to or greater than 50 μCi (1.85 MBq). It is highly probable that these breast cancers resulted from external exposure to radium placed in the small paint can, some 30 cm from the person. The high level of contamination is also evidence of long-term exposure.

The principle of lump-sum compensation for accidents at work set out in the law[5] of April 9, 1898, was extended to form the system for recognizing ODs, created by the law of October 25, 1919, after 18 years of debate in the "Chamber of Deputies" in France.

French legislation lagged behind that of neighboring countries which had set up compensation systems, such as Germany, United Kingdom, Austria, Belgium, Italy, Norway and Switzerland (law of 1881, improved in 1887 and 1901) [BON 05, MAR 18].

### 10.2.1. *The long beginnings of the law recognizing occupational diseases in France*

In 1901, MP Jules-Louis Breton (a chemical engineer) tabled a bill in the Chamber of Deputies[6], which equated ODs with accidents at work (law of April 9, 1898) [MAR 18]. The bill listed 11 substances, whose manufacture or use caused acute or chronic poisoning, as ODs. Although passed by 461 votes to 102, the bill was not voted on by the Chamber of Deputies until 1913, according to Anne Marchand, against a backdrop of considerable parliamentary instability [MAR 18]. The bill was again blocked for six years by the Senate before being adopted[7] on October 25, 1919. Its title stated that it was a law extending the law of April 9, 1898 on industrial accidents to diseases of occupational origin.

The scope of this law is disappointing, as it only covers the first two toxic substances on the list of Jules-Louis Breton's bill: lead and its compounds and mercury and its compounds. Article 10 of the law stipulates that a "Commission supérieure des maladies professionnelles" (Higher Commission on Occupational Illnesses) is to be set up, with special responsibility for advising on modifications to be made to the tables and on extensions to be made to the present law.

Anne Marchand investigated the work of this Commission. In the 20 years since the law was passed, the Commission met only 14 times between May 1920 and July 1938 (including a five-year break); a notorious lack of zeal resulted in a total absence of progress in the field of OD coverage.

---

5 Journal officiel of Sunday April 10, 1898, law of October 9, 1898: *Law concerning liability for accidents suffered by workers at work*.

6 Journal officiel of 6 December 1901, Chambre des députés, 2nd session of December 5, 1901, page 2543.

7 Law published in the JO of October 27, 1919, pp. 11973–11974, Journal officiel de la République française. Lois et décrets | 1919-10-27 | Gallica (bnf.fr).

### 10.2.2. *Contemporary knowledge of toxic substances and their deleterious effects*

Delays on the part of the Higher Commission were not, however, due to a lack of scientific and medical knowledge, for as early as 1905, Léon and Maurice Bonneff, two brothers united in a fight against various unacceptable working conditions, published their survey[8], *Les métiers qui tuent* (Occupations Which Kill). In this work, they describe in great detail work techniques and the various ailments affecting workers in many trades, caused by chemical toxins and work atmospheres [BON 05].

In 1927, the physician Gustave Roussy, after recalling the discovery in 1775 by the English surgeon Percival Pott (1714–1788) of the occupational cancer[9] of very young chimney sweeps (four to five years old) who penetrated chimneys to remove carcinogenic soot, a job carried out under very poor hygiene conditions, described the cancers associated with different professions exposed to various chemical or physical toxins: cancers of tar workers, kerosene workers, cotton weavers, oil refiners, aniline dyers, miners, X-ray and radium users [ROU 27]. Detailed and precise data on the toxic substances and occupational injuries affecting several trades were available, but what was missing was the social will of the Chamber of Deputies to protect the victims.

### 10.2.3. *Implementation of the law on the recognition of occupational diseases in France*

Since in almost all cases it is difficult to establish an unquestionable causal link between a disease and the work performed to cause it, and since the latency period for cancers can be several years, or even decades, making it even more difficult to identify the pathological factors, a principle of "presumption of origin" has been adopted. This presumption applies as soon as a series of conditions are met by the victim.

8 These include the health effects of industrial toxins (lead, ceruse (white lead), minium, mercury, arsenic, carbon sulfide, tar distillation products, benzine, etc.), ailments caused by dust in the working atmosphere, tuberculosis in sewage workers and laundry workers, etc.

9 Cancer of the scrotum, caused by carcinogenic soot, appeared on benign skin lesions. Depending on the study, they occurred between 20 and 25 years or 35–40 years after exposure.

The elements that enable this presumption of origin to be applied are described in a "table". The 1919 law contained only two tables[10]. It was not until 11 years later that a new law (January 1, 1931) amended the 1919 law, creating four other tables, including Table no. 6 on the toxic effects of X-rays and several radioelements.

This new law transferred the creation and modification of tables from public debate (Assemblée nationale) to an administrative procedure by decree (Ministry of Labor). These provisions were incorporated into the Social Security Code by the Ordinance of August 2, 1945.

At the same time, the United Kingdom was compensating 32 illnesses and Germany 21 illnesses. This expansion accelerated under the Popular Front: 14 new tables were created between 1936 and 1938 [MAR 18]. This rate of creation in two years has not been equaled since.

### 10.2.4. *Treatment of radiation-induced diseases*

The illnesses suffered by technicians and engineers working at the Radium Institute, directed by Marie Curie, were the starting point for the creation of an OD table. This was the case of the sudden and unexplained death of Mrs. Artaud, who lay in bed wasting away and dying, without being able to be treated, for lack of a precise diagnosis [LOR 92].

Then, it was Maurice Démenitroux's turn, who experienced the same fatigue and weight loss. In January 1925, he was admitted to Tenon Hospital, where he died a short time later of pernicious anemia. He was 40 years old and had been working with radium for two decades. A few days later, another of Marie Curie's companions, Marcel Demalander, aged 34, died of myeloid leukemia [LOR 92, MAR 18]. In August 1927, Mrs. Cotelle, a collaborator of the Radium Institute, who had been working on polonium-210 slides, suffered from stomach pains and lost her hair[11].

Both were members of the Syndicat des ingénieurs chimistes (French Union of Chemical Engineers), and their deaths were widely publicized. In 1926, the Tribunal d'instance de la Seine (Seine District Court), called upon by the families who had received the union's support, issued a ruling refusing to treat the injuries as an industrial accident, in the absence of a table. That same year, probably as a result of

10 JO of 27/10/1919, p. 11975, Table no. 1, Saturnisme professionnel (*Maladies causées par le plomb et ses composés)*, Table no. 2, Hydrargisme professionnel (*Maladies causées par le mercure et ses composés*).

11 Since the polonium-210 assassination of Alexander Litvinenko, an opponent of the Russian regime, on November 1, 2006 in London, we have known that the incorporation of this radionuclide leads, in particular, to total hair loss within a week [IRS 06, BER 16].

media pressure from the Syndicat des ingénieurs chimistes and former students of the Ecole de physique et chimie industrielle de Paris (Paris School of Physics and Industrial Chemistry), André Fallières, the Minister of Labor, tabled a bill to extend the scope of compensation limited to the two tables of lead and mercury. The bill was submitted to the Senate in 1929 and adopted on December 19, 1930 [MAR 18].

The law was promulgated on January 1 1931. In particular, it created Table no. 6: Intoxications caused by the action of X-rays or the following noxious radioactive substances: uranium and its salts, uranium X, radium and its salts, radon, polonium, mesothorium, radiothorium, thorium X, thoron, actinium.

## 10.3. Radiation-induced occupational diseases recognized in France since 1948

The first version of Table no. 6 included six radiation-induced conditions, two of which were cancers: leukemia and "radiologists' cancer" (skin cancer). Since 1950, the number of diseases has been increased to 12, and two new "historical" cancer diseases have been added to leukemia: bone sarcoma and bronchopulmonary cancer due to inhalation[12], but radiologists' cancer has been withdrawn. Since June 1984, this table has remained unchanged, due to a lack of consensus in the *Commission des maladies professionnelles* (Occupational Diseases Commission), which proposed new or modified tables to the Ministry of Labor. It now includes only the three historical cancers: bronchopulmonary cancer caused by inhalation in uranium miners, leukemia in the first radiologists who practiced radioscopy and bone sarcoma, which led to the death of several young women, notably those known in the United States as radium girls.

### 10.3.1. *The first statistical data on recognized radiation-induced occupational diseases*

For Table 6, on radiation-induced illnesses, statistics from the French Social Security system[13] from 1948 to 1955 were very poor, as only the total number of ODs recognized was provided. Then, from 1956 to 1967, the number of deaths was added. For this 12-year period, a total of 241 ODs were recognized, resulting in 23 deaths: an

12 The risks of industrial accidents and occupational illnesses covered for different categories of workers come under different social security schemes. The largest of these is the *Régime général*, which covers over 19 million workers.

13 The occupational injury and illness risks covered for different categories of workers fall within the remit of different organizations. The most important is the *Régime général*, which covers over 19 million workers. This scheme's National Health Insurance Fund (*Caisse nationale d'assurance maladie*) is responsible for managing occupational injuries and diseases.

average of 20.1 ODs/year, including 1.9 deaths per year, i.e. around one death for every 10 radiation-induced diseases recognized, which is a high mortality rate.

For the 2010–2021 period, some 50,000 occupational illnesses were recognized annually, resulting in between 300 and 570 deaths, or around one death for every 100 ODs. This general risk is therefore 10 times lower than that of ionizing radiation. However, analyses of the risk posed by ionizing radiation are few and far between, as the illnesses associated with this risk account for an average of just 1 per year out of 2,500 ODs.

The mortality figures we give are in fact very much underestimated, as they only concern deaths that have already occurred at the time of the claim for recognition of the OD and those that occur during the year in which the claim is processed. Deaths occurring after the year of recognition disappear from the statistics of the Caisse nationale d'assurance maladie (CNAM), and this applies to all tables.

The mortality data associated with all the OD tables are therefore all underestimated, due to the criteria used by the CNAM. If we add deaths due to recognized ODs in years N-1 and N-2 to deaths occurring in year N, the total number of deaths is multiplied on average by a factor of 2.5.

### 10.3.2. *Detailed data on recognized radiation-induced occupational diseases*

From 1968 to 1993, data published by the CNAM provided, in addition to the number of recognized ODs and associated fatalities, the number of leukemias and cases of radiodermatitis. Total and average values for this period are shown in Table 10.1 [ZER 18].

| | **Recognized conditions** | **Number of leukemias** | **Number of radiodermatitis cases** | **Number of deaths** |
|---|---|---|---|---|
| Total (1968–1993) | 452 | 174 | 92 | 125 |
| Annual average | 17.4 | 6.7 | 3.5 | 4.8 |

**Table 10.1.** *Relative balances to recognized occupational diseases from 1968 to 1993*

Since 1994, the statistics published in France by the CNAM provide information on the number of recognized illnesses for all conditions listed in Table no. 6. However, since 2017, the numbers of recognized ODs and associated deaths are no longer provided; only data related to the numbers of radiation-induced cancers are published.

| Year | Monitored workforce | Recognized conditions | Leukemias | Lung cancers | Bone sarcomas | Total cancers | Deaths |
|---|---|---|---|---|---|---|---|
| 2010 | 330,618 | 23 | 6 | 10 | 2 | 18 | 5 |
| 2011 | 343,988 | 20 | 7 | 9 | 1 | 17 | 2 |
| 2012 | 354,665 | 16 | 7 | 6 | 0 | 13 | 2 |
| 2013 | 352,082 | 25 | 9 | 8 | 1 | 18 | 6 |
| 2014 | 359,646 | 26 | 11 | 7 | 2 | 20 | 5 |
| 2015 | 365,830 | 24 | 9 | 6 | 0 | 15 | 2 |
| 2016 | 372,262 | 24 | 8 | 15 | 0 | 23 | 7 |
| 2017 | 360,694 | NP | 14 | 7 | 0 | 21 | NP |
| 2018 | 365,980 | 21 | 8 | 8 | 0 | 16 | 1 |
| 2019 | 395,040 | NP | 9 | 9 | 0 | 18 | NP |
| 2020 | 387,452 | NP | 8 | 9 | 0 | 17 | NP |
| 2021 | 392,180 | NP | 12 | 9 | 0 | 21 | NP |
| Average | 365,037 | 22.4 | 9 | 8.6 | 0.5 | 18.1 | 3.8 |

**Table 10.2.** *Change in the number of recognized radiation-induced cancers from 2010 to 2020 (based on [CNA 17] to [CNA 21]; [EUR 18]; [IRS 11] to [IRS 22]; [ZER 18]). NP = not provided*

It took a long time to compile this balance sheet, as the data for the last three years change from year to year in the published balance sheets, due to the fact that not all accepted OD cases have been settled administratively, and that legal challenges, whether positive by the employer or negative by the victim, delay the final outcome of the case. Figures for year N are not consolidated until year N+3.

We do not know the annual recognition rate for each Table no. 6 disease, but a publication on the volume of recognition of ODs provides us with average rates. For the years 2004–2013, 112 leukemias were recognized out of 130 claims (86%), 15 bone sarcomas out of 18 (83%) and 115 bronchopulmonary cancers out of 158 (73%). Taking all Table no. 6 diseases together, 364 ODs were recognized out of 470 claims, giving a recognition rate of 77% [CNA 15a, b]. This recognition rate is higher than that calculated for all ODs, which is 69.5%.

The 2010–2021 review shows that the average annual number of recognized radiation-induced cancers was close to 18, with leukemia (9/year) and bronchopulmonary cancer (8.6/year) almost equal. Bone sarcoma remained a rare cancer, with only 15 cases recognized in 28 years between 1994 and 2021, an average of one case every two years.

In terms of occupational risk, it is customary to assess the risk rate per 100,000 workers exposed to the risk in question. The average risk (for 2010–2021) of developing radiation-induced occupational cancer in France is 6.14 per hundred thousand workers exposed to the risks presented by ionizing radiation.

### 10.3.3. *Comparison of the risk of radiation-induced cancers with the risk of all occupational cancers in France*

We can compare this radiation-induced cancer risk rate with that for all occupational cancers recorded in France by the CNAM (rates calculated for the 2010–2021 period), to assess its relative importance.

| Year | Number of employees | Reported ODs | All recognized cancers (asbestos) | Recognized cancers per 100,000 employees | Radiation-induced cancers recognized per 100,000 employees |
|---|---|---|---|---|---|
| 2010 | 18,299,717 | 110,828 | 1,739 (1,473) | 9.50 (8.05) | 5.44 |
| 2011 | 18,492,444 | 121,410 | 1,812 (1,535) | 9.80 (830) | 4.94 |
| 2012 | 18,296,201 | 110,357 | 1,902 (1,579) | 10.4 (8.63) | 3.67 |
| 2013 | 18,649,775 | 110,388 | 1,707 (1,415) | 9.15 (7.59) | 5.11 |
| 2014 | 18,615,687 | 112,245 | 1,685 (1,363) | 9.05 (7.32) | 5.56 |
| 2015 | 18,775,282 | 107,889 | 1,879 (1,469) | 10.0 (7.82) | 4.10 |
| 2016 | 18,653,985 | 110,141 | 1,872 (1,409) | 10.0 (7.55) | 6.18 |
| 2017 | 19,250,334 | 112,614 | 1,940 (1,392) | 10.1 (7.23) | 5.54 |
| 2018 | 18,875,562 | 130,512 | 1,946 (1,424) | 10.3 (7.54) | 4.37 |
| 2019 | 19,649,134 | 113,187 | 1,790 (1,352) | 9.11 (6.88) | 4.56 |
| 2020 | 19,344,473 | 97,325 | 1,534 (1,210) | 7.93 (6.28) | 4.39 |
| 2021 | 19,283,655 | 118,082 | 1,633 (1,282) | 8.47 (6.65) | 5.35 |
| **Average** | **18,848,845** | **112,914** | **1,803 (1,409)** | **9.57 (7.48)** | **4.96** |

**Table 10.3.** *Comparison of risk rates for all occupational cancers, asbestos-related cancers and radiation-induced cancers (based on [CNA 17] to [CNA 22]; [EUR 12] to [EUR 18])*

Table 10.3 clearly shows the preponderance of risk presented by asbestos-induced cancers, a risk that affects virtually all occupations. We also note that the average risk rate associated with radiation-induced cancers is high. It is equal to 2/3 of the asbestos risk (66.3%).

## 10.4. Diseases recognized in Europe

Comparisons between European countries are difficult to make, as lists of ODs differ from one country to another, and some countries have no list of ODs at all. Reporting procedures and recognition criteria for the same disease may also differ. Financial compensation for ODs does not follow a single European model [EUR 99].

The European list of ODs, which dates back to 2003 [COM 03], was not updated until 2022, following the Covid-19 pandemic which affected all Member States and necessitated harmonized procedures for the treatment of health problems resulting from occupational viral contamination.

However, these recommendations are not binding. As stated in Article 8 of the Recommendation, the Commission encourages Member States to recognize the ODs included in the European list.

The list includes five families of diseases caused by: chemical agents (54 products or families of products), skin-attacking substances (10 substances, mainly carcinogens), inhalation of substances (20 substances), infectious and parasitic diseases (8 diseases) and diseases caused by physical agents (18, including ionizing radiation).

In France, occupational injury and disease risks for different categories of workers are covered by different organizations. The largest of these is the *Régime général*, which covered more than 19 million workers in 2022. This is the scheme used in Eurogip's European comparisons, including the latest one carried out in 2018.

| Country | Recognition requests | Unknown cancers | Ratio* of recognized cancers to 100,000 workers | Proportion of off-list cancers (%) | Share of cancers in recognized ODs (%) |
|---|---|---|---|---|---|
| Germany | 15,234 | 6,559 (1,910) | 15.1 | 0.43 (28 cases) | 32 |
| Austria | 148 | 129 (106) | 3.73 | 0 | 11 |
| Belgium | 344 | 181 (165) | 4.69 | 0.55 (1 case) | 7 |
| Denmark | 688 | 194 | 6.93 | 0 | 5 |
| France | 2,679 | 2,118 | 11.39 | 4.44 (94 cases) | 3 |
| Italy | 2,642 | 1,033 | 6.31 | NC | 6 |
| Sweden | 56 | 27 | 0.5 | NA | 5 |
| Switzerland | 253 | 177 | 4.41 | 0 | 7 |

*Population insured by each national organization, including the public sector (except in France, private sector only).

**Table 10.4.** *General data on occupational cancers in Europe in 2016 (from [EUR 18a]). NC = not communicated, NA = not applicable*

NOTE ON TABLE 10.4.– *Figures in brackets are for cancers caused by asbestos (bronchopulmonary and pleural cancers). The high number of cancer cases recognized in Germany is due to the inclusion (since 2015) of UV-induced skin cancers on the list of ODs (3,782/6,559).*

Radiation-induced diseases, which account for a very small proportion of OD statistics in Europe, have not been the subject of any published work.

## 10.5. Conclusions

Occupational diseases have a major impact on the workforce in France, Europe and the rest of the world. In 2015, the number of work-related deaths worldwide was estimated at 2.78 million, compared with 2.33 million in 2011. For fatal workplace accidents, there were 380,500 deaths, an increase of 8% in 2014 compared to 2010.

Fatal work-related illnesses were a concern of the International Labour Organization (ILO) in 1998, as their numbers were at least five times higher than deaths from work-related accidents. In 2015, there were 2.4 million deaths due to fatal occupational diseases, an increase of 0.4 million on 2011.

In total, it is estimated that over 7,500 people die every day: 1,000 from accidents at work and 6,500 from occupational diseases. The fatal accident rate has risen slightly. The number of non-fatal occupational accidents has been estimated at 374 million, a significant increase on 2010 [HÄM 17].

In France in 2021, 1.227 million accidents at work, 176,000 commuting accidents and 118,000 occupational diseases were reported, for a total of 1.4 million. While 94.5% of accidents at work and 97.1% of commuting accidents were recognized, only 64.4% of occupational diseases were recognized.

Radiation-induced occupational diseases were treated late, particularly in France. In the United States, the induction of radium-induced bone sarcomas led to the death of a dozen young women; in Europe, over a hundred cases of skin cancer, some of which metastasized and led to the death of the victims; and in France, the death of professionals barely forty years of age, caused by internal contamination by radium-226 and its descendants. There was a delay of a few more years before a law in France produced the first table of radiation-induced diseases, in 1931. From 1931 to 1984, the last version of this table, only minor advances were made, but in the end, the only recognized diseases were the historical radiation-induced cancers known in the 1920s-1930s. The gap, sometimes of a decade, between medical knowledge of the impact of internal and external irradiation and the translation of this knowledge into the treatment of occupational injuries is obvious. The same applies to the recognition of occupational diseases, particularly cancers.

It is a double penalty for victims of occupational diseases, who suffer not only from their injury but also from the non-recognition of the occupational nature of their condition.

## 10.6. References

[ADA 80] Adams E.E., Brues A.M, "Breast cancer in female radium dial workers first employed before 1930", *Journal of Occupational Medicine*, vol. 22, no. 9, pp. 583–587, September 1980.

[BEC 01] Becquerel H., "Action physiologique des rayons du radium", *CR des Séances de l'Académie des sciences*, vol. 132, pp. 1289–1291, 3 June 1901.

[BER 16] Bernard P., Mandraud I., "Affaire Litvinenko : Poutine en accusation, Londres veut éviter l'affrontement", *Le Monde*, 22 January 2016.

[BON 05] Bonneff L., Bonneff M., *Les métiers qui tuent : enquête auprès des syndicats ouvriers sur les maladies professionnelles*, Bibliographie sociale, Paris, 1905.

[CNA 15a] CNAM, Volumétrie de la reconnaissance des maladies professionnelles par tableau de MP sur la décennie 2004-2013, Study 2015-014vlog, Paris, 30 January 2015.

[CNA 15b] CNAM, Rapport de gestion 2014, L'Assurance maladie, Risques professionnels, Paris, July 2015.

[CNA 16] CNAM, Rapport de gestion 2015, L'Assurance maladie, Risques professionnels, Paris, December 2016.

[CNA 17a] CNAM, Rapport annuel 2016 de l'Assurance maladie, Risques professionnels, éléments statistiques et financiers, Paris, 1 December 2017.

[CNA 17b] CNAM, Statistiques de sinistralité 2016 tous CTN et par CTN, Accidents du travail, Accidents de trajet, Maladies professionnelles, Study 2017-197-CTN, Paris, February 2017.

[CNA 18] CNAM, Rapport annuel 2017 de l'Assurance maladie, Risques professionnels, éléments statistiques et financiers, Paris, 1 December 2018.

[CNA 19] CNAM, Rapport annuel 2018 de l'Assurance maladie, Risques professionnels, éléments statistiques et financiers, Paris, December 2019.

[CNA 20] CNAM, Rapport annuel 2019 de l'Assurance maladie, Risques professionnels, éléments statistiques et financiers, L'assurance maladie, Risques professionnels, Paris, December 2020.

[CNA 21] CNAM, Rapport annuel 2020 de l'Assurance maladie, Risques professionnels, éléments statistiques et financiers, L'assurance maladie, Risques professionnels, Paris, November 2022.

[CNA 22] CNAM, Rapport annuel 2021 de l'Assurance maladie, Risques professionnels, éléments statistiques et financiers, L'Assurance maladie, Risques professionnels, Paris, 2022.

[COM 03] COMMISSION UE, "Recommandation (CE) 2003/670/CE de la Commission du 17 septembre 2003 concernant la liste européenne des maladies professionnelles", *Journal officiel de l'Union européenne*, p.28, 25 September 2003.

[COM 22] COMMISSION UE, "Recommandation (UE) 2022/2337 de la Commission du 28 novembre 2022 concernant la liste européenne des maladies professionnelles", *Journal officiel de l'Union européenne*, pp. 12–21, 30 November 2022.

[COS 11] COSSET J.-M., HUYNH R., *La fantastique histoire du Radium, Quand un élément radioactif devient potion magique*, Editions Ouest-France, April 2011.

[CUR 18] CURIE E., *Madame Curie*, Gallimard, 2018 [Original published 1938].

[DAV 12] DAVAL M., La problématique de la radioactivité dans les objets patrimoniaux, Master's thesis, 2012.

[EUR 99] EUROGIP, Etude sur les maladies professionnelles en Europe, Chiffres et tendances, Paris, June 1999.

[EUR 09] EUROGIP, Les maladies professionnelles en Europe, Statistiques 1990-2006 et actualité juridique, Paris, 16 February 2009.

[EUR 12a] EUROGIP, Point statistique AT-MP France, Données 2011, Eurogip-79F report, Paris, December 2012.

[EUR 12b] EUROGIP, Point statistique AT-MP France, Données 2010, Eurogip-76F report, Paris, October 2012.

[EUR 13] EUROGIP, Point statistique AT-MP France, Données 2012, Eurogip-90F report, Paris, December 2013.

[EUR 14] EUROGIP, Point statistique AT-MP France, Données 2013, Eurogip-104F report, Paris, December 2014.

[EUR 15] EUROGIP, Point statistique AT-MP France, Données 2014, Eurogip-112F report, Paris, December 2015.

[EUR 18a] EUROGIP, Sinistralité et repérage des cancers professionnels dans neuf pays européens Allemagne, Autriche, Belgique, Danemark, Finlande, France, Italie, Suède et Suisse, Study report, Réf. Eurogip-141/F, Paris, December 2018.

[EUR 18b] EUROGIP, Point statistique AT-MP France, Données 2016, Eurogip-135/F, Paris, July 2018.

[FEY 14] FEYGIN S., Du cancer radiologique, Thesis, Bibliothèque interuniversitaire de santé, Paris, 1914.

[HÄM 17] HÄMÄLÄINEN P., TAKALA J., BOON KIAT T., "Global estimates of occupational accidents and work related illnesses", *21st World Congress on Safety and Health at Work*, Singapore, 2017.

[HES 11] HESSE O. *Symptomatologie, Pathogenese und Therapie des Röntgenkarzinoms*, V. von A. Barth, Leipzig, 1911.

[IRS 06] IRSN, Impact du polonium 210 sur l'Homme, 28 November 2006.

[IRS 11] IRSN, La radioprotection des travailleurs, Exposition professionnelle aux rayonnements ionisants en France : Bilan 2010, Fontenay aux Roses, 2011.

[IRS 12] IRSN, La radioprotection des travailleurs, Exposition professionnelle aux rayonnements ionisants en France : Bilan 2011, Fontenay aux Roses, 2012.

[IRS 13] IRSN, La radioprotection des travailleurs, Exposition professionnelle aux rayonnements ionisants en France : Bilan 2012, Fontenay aux Roses, 2013.

[IRS 14] IRSN, La radioprotection des travailleurs, Exposition professionnelle aux rayonnements ionisants en France : Bilan 2013, Fontenay aux Roses, 2014.

[IRS 15] IRSN, La radioprotection des travailleurs, Exposition professionnelle aux rayonnements ionisants en France : Bilan 2014, Fontenay aux Roses, 2015.

[IRS 16] IRSN, La radioprotection des travailleurs, Exposition professionnelle aux rayonnements ionisants en France : Bilan 2015, Fontenay aux Roses, 2016.

[IRS 17] IRSN, La radioprotection des travailleurs, Exposition professionnelle aux rayonnements ionisants en France : Bilan 2016, Fontenay aux Roses, 2017.

[IRS 18] IRSN, La radioprotection des travailleurs, Exposition professionnelle aux rayonnements ionisants en France : Bilan 2017, Fontenay aux Roses, 2018.

[IRS 19] IRSN, La radioprotection des travailleurs, Exposition professionnelle aux rayonnements ionisants en France : Bilan 2018, Fontenay aux Roses, September 2019.

[IRS 20] IRSN, La radioprotection des travailleurs, Exposition professionnelle aux rayonnements ionisants en France : Bilan 2019, Fontenay aux Roses, June 2020.

[IRS 21] IRSN, La radioprotection des travailleurs, Exposition professionnelle aux rayonnements ionisants en France : Bilan 2020, Fontenay aux Roses, June 2021.

[IRS 22] IRSN, La radioprotection des travailleurs, Exposition professionnelle aux rayonnements ionisants en France : Bilan 2020, Fontenay aux Roses, June 2022.

[LAB 37] LABORDE S., LECLERC J., "Les maladies professionnelles causées par les rayons X et par les substances radioactives", *Journal belge de radiologie*, no. 148, pp. 201–253, May–June 1937.

[LAG 74] LAGROT F., *Radiodermites des mains*, Editions Douin, Paris, 1974.

[LOR 92] LORIOT N., *Irène Joliot-Curie*, Livre de Poche, Paris, 1992.

[MAR 18] MARCHAND A., Reconnaissance et occultation des cancers professionnels le droit à réparation à l'épreuve de la pratique (Seine-Saint-Denis), Technical PhD Thesis, Université Paris-Saclay, 22 March 2018.

[MEN 12] MENGEOT M.-A., ZERBIB J.C., *L'atome, au jour le jour. Les travailleurs faces aux risques de la technologie nucléaire*, European Trade Union Institute, Brussels, 2012.

[MIN 21] MINISTERE DU TRAVAIL, Conditions de travail, Bilan 2020, Ministère du Travail, de l'Emploi et de l'Insertion, Direction générale du travail, Conseil national d'orientation des conditions de travail, Paris, 2021.

[MOO 17] MOORE K., L'histoire oubliée des "radium girls", dont la mort a sauvé la vie à des milliers d'ouvrières, 2017. Available at: https://www.buzzfeed.com/fr/authorkatemoore/lhistoire-oubliee-des-radium-girls-dont-la-mort-a-sauve-la.

[RAD 17] RADIOPROTECTION, "Notes recueillies au cours de la conférence du professeur Lagrot", *Radioprotection*, vol. 6, no. 3, 2017. Available at: radioprotection.org.

[RAM 00] RAMAZZINI B., *Traité des maladies des artisans et de celles qui résultent des diverses profession*, first published in Modena, Italy, ca. 1700.

[ROU 27] ROUSSY G., HERAUX A., "Les cancers professionnels", *La presse médicale*, no. 103, 24 December 1927.

[ZER 18] ZERBIB J.C., "Les maladies professionnelles radioinduites reconnues en France : bilan et évolutions depuis 1956", *Radioprotection*, vol. 53, no. 1, p. 53, Société Français de Radioprotection, 2018.

# 11

# The Dose–Response Relationship to the Effects of Ionizing Radiation

## 11.1. Introduction: the dose–response relationship

To estimate the radioactive risk, it is necessary to carry out a good assessment of the relationship between the dose received and the harmful effect. For all living organisms except humans, the dose estimate stops at the absorbed dose and is expressed in gray (Gy).

The linear non-threshold (LNT) model has been used in radiation protection for over 40 years, and is the subject of lively debate. It relies heavily on human epidemiology, with support from radiobiology. Scientific foundations include NCRP Report No. 136 [NCR 01], UNSCEAR (United Nations Scientific Committee on the Effects of Atomic Radiation) Synthesis [UNS 00a,b], ICRP Publication 99 [ICR 05] and BEIR VII [NRC 06].

Recent studies following the BEIR VII report are under critical review and include atomic bomb survivors, workers at the Russian Mayak site, Tetcha River populations, atomic veterans, US radiological technicians, the US million-person study, international workers (INWORKS), Chernobyl decontaminators and firefighters (known as liquidators), children who underwent computerized tomography scans and tuberculosis patients under fluoroscopy [BOI 17].

## 11.2. Dose–response relationships for high doses

Two events can lead to high doses of radiation in humans: an accident or an atomic explosion. In nuclear history, a number of accidents have resulted in high doses of radiation for dozens of individuals, due to irradiator handling errors or a major nuclear accident such as a criticality event. However, it is above all the two atomic weapon blasts in Hiroshima and Nagasaki that were responsible for a very large number of severe irradiations.

For high doses, the main data come from the effects observed following accidents or the Hiroshima and Nagasaki explosions. These are described in detail in Volume 1 of the "Radioactive Risk" series [AMI 19a].

### 11.2.1. *Survivor studies*

In 1947, the Atomic Bomb Casualty Commission (ABCC) was established by the US National Academy of Sciences (NAS) with funding from the US Atomic Energy Commission. This commission launched extensive studies on the health of bomb survivors in cooperation with the Japanese National Institute of Health of the Ministry of Health and Welfare, which joined the research program in 1948. A thorough review of the ABCC's work by the Francis Committee (https://www.rerf.or.jp/en/glossary/francomm-en/) in 1955 led to a wide-ranging revision of the research design and laid the foundations for the future population studies that continue today. The Commission was replaced in April 1975 by the Radiation Effects Research Foundation (RERF). This non-profit foundation, governed by Japanese civil law, comes under the jurisdiction of the Japanese Ministries of Foreign Affairs and Health and Welfare, and benefits from an agreement between the governments of Japan and the USA.

One of the first problems to be solved was quantifying the radiation dose suffered by each individual. Numerous studies provided increasingly precise estimates. For a long time, the main system was DS86 (Dosimetry System of 1986). The most recent is DS02 (2002 estimate of the dose received) [YOU 05].

Table 11.1 shows colon dose values as a function of distance from the hypocenter. This radiation dose is halved for every 200-meter increase in distance from the bomb's plumb line. The average dose received by survivors was around 0.2 Gy (see Chapter 7).

| Weighted colon dose (Gy) | Approximate distance from hypocenter (m) | |
|---|---|---|
| | Hiroshima | Nagasaki |
| 0.005 | 2,500 | 2,700 |
| 0.05 | 1,900 | 2,050 |
| 0.1 | 1,700 | 1,850 |
| 0.5 | 1,250 | 1,450 |
| 1 | 1,100 | 1,250 |

**Table 11.1.** *Weighted colon dose for subjects in the LSS (Life Span Study) cohort as a function of average distance from the hypocenter. (Different protection conditions among survivors mean that this radiation dose/distance relationship does not apply to all survivors)*

The most recent report on the long-term effects of radiation on cohorts of survivors of the Hiroshima and Nagasaki atomic bombs (LSS, Life Span Study) provides observations for the period 1950–2003 [OZA 12]. During this period, 58% of the 86,611 members of the LSS cohort, where dose estimates with DS02 are available, died. The excess cancer cases are $10^4$ people per year and per Gy.

A formal analysis of the dose-threshold response indicates that there is no threshold. This means that the zero dose was the best estimate of the threshold (this zero dose is purely theoretical, since there is natural radioactivity everywhere on Earth, and some of it comes from the cosmos). The risk of cancer mortality increased significantly for most major organs, including stomach, lung, liver, colon, breast, gallbladder, esophagus, bladder and ovary.

### 11.2.2. *Other studies*

More recently, studies have focused on other harmful effects of ionizing radiation besides cancer and the resulting mortality. These include heart disease and stroke.

In one study, Shimizu et al. (2018) measured reaction time (RT) as a cognitive function and grip strength (GS) as a physical function and attempted to demonstrate that they were associated, alone or in combination, with mortality from heart disease or stroke. RT was positively correlated and GS was negatively associated with heart disease and stroke mortality [SHI 18b].

Yamada et al. [YAM 04] found a significant positive linear dose response for cataract ($P = 0.026$), a negative linear dose response for glaucoma ($P = 0.025$) and a significant quadratic dose response for hypertension ($P = 0.028$) and myocardial infarction in survivors exposed to less than 40 years ($P = 0.049$). The effects of radiation on cataracts, glaucoma, hypertension and renal and ureteral calculi in humans are new findings for Hiroshima and Nagasaki survivors [YAM 04].

The radiation-related risk of developing ovarian cancer was estimated between 1958 and 2009 in 62,534 female survivors of the two atomic bomb firings. Radiation dose was positively, though not significantly, associated with the risk of total ovarian cancer (ERR/Gy (excess relative risk per gray) of 0.30, 95% confidence interval (CI) -0.22 to 1.11) [UTA 21].

## 11.3. Dose–response relationships for low doses

For a long time, scientists believed that no harmful effects were observed in humans at low doses. Since then, there has been increasing evidence of harmful effects at doses of less than 100 mSv, or at dose rates of less than 6 $mSv.h^{-1}$.

What is this evidence? Is the relationship to low doses always linear? Is there a threshold below which no adverse effects of ionizing radiation can be observed? Is there an adaptive human response to low dose rates?

### 11.3.1. *Epidemiological studies*

In the case of medium and low doses, the effects are stochastic and can only be apprehended by epidemiological studies. In epidemiology, there are three main types of study: aggregate studies, case-control studies and cohort (or prospective) studies. The last two types are analytical studies. Aggregate studies are the most frequent, but generally provide less information. Cohort studies are more powerful because the population studied is more homogeneous. Details are given in [AMI 17a].

#### 11.3.1.1. *The general principle of epidemiological studies*

The power (or resolution) of epidemiological studies depends on a number of parameters, the most important of which is the size of the populations studied. This is the reason for the emergence of multiple meta-analyses, in which several epidemiological studies are merged. There are also many possible biases in epidemiological studies. One of the most frequent is the poor choice of control group or unexposed population, which is compared with the exposed population.

Confounding factors (or cofactors) are also numerous, especially if the study is conducted over the long term. For example, the effects of irradiation and smoking are linked for a large number of pathologies. Corrective factors must therefore be applied to eliminate the influence of confounding factors.

The analysis of epidemiological studies uses a large number of statistical tools such as risk ratio (RR), excess relative risk (ERR) or odds ratio (OR), standardized mortality ratio (SMR) or standardized incidence ratio (SIR). Further information can be found in [AMI 17a][1].

The fact that the correlation is positive in no way means that there is a real cause-and-effect relationship between these two parameters. This is an all-too-frequent shortcut, and as the saying goes, "*Cum hoc ergo propter hoc*" (Latin for with this, therefore because of this). Indeed, it is common to claim that if two events are correlated, then there is a causal link between them. This confusion between correlation and causation should absolutely be avoided.

In the nuclear field, the main epidemiological studies have focused on populations of nuclear workers, including uranium miners; populations living in the vicinity of nuclear power plants and other nuclear facilities, particularly children; military personnel and populations living in the vicinity of atomic test sites; populations living on primary massifs, notably granite, where houses are rich in radon; populations living in areas with high levels of natural radioactivity. The main results of epidemiological studies following military and civilian nuclear accidents are given in the books by Amiard [AMI 19a, AMI 19b].

It is becoming increasingly important to estimate dose as accurately as possible when assessing low dose and low dose rate, as the calculation of ERR/Gy in an epidemiological study is very sensitive to the number of excess cancers observed, which can lead to errors if dosimetry is of poor quality [TIL 17].

#### 11.3.1.2. *Exposure due to atomic blasts in Japan*

The first comprehensive report on leukemia mortality among Japanese atomic bomb survivors from the LSS program provides clear results. Overall, 103 of the 310 leukemia deaths observed were estimated to be excess deaths due to exposure to ionizing radiation. In the last decade of observation (1991–2000), the estimated

1 The ratio O/A of the number of cases observed in the study population (O) to the calculated number of cases, known as the expected number of cases (A), is called the standardized mortality ratio (SMR) or the standardized incidence ratio (SIR). The RR is obtained by dividing $R^1$ (the risk of disease in exposed subjects) by $R^0$ (the risk of disease in unexposed subjects). The ERR is equal to RR - 1. The OR is the ratio of the disease risk R divided by 1 - R (R/(1 - R). The hazard ratio (HR) is the ratio of risk at a given time between the exposed and the unexposed.

attributable fraction of leukemia deaths among survivors exposed to more than 0.005 Gy was 0.34, suggesting that the effect of atomic bombing on leukemia mortality persisted in this cohort for more than five decades [RIC 09].

Atomic bomb survivors in Japan had ERRs of 14% per Gy for heart disease and 9% per Gy for stroke [SHI 10]. These two risks have only recently been taken into account by the ICRP.

Grant et al. [GRA 17] carried out the third analysis of solid cancer incidence among the LSS cohort of Hiroshima and Nagasaki atomic bomb survivors. The follow-up period was from 1958 to 2009, providing 3,079,484 person-years of follow-up. ERR and excess absolute risk (EAR) models were adjusted for smoking. In the cohort, 22,538 incident cases of first primary solid cancer were identified. For women, dose response was consistent with linearity, with an estimated ERR of 0.64 per Gy (95% CI 0.52 to 0.77). For men, a linear quadratic model was used, resulting in an ERR of 0.20 (95% CI 0.12 to 0.28) at 1 Gy and an ERR of 0.010 (95% CI 0.0003 to 0.021) at 0.1 Gy. A statistically significant dose response using the gender-averaged linear ERR model was obtained from 0 to 100 mGy (P = 0.038).

#### 11.3.1.3. *Exposure due to atmospheric atomic testing*

A group of experts from INSERM (Institut national de la santé et de la recherche médicale; French National Health and Medical Research Institute) has summarized the work on the health effects of atmospheric atomic testing. The few epidemiological studies carried out in French Polynesia show no major impact of fallout from nuclear testing on the health of the Polynesian population. A case-control study carried out on the local population shows a slight increase in the risk of thyroid cancer, linked to an increase in the dose received by the thyroid gland before the age of 15. A mortality study carried out on military personnel revealed an increased risk of mortality from hematological malignancies in individuals with non-zero dosimetry. This epidemiological cohort study, conducted to assess the health consequences in French Polynesia, was complex to implement due to four major difficulties: gaps in knowledge about the health of the population, the low-dose levels of ionizing radiation received, the small size of the Polynesian population and the time elapsed since exposure.

The group of experts made three main recommendations: 1) improve, develop and perpetuate the health monitoring system for non-communicable diseases (cancers, cardiovascular diseases, congenital anomalies); 2) refine estimates of doses received by local populations and by civilian and military personnel; and 3) carry out a careful and rigorous review of the international scientific literature on the effects of low doses of ionizing radiation, in particular for certain cancers (not yet recognized by international bodies as radiation-induced), cardiovascular diseases and effects on offspring [INS 20].

#### 11.3.1.4. *Medical exposure*

Ionizing radiation is widely used in medicine, either as radioactive tracers for diagnostic purposes, or to destroy cancerous tumors using the ionizing properties of radionuclides. Low doses concern exposure during diagnostic examinations (X-rays and CT scans), for which the dose rate may be high, but the dose will be fractionated in the case of multiple examinations. Recent studies have focused on the consequences of exposure in utero or during childhood and adolescence [CEA 16].

Therefore, while CT scans are very useful clinically, there are potential cancer risks associated with the associated ionizing radiation, particularly for children who are more radiosensitive than adults.

In the UK, children who underwent CT scans before the age of 22 were subsequently followed up. Of 178,604 patients, 74 were diagnosed with leukemia, and 135 of 176,587 with brain tumors. Pearce et al. [PEA 12] found a positive association between CT radiation dose and leukemia ([ERR] per mGy 0.036, 95% CI 0.005–0.120; p=0.0097) and brain tumors (0.023, 0.010–0.049; p<0.0001). In comparison to patients who received a dose of less than 5 mGy, the relative risk of leukemia for patients who received a cumulative dose of at least 30 mGy (mean dose 51.13 mGy) was 3.18 (95% CI 1.46–6.94), and the relative risk of brain cancer for patients who received a cumulative dose of 50–74 mGy (mean dose 60.42 mGy) was 2.82 (1.33–6.03). The authors conclude that using CT scans in children to deliver cumulative doses of around 50 mGy could almost triple the risk of leukemia, and doses of around 60 mGy could triple the risk of brain cancer [PEA 12].

Similarly, an Australian study showed that the number of scans undergone by patients under the age of 20 increased the risk of developing several types of cancer [MAT 13]. In 11 million Australian young people and adolescents, 60,674 cancers were identified, including 3,150 in 680,211 people exposed to a CT scan at least one year before any cancer diagnosis. Overall cancer incidence was 24% higher in those exposed than in those unexposed, with an incidence rate ratio of 1.24 (95% CI 1.20–1.29; p < 0.001). The incidence rate ratio was higher after exposure at younger ages (p<0.001 for trend). At 1–4, 5–9, 10–14 and 15 years or more since first exposure, incidence rate ratios were 1.35 (1.25 to 1.45), 1.25 (1.17 to 1.34), 1.14 (1.06 to 1.22) and 1.24 (1.14 to 1.34), respectively. The absolute excess incidence rate for all cancers combined was 9.38 per 100,000 person-years at risk. The mean effective radiation dose per scan was estimated at 4.5 mSv [MAT 13].

In Germany, a cohort was set up to assess the risk of developing cancer, in particular leukemia, central nervous system tumors and lymphoma, before the age of 15, in children previously exposed to CT scans. The cohort included information on 71,073 CT examinations in 44,584 children, representing 161,407 person-years. For leukemia, the SIR (ratio of the number observed to the number expected) was

1.72 (95% CI 0.89–3.01, for 12 observed), and for CNS tumors, the SIR was 1.35 (95% CI 0.54–2.78, for 7 observed) [KRI 15].

In Taiwan, the increased risk of subsequent malignancy and benign brain tumors was quantified following a pediatric CT scan of the head. In the exposed cohort, 24,418 participants were under 18 years of age. The overall risk was not significantly different in the two cohorts (incidence rate 36.72 per 100,000 person-years in the exposed cohort, 28.48 per 100,000 person-years in the unexposed cohort, hazard ratio (HR)) was 1.29 (95% CI 0.90–1.85). The risk of benign brain tumors was significantly higher in the exposed cohort than in the unexposed cohort (HR 2.97, 95% CI 1.49–5.93) [HUA 14].

In France, recent epidemiological findings have suggested an increased risk of cancer after receiving CT scans in childhood or adolescence. The cohort included 67,274 children who had their first CT scan before the age of 10 from 2000 to 2010 in 23 French departments. During an average follow-up of four years, 27 cases of central nervous system tumors, 25 of leukemia and 21 of lymphoma were diagnosed; 32% of these among children with predisposing factors[2] [JOU 14].

#### 11.3.1.5. *Occupational exposure*

There are a large number of people working in the nuclear industry, and they are spread all over the world. Some workers, such as uranium miners, are more exposed to ionizing radiation than others. The vast majority of these workers have regular medical check-ups, which is an advantage when it comes to carrying out epidemiological studies.

##### Soviet and Russian workers

Historically, the Soviet nuclear industry developed mainly around the Mayak nuclear complex, where the first plutonium reactor was built in January 1947. Mayak is located east of the Urals, some 1,500 km from Moscow. The site discharged liquid radioactive waste into the Tetcha River, a right-bank tributary of the Isset and a sub-tributary of the Ob.

The incidence and mortality of cerebrovascular disease (CVD) were studied in a cohort of 22,377 workers, first employed at the Mayak Production Association (Mayak PA) in 1948–1982 and followed up until the end of 2008. The ERR/Gy were 0.46 (95% CI 0.37, 0.57) and 0.28 (95% CI 0.16, 0.42), respectively, based on a linear dose–response model for total absorbed external gamma radiation doses above 0.1 Gy and total absorbed liver doses above 10 mGy, delivered by alpha particles emitted by incorporated plutonium. These updated results provide good

---

2 Predisposing factors are those factors that increase a person's risk of developing a given disease.

evidence of a linear trend in the risk of incidence of cerebrovascular disease (CVD) with an external dose of gamma radiation [AZI 14].

The new Mayak Worker Dosimetry System 2008 (MWDS-2008) was published in 2013. A cohort of 18,856 workers, first employed at one of the main Mayak PA plants between 1948 and 1972 and followed up until 2005, was identified. Significant associations were observed between external gamma radiation doses and the incidence of ischemic heart disease (IHD) and CVD, as well as between internal alpha radiation doses and IHD mortality and CVD incidence [MOS 14].

## Uranium miners

Mortality and cancer incidence in a cohort of 17,660 Eldorado uranium miners employed for the first time in 1932–1980 and followed up until 1999 have been studied. The results are homogenized in terms of working level months (WLM). Individual cancer rates were also reduced, with the exception of lung cancer mortality (SMR = 1.31, $P < 0.001$) and incidence (SIR = 1.23, $P < 0.001$). The ERR per 100 WLM (ERR/100 WLM) for lung cancer mortality (N = 618, ERR/100 WLM = 0.55, 95% CI 0.37, 0.78, $P < 0.01$) and incidence (N = 626, ERR/100 WLM = 0.55, 95% CI 0.37, 0.81, $P < 0.001$) increased linearly with increasing exposure to radon decay products (RDPs) [LAN 10].

It is well established that high radon exposure increases the risk of lung cancer mortality. The risk of lung cancer mortality at low radon exposures (< 100 working months) was assessed in a joint cohort analysis of Czech, French and Canadian uranium miners, employed in 1953 or later. A statistically significant linear relationship between radon exposure and lung cancer mortality was found. The ERR per month of work level was 0.022 (95% CI 0.013 to 0.034), based on 408 lung cancer deaths and 394,236 person-years of risk [LAN 19].

A two-stage biological model of carcinogenesis was applied to epidemiological data on lung cancer mortality in a large cohort of uranium miners from the WISMUT company (Germany). This cohort comprised 35,084 workers, 461 of whom died of lung cancer during the follow-up period 1955–1998. The ERR model calculations for lung cancer deaths are in good agreement with those of the BEIR VI-type descriptive exposure-age-concentration model for the Wismut miners [VAN 11].

The results obtained with the Wismut cohort were analyzed using the two-stage clonal expansion (TSCE) model. A total of 2,996 lung cancer deaths among the 58,695 male workers were observed during the follow-up period between 1946 and 2003, which is a high mortality. In comparison to the ERR model, the TSCE model predicts a significantly greater risk for low exposure rates below 50 WLM per year. This suggests that a nonlinear response mechanism in clonal expansion, with some level of saturation at high exposure rates, could play a crucial role in the development of lung cancer after alpha particle irradiation [ZAB 16]. Similarly, in French uranium miners, a relationship between radon exposure and lung cancer risk has been observed [VAC 09].

### The INWORKS cohort

Nuclear workers chronically exposed to low doses of ionizing radiation are subject to epidemiological monitoring, the results of which are regularly updated. The "15 countries" study [CAR 07], involving over 400,000 workers who received an average cumulative whole-body dose of 19.4 mSv, showed a non-significant linear relationship between dose received and mortality from leukemia (excluding chronic lymphocytic leukemia, CLL). A significant excess risk was found for all-cause mortality.

INWORKS (International Nuclear Workers Study), which combines cohorts of French, British and American workers, has a longer follow-up than the "15 countries" study (27 years on average, compared with 13 years). However, information on co-factors and other exposures to ionizing radiation (medical, etc.) is not specified, and workers exposed to neutrons or through internal contamination have not been excluded (unlike in the "15 countries" study). The characteristics of the INWORKS cohort are provided by Hamra et al. [HAM 15].

The cohort comprised a total of 308,297 workers, including 40,035 women. The collaborating industrial establishments were the *Commissariat à l'énergie atomique*, *AREVA Cycle nucléaire* and *Compagnie nationale d'électricité* in France, the Departments of Energy and Defense in the USA, and nuclear industry employers included in the National Register of Radiation Workers in the UK. The number of workers reported as exposed to neutrons was 13% for the whole cohort, with 15% of the cohort in France, 12% of the cohort in the UK and 14% in the USA. The overall mean annual individual dose equivalent, including zero doses, was 1.73 mSv (with the median at 0.42 and an interquartile range (IQR) of 0.07, 1.59) [THI 15].

The INWORKS cohort provided 8.2 million person-years of observation from a combined follow-up period (earliest in 1944 and latest in 2005) of workers from France, the UK and the USA. The average absorbed dose to the colon was 17.4 mGy. Significant risk heterogeneity was evident in chronic myeloid leukemia with time since exposure, where Daniels et al. [DAN 17] observed an increase in estimates of ERR per Gy soon after exposure (2 to 10 years) and again later (20 to 30 years).

Doses were accumulated at very low rates (mean 1.1 mGy per year, SD 2.6). The ERR of leukemia mortality (excluding CLL) was 2.96 per Gy (90% CI 1.17–5.21; lagged two years), notably due to an association between radiation dose and chronic myeloid leukemia mortality (ERR/Gy 10.45, 90% CI 4.48–19.65) [LEU 15]. This study is the first to report late-onset leukemia due to prolonged occupational exposure to radiation.

Of 66,632 known deaths at the end of follow-up, 17,957 were due to solid cancers. The estimated mean cumulative dose to the colon in exposed workers was 20.9 mGy (median 4.1 mGy). The estimated mortality rate for all cancers except leukemia increased with cumulative dose by 48% per Gy (90% CI 20–79%), with a 10-year lag. Similar associations were observed for mortality from all solid cancers (Average 47%, Range 17 to 79%) and within each country [ICR 15].

A statistically significant association between radiation dose and all non-cancer causes of death was observed. The ERR per sievert ($ERR.Sv^{-1}$) was 0.19 (90% CI 0.07, 0.30). This was largely due to the association between radiation dose and circulatory disease mortality ($ERR.Sv^{-1}$ 0.22; 90% CI 0.08, 0.37). Within circulatory disease subtypes, associations with dose were observed for cerebrovascular disease mortality ($ERR.Sv^{-1}$ 0.50; 90% CI 0.12, 0.94) and IHD mortality ($ERR.Sv^{-1}$ 0.18; 90% CI 0.004, 0.36) [GIL 17].

#### French nuclear workers

Various sources of bias appear in the analysis of cancer risk associated with external exposure to ionizing radiation in the CEA–AREVA–EDF cohort of nuclear workers. This cohort included 59,004 employees. The first difficulty was estimating internal contamination. A level of internal contamination risk for each worker was therefore incorporated into the analyses of cancer mortality risk linked to external exposures. Non-occupational exposure of workers to ionizing radiation, notably environmental exposure of natural origin and medical exposure due to radiological examinations carried out as part of occupational medicine, was poorly assessed. The third bias concerned dosimeter registration thresholds, as doses below these are poorly quantified [FOU 17].

### 11.3.1.6. *Young people's exposure to nuclear facilities*

Young people are more radiosensitive than adults to the harmful effects of ionizing radiation. This is particularly true of leukemias. Although the overall incidence is rare, leukemia is the most common type of childhood cancer. It accounts for 30% of all cancers diagnosed in children under 15. Within this population, acute lymphocytic leukemia (ALL) occurs about five times more frequently than acute myelogenous leukemia (AML) and accounts for around 78% of all childhood leukemia diagnoses. Only one environmental risk factor (ionizing radiation) has been significantly linked to ALL or AML [BEL 07].

## Excessive childhood leukemia in the Beaumont-Hague area

Ionizing radiation has long been known to cause cancers, and leukemias in particular [UNS 00a]. Leukemias have shorter latency periods than most cancers, with the first appearances occurring two to three years after irradiation, and the maximum occurring after seven to eight years. The younger the age of exposure, the greater the relative risk. Every year in France, 470 new cases of leukemia are detected, and 75 deaths due to leukemia occur in children, for a population of around 12 million children (0–14 years) [BER 08]. There are many potential risk factors, including ionizing radiation (external or internal exposure), but the cause is not certain. Numerous epidemiological studies have been carried out in the vicinity of nuclear facilities. Bernier et al. [BER 08] analyzed the results from 198 sites where a result was available. In 179 sites, no excess of leukemia was observed. In 13 sites, an excess of leukemias was reported but not confirmed. In three sites, excesses were possible: in France at La Hague and in the UK at Aldermaston and Burghfield. Excesses were confirmed at three sites: Sellafield and Dounreay in the UK, and Kruemmel in Germany.

In the Nord-Cotentin region of France, four cases of leukemia in young people (aged 0 to 24) were recorded between 1978 and 1992, compared with the French average of 1.4 cases. This difference is borderline significant. These cases of leukemia have been attributed to the radioactivity released into the environment by the region's nuclear facilities, which contaminate beaches and seafood [VIE 90, VIE 95, GUI 97, POB 97]. The Ministers of the Environment and Health therefore decided to set up a group of experts to assess the risk of leukemia and cancer due to nuclear facilities, called the *Groupe Radioécologie Nord-Cotentin* (GRNC).

The Nord-Cotentin region is home to four facilities with nuclear activities. Firstly, the COGEMA (now AREVA NC, then ORANO) spent fuel reprocessing plants at La Hague, 20 km west of Cherbourg, at the north-western tip of the Cotentin peninsula. Then there is ANDRA's Manche disposal center, located at the tip of the Cotentin peninsula (Beaumont-Hague administrative district) on a surface area of around 15 ha. Commissioning began in 1991, was completed in 1997, and the center has now entered the monitoring phase (2003).

The region has one EDF power plant, Flamanville, located 21 km southwest of the port of Cherbourg, on the west coast of the Cotentin peninsula. At the time of the leukemia outbreak in young people, the plant comprised two pressurized water nuclear units, each with an electrical capacity of 1,300 MWe. The first unit was connected to the grid in December 1985, and Flamanville No. 2 in July 1986.

Lastly, *Direction des constructions navales* (DCN, now Naval Group) has a major nuclear submarine construction and maintenance business (including commissioning, dismantling and deconstruction) at the Cherbourg military port.

Around all nuclear facilities, regulations provide for the monitoring of radioactivity in the environment. The aim of these measurements is to verify a posteriori that the radioactivity released is dispersed in accordance with forecasts, and that there is no abnormal increase in environmental radioactivity levels over time. This regulatory monitoring of the environment is carried out on an essentially predictive basis. Indeed, in the event of an incident or accident, it is estimated that maximum radioactivity will be found at certain sites, and it is these that are monitored. In the interests of greater efficiency of the resources allocated to protection, this monitoring must evolve over the life of a nuclear facility. In addition, recent regulatory developments mean that environmental monitoring systems are increasingly confronted with the need to produce realistic assessments of the health impact of industrial releases, rather than simply monitoring them a posteriori. With this in mind, environmental monitoring measures are no longer just used to control releases, as well as to interpret the results of epidemiological studies. This increases the need for realistic data production. It is therefore no longer possible to limit ourselves to conservative assessments.

| Exhibition | Dose (h.Sv) | Percentage |
|---|---|---|
| Natural ($^{210}$Po) | 241 | 74 |
| Medical | 76 | 24 |
| Nuclear testing | 5 | 2 |
| BNI | 0.3 | 0.09 |

**Table 11.2.** *Comparison of ex utero exposure levels from different sources in the Beaumont-Hague area. BNI: basic nuclear infrastructure (according to [GRN 99])*

In the Nord-Cotentin region, the collective ex utero risk of leukemia due to BNIs (for the 43 generations from 1954 to 1996 aged between 0 and 24 from 1978 to 1996) was 0.0009 cases (87% due to marine exposure). This same risk due to all exposures was 0.835 cases (74% due to natural exposures, 24% to medical exposures and 0.1% to BNIs) (Table 11.2). Nuclear incidents at the La Hague plant increased the risk by 0.0001 cases for offshore pipe rupture and 0.0004 cases for silo fire. The collective in utero risk of leukemia was estimated at 0.0003 cases. These values can be compared with those obtained in geographical areas subject to the same type of nuclear pollution (Table 11.3).

| | Thurso (Dounreay) | Seascale (Sellafield) | Beaumont-Hague (Nord-Cotentin BNI)* |
|---|---|---|---|
| Documentary source | Dionan et al. [DIO 86] | COMARE [COM 96] | GRNC [GRN 99] |
| Period | 1950–1984 | 1945–1992 | 1976–1996 |
| Workforce | 4,550 | 1,348 | 6,656 |
| Person-years | 74,750 | 25,300** | 69,308 |
| Estimated number of radiation-induced cases in the cohort | 0.004[a] | 0.020[a] | 0.0012b |
| Risk of leukemia per 100,000 person-years | 0.005[a] | 0.079[a] | 0.002[b] |

**Table 11.3.** *Comparison of average risk estimates attributed to routine releases from facilities (in utero and ex utero doses) with those of previous British radioecological studies. *La Hague reprocessing plant, Flamanville power plant, Cherbourg arsenal. **approximation. a: leukemia + non-Hodgkin's lymphoma, relative risk model. b: leukemia, additive risk model*

Other examples of recognized excess cases of childhood leukemia exist in Great Britain near the Sellafield and Dounreay reprocessing plants, and in Germany near the Kruemmel power plant. Nevertheless, all the multi-site studies currently available, including those in France, show no overall increase in the frequency of leukemia in 0–14 or 0–24-year-olds in the vicinity of nuclear sites [BER 08].

The reporting of spatio-temporal clusters of non-infectious diseases is a long-established phenomenon. The seriousness of the diseases reported (often cancers) and the presence of potential risk factors in the environment (nuclear or industrial facilities, etc.) generate concern among the appropriate populations.

The number of aggregate reports is rising steadily in many countries (USA, France). However, for the majority of cases, no explanatory environmental hypothesis has been identified. This is mainly due to a statistical bias linked to the fact that a local population contains a limited number of individuals and that, on the contrary, the numbers of observed and expected cases are also very small [GER 05].

In the nuclear sector, 11 in-depth investigations of spatio-temporal clusters of childhood leukemias (La Hague, Sellafield, Woburn, Dounreay) failed to produce any convincing etiological indications. However, a few investigations of clusters have been fruitful. These include the phocomelia epidemic in Germany (Contergan:

thalidomide), adenocarcinoma of the vagina associated with estrogen stilbestrol therapy in the mother, respiratory cancers and tremolites in New Caledonia, angiosarcomas of the liver in vinyl chloride production plant workers, germ cell tumors of the testicles in aircraft repair workers and lung cancer in chloromethyl methyl workers [GER 05]. Since then, other relationships between a pollutant and the appearance of various cancers have been highlighted (chlordecone and prostate cancer, asbestos and various lung cancers such as pleural cancer, etc.).

#### 11.3.1.7. *Environmental exposure*

Populations can be exposed to ionizing radiation from environmental sources at higher levels than the global average. This is mainly due to the geological composition of soils, where primary rocks contain higher concentrations of uranium or thorium. This means that there are naturally highly radioactive areas, and to a lesser extent, radon emanating from granitic mountain regions.

##### Naturally highly radioactive areas

Inhabitants of global high background natural radiation areas (HBNRAs or HLNRAs, High-Level Natural Radiation Areas), such as Ramsar (in Iran), Guarapari (in Brazil), Orissa and Kerala (in India) and Yangjiang (in China), have lived in these areas for generations under extraordinary radiation field conditions. The failure of previous epidemiological studies to report a substantial increase in cancer incidence in HBNRAs has raised some controversy as to the validity of the LNT hypothesis [ALI 15]. It should be noted, however, that the majority of these studies involve relatively small numbers of individuals, and that numerous biases mar these studies.

The effective dose equivalents in these areas of very high Ramsar background radiation, particularly at Talesh Mahalleh, are several times higher than the dose limits for radiation workers. Inhabitants living in this area receive annual doses of up to 132 mSv from external terrestrial sources and show no health drift, as if they were developing a radio-adaptive response to high levels of background radiation. The authors wrote: “Futility of any urgent limiting regulations is discussed” [MOR 02]. This is an unscientific sentence. Curiously, the second author of this publication signed another scientific article two years later, contradicting the first. Thus, in Ramsar, a coastal town in northern Iran overlooking the Caspian Sea, the average whole-body dose received by the population in these areas is around 5 times higher than the normal level of background radiation. The results of Gnassi-Nejad et al.’s study [GHI 04] showed a significant increase in CD69 expression on TCD4+-stimulated cells ($p < 0.004$) and a significant increase in total serum IgE ($p < 0.05$), as well as a higher incidence of stable and unstable chromosomal aberrations in the HLNRA group compared with the control group.

### Sites radioactively contaminated by human activities

Near the Mayak nuclear complex (former USSR), the Tetcha River cohort, composed of individuals residing in riverside villages between 1950 and 1961, when discharges from the Mayak plutonium production complex contaminated the river, nearly half of the 72 cases of leukemia (excluding CLL) were associated with radiation exposure. These data are consistent with a linear dose response. The ERR estimate was 0.22 per 100 mGy [KRI 15]. Solid cancer mortality in a cohort of 29,730 people living along the Tetcha River between 1950 and 1960 showed a linear ERR/Gy of 0.61 (95% CI 0.04–1.27, p = 0.03) [SCH 13].

### Radon-enriched areas

In a review, the WHO [WHO 09] noted that the number of lung cancer cases increased proportionally with increasing radon exposure. Several studies have been carried out on the associations between childhood cancer and natural background radiation. This organization reported on three epidemiological studies (Chinese, North American and European) that had focused on residential radon exposure, and were analyzed together. Overall, they showed a tendency for the risk of developing lung cancer to increase with increasing radon concentration in the home, although the results were not always significant. The weighted average increase in the risk of developing lung cancer per 100 $Bq.m^{-3}$ of radon was estimated at 10% [WHO 09].

Since then, two case-control studies in England [KEN 13] and Switzerland [HAU 13, SPY 15] have investigated the relationship between the occurrence of childhood cancers and the level of natural domestic radioactivity (cosmic and terrestrial, including radon). In these studies, exposure is estimated on the basis of national surveys, using an average regional gamma dose rate and radon concentration. Co-exposures and risk co-factors are not reported, with the exception of socio-economic status. In England, a significant dose–response relationship was found between bone marrow dose from gamma radiation and the risk of developing leukemia. The same is true in Switzerland for leukemia, as well as for lymphomas, tumors of the central nervous system and all types of cancer combined. This study used the external dose rate as the exposure variable, with an average of 109 $nSv.h^{-1}$ [SPY 15]. No relationship between radon exposure and childhood-onset cancers was found in Switzerland [HAU 13], in contrast to England, where there was a 12% (95% CI 3, 22; two-tailed p = 0.01)[3] ERR of childhood leukemia per millisievert of cumulative red bone marrow dose from gamma radiation [KEN 13].

Radon is the second leading cause of lung cancer, ranked by the World Health Organization as the fifth leading cause of death in 2010. An updated database of

3 A two-tailed test is associated with the alternative hypothesis that the sign of the potential difference is unknown.

national radon exposures for 66 countries is used to estimate the global burden (population attributable risk, PAR) of lung cancer mortality attributable to radon. These estimates in 2012 ranged from 4.2% (95% CI 0.9, 11.7) for Japan, to 29.3% (95% CI 22.9, 35.7) for Armenia, with a median for the 66 countries of 16.5%. Lung cancer deaths attributable to radon for the 66 countries totaled 226,057 cases in 2012 [GAS 18].

### 11.3.1.8. *Cardiovascular disease and cataracts*

The association between lower-dose exposures and late-onset cardiovascular disease has only recently begun to emerge and remains controversial. One possible biological mechanism is endothelial cell damage and the subsequent induction of an inflammatory response. An arguably more plausible mechanism for fractional effects at low doses, based on the destruction of monocyte cells in the intima, is proposed by Little et al. [LIT 10].

Estimated population excess risks for all circulatory diseases combined ranged from 2.5%/Sv (95% CI 0.8, 4.2) for France, to 8.5%/Sv (95% CI 4.0, 13.0) for Russia. Radiation-related mortality is roughly double that currently estimated on the basis of cancer assessment criteria alone (which range from 4.2% to 5.6%/Sv for these populations) [LIT 12]. This doubling of mortality is due to the onset of circulatory diseases.

Exposure to therapeutic doses of ionizing radiation is associated with damage to the heart and coronary arteries. The results of a meta-analysis suggest that the ERRs per unit dose for various types of heart disease do not show statistically significant heterogeneity ($p > 0.2$) between studies. The review by Little [LIT 16] provides strong evidence in support of a causal association between both low- and high-dose radiation exposure and most types of circulatory disease.

It has long been known that exposure to ionizing radiation at relatively high doses (1 Gy) can induce cataracts, but few data existed for low doses (< 100 mGy). However, among American radiobiologists, an ERR per mGy of $0.69.10^{-3}$ (95% CI $0.27.10^{-3}$ to $1.16.10^{-3}$, $p = 0.001$) was observed [LIT 18].

In nuclear workers, there were 11,345 cataract diagnoses and 5,440 cataract surgeries, during 832,462 and 888,402 person-years of follow-up, respectively. Cumulative occupational radiation exposure was associated with self-reported cataracts, but not cataract surgery, with an EAR/$10^4$ person-year per Gy of 94 (95% CI 47 to 143, $p < 0.001$) and an EAR/$10^4$ person-year per Gy of 13 (95% CI < 0 to 57, $p = 0.551$) [LIT 20].

#### 11.3.1.9. *Illnesses following a nuclear accident*

Epidemiological studies monitoring the consequences of major nuclear accidents were reviewed in volume 2 of this series [AMI 19b]. One of the consequences is an increased risk of early-onset and late-onset thyroid cancer in children who encountered the radioactive cloud in the early days of the nuclear crisis. This consequence is mainly linked to the release of several radioactive iodine isotopes in the atmospheric fallout. Large-scale ultrasound screening in Japan's Fukushima prefecture showed a high detection rate of thyroid cancer in young people, revealing 116 and 71 cases, respectively, at the first and second thyroid ultrasound series, among the same cohort of around 300,000 subjects [YAM 18].

### 11.3.2. *Controversies over responses to low doses*

According to the UNSCEAR [UNS 93], exposure to ionizing radiation is classified as low dose (0 to 0.2 Gy), medium dose (0.2 to 2 Gy), high dose (2 to 10 Gy) and very high dose (over 10 Gy). Other authors have proposed other limits between low and medium doses, such as 0.5 or 0.1 Sv.

Due to their very nature, stochastic (random) radiation-induced effects such as carcinogenic and hereditary effects occurring at lower radiation doses are more difficult to control, and protection is less obvious [AVE 99]. In the case of low doses of radiation, epidemiological studies cannot in some cases provide any usable results, as they are at the limit of the significance of excess damage due to irradiation compared with naturally occurring damage [ANC 12], notably because of the problem of cofactors.

The effects of low doses can be approached with the help of results from experimental disease models (animal studies and cell cultures) providing information on the biological mechanisms responsible. Molecular biology makes it possible to identify DNA damage, and repair mechanisms (hormesis and apoptosis) exist, although there is no guarantee of 100% repair.

Based on the morbidity results recorded in populations generally irradiated at moderate to high doses (epidemiological studies), it is possible to extrapolate the dose–response relationship towards low doses according to four curves (Figure 11.1). The four scenarios are (1) linear without threshold, (2) infra-linear, (3) supra-linear and (4) linear with threshold. Scenario 2 involves hormesis and is optimistic. Scenario 3 involves the proximity effect and is pessimistic. Scenario 4 involves a threshold and is also optimistic. To be on the safe side, the responsible bodies (ICRP, WHO, etc.) have decided to extrapolate the dose–response relationship linearly.

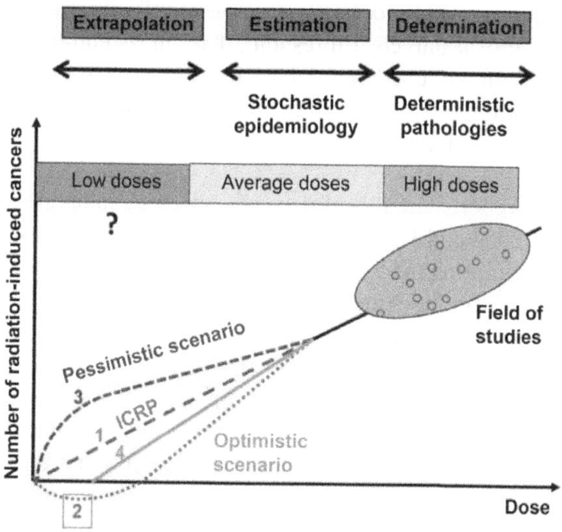


**Figure 11.1.** *Diagram illustrating the principle of dose–response extrapolation from high-dose data (modified from [AMI 13]). 1. Linear without threshold; 2. infra-linear (hormesis); 3. supra-linear (proximity effect); 4. linear with threshold*

The relationship between low doses of radiation and their consequences in terms of morbidity and mortality has been the subject of much controversy, as many questions remain unanswered. What is the relationship between adverse effects and low levels of radiation? Is there a threshold or not? Is the dose-effect relationship linear or not? Can the hormesis phenomenon be generalized? What are the consequences of bystander effects, i.e. stress on cells adjacent to the irradiated cell, on the onset of morbidity? These essential questions will be developed in the following sections.

#### 11.3.2.1. *Responses with or without thresholds: controversy*

Masse [MAS 00] considers that the choice of linearity without threshold is mainly due to a weakness in dosimetric estimates and that once this poor quantification of doses has been overcome, another alternative will be possible. In 2002, Masse [MAS 02] emphasized that no pathogenic effects are known to be induced in humans by low dose rates up to 100 mSv per year, and that the effects attributed to low exposures are the result of extrapolations. Moreover, many adverse effects are attributed to exposure to ionizing radiation when they are most likely due to real health effects in distressed populations, and are attributable to social disorganization, deficiencies and possibly other environmental factors. The ICRP report [ICR 05] concludes that, while the existence of a threshold at a low radiation dose does not seem improbable for cancers of certain tissues, the evidence does not support the existence of a universal threshold. The linear non-threshold hypothesis (LNT), combined with an uncertain DDREF (dose and dose-rate effectiveness factor) for extrapolation from high doses, remains a cautious basis for radiation protection at low doses and low dose rates.

However, three reports on the effects of low-dose radiation were published between December 2004 and July 2005 [TUB 06a, TUB 06b]. These were the ICRP report [ICR 05], the joint report by the French Academies of Science and Medicine [TUB 05] and a report by the American Academies of Science [NRC 05]. Despite the fact that these reports are based on the same studies of the biological effects of low doses, their conclusions differ significantly. The French report concludes that recent biological data show that the efficacy of defense systems is modulated by dose and dose rate, and that the LNT relationship is not plausible. Conversely, the ICRP and BEIR VII (NRC) reports, while acknowledging that there are biological arguments against the LNT relationship, consider that there is insufficient biological evidence to counter and change the current risk assessment methodology and regulatory policy based on the LNT.

Tubiana et al. [TUB 07] believe that multi-cellular organisms are more effective in their defenses against ionizing radiation at low doses (oxidative stress, apoptosis, etc.) than at high doses. However, they acknowledge the lack of mechanistic explanations for phenomena such as hypermortality at low doses or adaptive response (i.e. better tolerance to irradiation doses). These authors consider that epidemiological and experimental data challenge the validity of the LNT hypothesis for assessing the carcinogenic effect of low doses, but do not rule it out. Therefore, the main criteria for choosing the most scientifically reliable dose–response relationship should be based on biological data. Their analysis should help to understand the current controversy.

However, the controversy is becoming increasingly fierce. In February 2015, for example, a petition was submitted to the US Nuclear Regulatory Commission (NRC) to reject the LNT hypothesis as the basis for radiation protection regulation and to promote the use of threshold and hormesis evidence [GLA 18]. There are many authors who refute the LNT hypothesis (e.g. [CAR 18]). It should be noted that this controversy concerns only the appearance of cancers. However, the effects of irradiation are not limited to this type of morbidity. This is particularly true of cardiovascular disease, where the relationship between low-dose ionizing radiation (LDIR) and this pathology has not been fully elucidated [DIN 17]. These conflicts demonstrate that our knowledge of the effects of LDIR is far from complete [BAL 15].

#### 11.3.2.2. *Linear or linear-quadratic or nonlinear responses?*

The relationship between dose and effect at the cellular level can take various forms: linear, linear-quadratic or nonlinear. The linear response to stress assumes a constant increase in adverse effect as stress increases. However, many toxicological phenomena can be described by "S"-shaped dose–response curves (also known as sigmoidal curves), with a threshold of detectability at low stress levels, followed by a proportional increase in effect, before reaching a saturation point. Most of this curve can be described by an exponential of a second-degree polynomial. The formula relating effect to dose is the linear-quadratic model [BOD 16]. Recently, nonlinear dose–response relationships have been described mainly in the case of endocrine disruption [AMI 17b]. However, in the case of ionizing radiation stress, only linear and linear-quadratic relationships have been reported.

#### 11.3.2.3. *Radio-adaptation or hormesis*

The term hormesis is used to describe a stimulation of biological defenses in organisms exposed to low doses of various physical or chemical stressors, which is obviously likely to have an impact on the dose–response curve. However, it has not been fully established whether hormesis is a common and important phenomenon. Indeed, the idea that low-dose effects can be null is variously accepted, but that the effect of low doses is positive is for many authors doubtful. One of the fields in which hormesis is best studied is that of radiation effects [AGA 18].

Hormesis is based on the concept that a low dose of radiation received beforehand can protect the victim. This phenomenon corresponds to an "adaptive" response. Radiation adaptation occurs when cells subjected to a low prior irradiation dose subsequently withstand higher irradiation doses without developing harmful effects. A low dose (a few tens of mSv), delivered a few hours before a higher-dose irradiation, reduces the effect (particularly mutagenic) of this second dose, and could therefore reduce the risk of carcinogenesis. Rodgers and Holmes [ROD 08] observed this phenomenon of radio-adaptation in mice subjected in situ to various doses of irradiation in the Chernobyl exclusion zone. For their part, Møller and Mousseau [MOL 16] analyzed 17 studies and found that only one experimental study met the criteria for evolutionary adaptation. Finally, they concluded that no evidence of hormesis could be detected. The lack of evidence of adaptation stems mainly from the absence of replication and rigor in experimental design. In particular, most studies were based on transplants with organisms from a limited number of sites.

The mechanisms responsible for radiation adaptation are complex, involving more intense activation of repair systems, the action of certain cytokines (interleukins I, etc.) and slowing of the cell cycle. The interpretation would be that the initial low dose has triggered repair processes, thereby limiting the toxic effects of the high dose delivered later. To what extent can this phenomenon lead to "protection" in the human clinic? To date, the answer is not precisely known.

The literature on hormesis is abundant. However, the phenomenon is also controversial. Some authors consider the phenomenon to be important. Under certain conditions, according to Shibamoto and Nakamura [SHI 18a], low-dose radiation can have beneficial effects and the LNT theory may be outdated. Kudryashevaa and Rozhkob [KUD 15] studying the bioluminescent response of microorganisms to americium-241 or tritium irradiation observed three successive stages in this response: a first stage with no effects (stress recognition); a second stage with activation (adaptive response) and a third stage with inhibition (suppression of physiological function, i.e. irradiation-induced toxicity). If the second stage can be likened to hormesis, it is cancelled out by the third stage.

On the contrary, Sugie et al. [SUG 16] consider that they do not have sufficient data to assert a hormesis phenomenon on human salivary cells. Moreover, the concept of hormesis must be applied with caution, as the hormesis stimulus may act without threshold on pre-damaged or atrophic tissues, or in synergy with other noxious agents. Experimental evidence in favor of hormesis is considerable, but further studies are needed [JAR 17]. In a recent review, Beyea [BEY 17] considers the linear, no-threshold relationship to be the most plausible hypothesis.

In their opinions, the US National Research Council (a part of the NAS), the National Council on Radiation Protection and Measurements (NCRP, a body created by the US Congress) and the UNSCEAR believe that any stimulation of the hormesis phenomenon at low doses of ionizing radiation has a significant health benefit for the population, exceeding the potential harmful effects of irradiation, and is an unjustified assumption [NRC 06], and this position remains valid to this day.

However, the debates remain very lively and several scientists still support the hormesis hypothesis. For example, Vaiserman et al. [VAI 18] write that on the whole, although it probably has not yet been proven wrong, LNT certainly has not been proven right. At this point, given the high price (in economic and human terms) borne by LNT-inspired regulation, there can be little doubt that the current regulatory burden should be reduced. These economic aspects are obviously linked to the nuclear industries.

Various studies report adaptive responses to low doses of ionizing radiation. These highly conserved evolutionary defense mechanisms determine the cellular response to chemical and physical aggression. They include DNA damage repair (p53, ATM, PARP pathways), the antioxidant response (Nrf2 pathway), the immune/inflammatory response (NK-κB pathway), the cell survival/death pathway (apoptosis), the endoplasmic stress response (unfolded protein response) and other cytoprotective processes, including autophagy, cell cycle regulation and the unfolded protein response [GUÉ 19]. In mice contaminated with low doses of uranium or fluorine, Bontemps-Karcher et al. [BON 20] identify adaptive renal responses to these two elements at doses of 20 and 50 $mg.L^{-1}$, probably through reduced apoptosis and recruitment of inflammatory cells.

The conservative nature of DNA repair suggests that cells may have retained some ability to repair damage caused by higher levels of radiation than exist today. The existing evidence in favor of hormesis is substantial, with experimental data partly at odds with the results of epidemiological studies [JAR 18]. Many recent biological studies have suggested the induction of positive responses such as increases in immunity and antioxidants by LDIR [SHI 18a].

How can radiation hormesis be explained? While hormesis has beneficial effects on biological systems subjected to low doses of ionizing radiation, the molecular mechanisms are not understood [SHA 19, KAB 20].

In rats exposed to X-rays, levels of reduced glutathione (GSH) and total glutathione (TG) increased in liver, kidney, brain and blood at 6 hours and 24 hours post-exposure. In contrast, no significant changes in oxidized glutathione (GSSG) content were observed after X-ray irradiation in any of the organs. In addition, low-dose X-rays significantly decreased lipid peroxidation (LPO) in liver, kidney and brain, while increasing LPO levels in blood. The enzymatic activities of catalase (CAT) and glutathione-S-transferase (GST) also increased in various organs after X-ray exposure. In addition, low-dose X-ray irradiation significantly increased the number of total leukocytes, lymphocytes and eosinophils, while decreasing the number of neutrophils and monocytes. Consequently, our results clearly indicate that low-dose X-ray exposure stimulates the endogenous antioxidant defense machinery and also causes an increase in whole-blood lymphocytes and eosinophils responsible for providing key defenses [SHA 19].

Vuillez [VUI 19] reminds us of some important concepts that are not well known because they are recent, but which are very likely to shed useful light on the debate concerning the effects of low and very low doses, and above all to introduce coherence into it. The effect of ionizing radiation, at low doses, is ambivalent and must be interpreted through a dynamic balance between two phenomena acting in opposite directions: on the one hand, the contribution of radiation to potentially deleterious lesions, and on the other hand, the stimulation by this same radiation of defense mechanisms, notably immunological, against the appearance of tumors.

Adaptive responses of cells to stress such as ionizing radiation include damage prevention through increased antioxidants, DNA repair, damage suppression through apoptosis and immunological functions (immuno-surveillance) [SAN 10]. Sanders

[SAN 17] provides examples of hormesis for breast cancer mortality in Canadian tuberculosis patients, and for leukemia deaths in survivors of the Hiroshima and Nagasaki bombs.

The crucial question is whether the up-regulation of the immune system or other forms of adaptive response that can result from a radiation dose compensates for the undeniable carcinogenic damage that results. Although the adaptive response modulating the effect of relatively high radiation doses (of several Gy) following a lower initiation dose (usually of at least several tens of mGy) is well known experimentally (mainly in vitro), it is not universally observed in all experimental systems, nor does it persist for more than a few days, and there is little or no evidence of its involvement at low initiation doses [DOS 14].

In conclusion, the main limitations of hormesis stem from the biases of the studies supporting this phenomenon. This is due in particular to the fact that they do not concern humans, but other mammals, or cell cultures. They are therefore not representative of the actual reaction of an entire individual.

#### 11.3.2.4. *Pre-exposure*

It has been hypothesized that pre-exposure to low doses triggers various defense mechanisms and enables individuals to better resist higher doses of irradiation. Various experiments have been carried out on cells and mammals.

Human keratocyte cells (fragments of red blood cells) were pre-exposed to 5 mGy or 0.5 Gy for 24 h, then exposed to 0.5 Gy or 5 Gy. Their clonogenic survival indicated an adaptive response induced by irradiation [MAG 07]. Plateau-phase mouse embryo cells were pre-exposed to an adaptive dose of 0.1–1.5 Gy of low-dose-rate $\gamma$-radiation 3.5 h before an acute challenge dose of 4 Gy. Regardless of dose, no adaptation improved clonogenic survival detectably. However, all chronic adaptation doses resulted in both a reduction in the frequency of micronuclei in binucleated cells and an approximately twofold reduction in the frequency of neoplastic transformation per viable cell. In this case, low-dose-rate pre-exposure to ionizing radiation induces an adaptive response in cells [AZZ 94].

In mice, low-level radiation-induced hormetic and adaptive responses in the hematopoietic and immune systems were observed, as evidenced by stimulatory effects on cell growth and resistance to subsequent radiation-induced cytogenetic damage. However, in terms of cell death by apoptosis, the effects of low-intensity radiation are controversial. The effects of low-level radiation at doses up to 200 mGy on male mouse germ cells in terms of apoptosis and expression of apoptosis-related proteins, were examined at different time points, following whole-body exposure of mice to low levels of radiation. The results showed that low-level radiation in the dose range 25 to 200 mGy induced significant increases in apoptosis in spermatogonia and spermatocytes, with a maximal effect at 75 mGy. Increased apoptosis is most likely associated with Trp53 protein expression. These results suggest that low-level radiation induces increased apoptosis in male germ cells, as well as induces a significant adaptive response that decreases cell death after subsequent high-dose irradiation [LIU 06].

However, this is not confirmed by a study of an adaptive response in rats pre-irradiated with chronic contamination (tritiated water, 7 or 35 cGy), then irradiated with fast neutrons (1 Gy). In radiosensitive tissues, no protective response on the antioxidant defense system of chronic pre-exposure to low doses and low linear energy transfers (LETs) has been demonstrated. This type of response has, however, appeared in the most radioresistant tissues [PET 02].

Male mouse brains were irradiated with 0.05 Gy ($^{12}C^{6+}$ or $^{60}Co$ ion) as a pre-exposure dose, then irradiated with 2 Gy as a stimulating irradiation dose 4 h after pre-exposure. Serum pituitary gonadotropic hormones, follicle-stimulating hormone (FSH) and luteinizing hormone (LH), testosterone, testicular weight, sperm count and shape were measured on day 35 post-irradiation. The authors found a significant reduction in serum levels of FSH, LH, testosterone, testicular weight and sperm count, and a significant increase in sperm abnormalities by irradiation of the mouse brain with 2 Gy. Moreover, the effects were more evident in the group irradiated with the $C^{126+}$ ion than in that irradiated with $^{60}Co$ gamma rays. Pre-exposure therefore considerably attenuated the harmful effects induced by subsequent high-dose irradiation [ZHA 06].

In rhesus macaque survivors of 700 cGy total irradiation, 60 days after radiation exposure, the authors found transcriptional changes independent of sex and cytokine treatment. Females had almost twice as many genes deregulated and appeared more radiosensitive than males. Bioinformatics analysis revealed an association with immunological processes and WNT proteins involved in embryogenesis and cancer for both sexes [MAJ 21].

Ionizing radiation interacts with the immune system in numerous ways, a multiplicity that reflects the complexity of the immune system itself. As a result, ionizing radiation fundamentally and permanently reshapes the immune system [LUM 21].

#### 11.3.2.5. *Non-target effects*

Non-targeted effects (NTE) are mainly bystander effects, genomic instability and epigenetics.

Recently, studies have shown that living organisms respond to ionizing radiation not only through direct reactions at the DNA level but also indirectly with NTE.

Bright and Kadhim [BRI 18] review the different types of NTE and their potential implications for radiobiology research and applications. The best-known NTE are genomic instability, neighborhood effects and epigenetic phenomena. Non-target effects need to be taken into account, and modeling, experimental and epidemiological approaches could all be used to determine the impact of non-target effects on the LNT model currently used in radiation protection [MOT 18].

The bystander effect corresponds to ionizing radiation-induced responses in directly irradiated cells, as well as in neighboring cells that have not been affected by ionizing radiation and the corresponding energy deposition. These neighboring cells react to a soluble signal secreted by the irradiated cells [COA 04, MOR 07, ERM 11].

The proximity effect is well demonstrated and predominates at low doses (0.2 Gy). At higher doses, the response saturates. This effect is of obvious interest for estimating the risk due to ionizing radiation [PRI 03]. Examples of the neighborhood effect have recently been published. For example, depleted uranium (DU) plays a role in both toxic and neoplastic "bystander effects" on human osteoblast cells (HOS) [MIL 17]. The proximity effect has been demonstrated in fish. Moreover, this effect is triggered even if only one parent is irradiated, and persists two generations after the irradiated generation [SMI 16].

The radiation-induced bystander effect (RIBE) presents potential risks to normal tissues in radiotherapy and confers a higher risk of low radiation doses than previously thought [WAN 15].

Genomic instability is the loss of DNA damage detection and repair mechanisms, and is certainly the original mechanism of oncogenesis. Primordial alterations occur at genomic (mutations) and epigenomic (transcription) levels, but can also occur as a result of functional alterations [ROB 21].

Irradiation-induced genome instability is well documented for directly exposed cells and organisms [MOT 16]. This instability has also been observed in neighboring but unirradiated cells. Enigmatically, increased instability is even observed in the offspring of pre-conceptually exposed animals, as well as in humans [MOR 11]. Twenty years after the Chernobyl accident, liquidators showed high genomic instability. The effects of irradiation are still evident at cytogenetic and molecular levels [MEL 07]. Moreover, genomic instability is one of the major limitations limiting the efficacy of targeted therapies in oncology [BIL 20].

In recent years, the discovery of epigenetic chromatin modifications, such as DNA methylation, following exposure to very low doses has been well documented [SCH 18].

NTE represent a paradigm shift away from the "DNA-centric" view that ionizing radiation only leads to biological effects and health consequences arising from the deposition of energy in the cell nucleus [MOR 15].

## 11.4. Conclusions

The vast literature review of human epidemiological studies suggests that exposure to LDIR (≤100 mSv) or low-dose-rate ionizing radiation (LDRIR) (< 6mSv/H) may induce negative or sometimes positive health effects.

Epidemiological and clinical studies show that exposure to LDIR or LDRIR can induce cancers, birth defects, cardiovascular and cerebrovascular diseases, cognitive and other neuropsychiatric disorders, cataracts and other ocular and somatic pathologies (endocrine, bronchopulmonary, digestive, etc.) [TAN 18].

In the United States, the NCRP widely disseminates information, guidance and recommendations on radiation protection and measurements that represent the consensus of cutting-edge scientific thinking. In 2017, the NCRP carried out an updated review of the results of epidemiological studies. This was necessary because a considerable number of radiation epidemiology study reports based on new or updated data have been published since other important reviews were carried out by national and international scientific committees. The NCRP reviews the strengths and weaknesses of the epidemiology, dosimetry and statistics of 29 low-LET radiation studies in terms of quality and degree of support for the LNT model. The Committee judged that the available epidemiological data were largely supportive of the LNT model, and that to date no other dose–response relationship appears more pragmatic or conservative for radiation protection purposes [NCR 17, SHO 18].

Other groups include underground hard rock miners who inhaled radioactive radon and its decay products, a large number of patients irradiated for therapeutic purposes, and workers who received high doses as part of the former USSR's nuclear weapons program. The degree of carcinogenic risk resulting from low levels of exposure is more controversial, but the available evidence points to an increased risk that is approximately proportional to the dose received [WAK 04].

For a long time, many epidemiological studies had limitations, and some contained biases. The major methodological problem was bias due to selection effects [GOL 08]. For example, the control group was not similar to the test group. On the contrary, it was not always possible to take confounding factors into account, and in particular it was sometimes impossible to collect adjustment data (age, sex, professions and socio-professional categories) for all subjects [LAU 00, SAM 18]. The major weakness of epidemiological studies is frequently the small size of the population studied. In recent years, this has been compensated for by the merging of several studies (meta-analyses).

## 11.5. References

[AGA 18] AGATHOKLEOUS E., "Environmental hormesis, a fundamental non-monotonic biological phenomenon with implications in ecotoxicology and environmental safety", *Ecotoxicology and Environmental Safety*, vol. 148, pp. 1048–1053, 2018.

[ALI 15] ALIYU A.S., RAMLI A.T., "The world's high background natural radiation areas (HBNRAs) revisited: A broad overview of the dosimetric, epidemiological and radiobiological issues", *Radiation Measurements*, vol. 73, pp. 51–59, 2015.

[AMI 13] AMIARD J.C., *Le risque radioactif. Devenir des radionucléides dans l'environnement et impacts sur la santé*, Lavoisier, Tec&Doc, Paris, 2013.

[AMI 17a] AMIARD J.C., *Les risques chimiques environnementaux. Méthodes d'évaluation et impacts sur les organismes*, 2nd edition, Lavoisier, Tec&Doc, Paris, 2017.

[AMI 17b] AMIARD J.C., COUDERC M., XUEREB B., "Perturbations des fonctions endocrines : causes et conséquences", in AMIARD J.-C., AMIARD-TRIQUET C. (eds), *Biomarqueurs en écotoxicologie aquatique*, Lavoisier, Tec&Doc, Paris, pp. 163–204, 2017.

[AMI 19a] AMIARD J.C., *Military Nuclear Accidents: Environmental, Ecological, Health and Socio-economic Consequences*, vol. 1. ISTE, London, and John Wiley & Sons, New York, 2019.

[AMI 19b] AMIARD J.C., *Industrial and Medical Nuclear Accidents: Environmental, Ecological, Health and Socio-economic Consequences*, vol. 2, ISTE, London, and John Wiley & Sons, New York, 2019.

[ANC 12] ANCCLI-InVS-IRSN, Santé publique à proximité des installations nucléaires : Comment aborder les questions posées, January 2012.

[AVE 99] AVERBECK D., "Les effets des faibles doses et des faibles débits de doses de rayons gamma", *SFRP National Conference*, Angers, 2–4 June 1999.

[AZI 14] AZIZOVA T.V., HAYLOCK R.G.E., MOSEEVA M.B. et al., "Cerebrovascular diseases incidence and mortality in an extended Mayak Worker Cohort 1948–1982", *Radiation Research*, vol. 182, no. 5, pp. 529–544, 2014.

[AZZ 94] AZZAM E.I., RAAPHORSTT G.P., MITCHEL R.E.J., "Radiation-induced adaptive response for protection against micronucleus formation and neoplastic transformation in C3H 10T1/2 mouse embryo cells", *Radiation Research*, vol. 138, pp. S28–S31, 1994.

[BAL 15] BALDWIN J., GRANTHAM V., "Radiation hormesis: Historical and current perspectives", *Journal of Nuclear Medicine Technology*, vol. 43, pp. 242–246, 2015.

[BEL 07] BELSON M., KINGSLEY B., HOLMES A., "Risk factors for acute leukemia in children: A review", *Environmental Health Perspectives*, vol. 115, no. 1, pp. 138–145, 2007.

[BER 08] BERNIER M.O., GRÉGOIRE E., JACOB S. et al., Les études épidémiologiques des leucémies autour des installations nucléaires chez l'enfant et le jeune adulte : revue critique, IRSN, DRPH/SRBE report, no. 2008-001, 2008.

[BEY 17] BEYEA J., "Lessons to be learned from a contentious challenge to mainstream radiobiological science (the linear no-threshold theory of genetic mutations)", *Environmental Research*, vol. 154, pp. 362–379, 2017.

[BIL 20] BILLAUD A., CHEVALIER L.M., CAMPONE M. et al., "L'instabilité génomique, paramètre limitant l'efficacité des thérapies ciblées en oncologie", *Bulletin du cancer*, vol. 107, no. 11, pp. 1161–1170, 2020.

[BOD 16] BODGI L., CANETA A., GRANZOTTO A., et al., "L'énigme de l'interprétation biologique du modèle linéaire-quadratique enfin résolue ? Une synthèse pour les non-mathématiciens", *Cancer/Radiothérapie*, vol. 20, pp. 314–321, 2016.

[BOI 17] BOICE, J.D., Jr., "The linear nonthreshold (LNT) model as used in radiation protection: An NCRP update", *International Journal of Radiation Biology*, vol. 93, no. 10, pp. 1079–1092, 2017.

[BON 20] BONTEMPS-KARCHER A., MAGNERON V., CONQUET L. et al., "Renal adaptive response to exposure to low doses of uranyl nitrate and sodium fluoride in mice", *Journal of Trace Elements in Medicine and Biology*, doi: 10.1016/j.jtemb.2020.126708, 2020.

[BRI 18] BRIGHT S., KADHIM M., "The future impacts of non-targeted effects", *International Journal of Radiation Biology*, vol. 94, no. 8, pp. 727–736, 2018.

[CAR 07] CARDIS E., VRIJHEID M., BLETTNER M. et al., "The 15-country collaborative study of cancer risk among radiation workers in the nuclear industry: Estimates of radiation-related cancer risks", *Radiation Research*, vol. 167, no. 4, pp. 396–416. 2007.

[CAR 18] CARDARELLI J.J., ULSH B.A., "It is time to move beyond the linear no-threshold theory for low-dose radiation protection", *Dose-Response*, vol. 16, no. 3, pp. 1–24, July–September 2018.

[CEA 16] CEA, "Les effets des faibles doses de rayonnements ionisants. Où en est-on de la compréhension des mécanismes ?", *ProsInfo*, no. 22, pp. 4–10, October 2016.

[COA 04] COATES P.J., LORIMORE S.A., WRIGHT E.G., "Damaging and protective cell signalling in the untargeted effects of ionizing radiation", *Mutation Research*, vol. 568, pp. 5–20, 2004.

[COM 96] COMARE (COMMITTEE ON MEDICAL ASPECTS OF RADIATION IN THE ENVIRONMENT). The incidence of cancer and leukaemia in young people in the vicinity of the Sellafield site, West Cumbria, Further studies and an update of the situation since the publication of the report of the Black Advisory Group in 1984, London, Department of Health (COMARE IV), 1996.

[DAN 17] DANIELS R.D., BERTKE S.J., RICHARDSON D.B. et al., "Examining temporal effects on cancer risk in the International Nuclear Workers' Study (INWORKS)", *International Journal of Cancer*, vol. 140, no. 6, pp. 1260–1269, 2017.

[DIN 17] DING C.Y., GAO L., "Low dose radiation exposure and cardiovascular diseases: A review", *International Journal of Cardiovascular Practice*, vol. 2, no. 4, pp. 76–79, 2017.

[DIO 86] DIONAN J., MUIRHEAD C.R., WAN S.L. et al., The risks of leukaemia and other cancers in Thurso from radiation exposure, Report NRPB-196, HMSO, London, 1986.

[DOS 14] DOSS M., LITTLE M.P., ORTON C.G., "Point/counterpoint: Low-dose radiation is beneficial, not harmful", *Medical Physics*, vol. 41, no. 7, p. 070601, 2014.

[ERM 11] ERMAKOV A.V., KONKOVA M.S., KOSTYUK S.V. et al., "An extracellular DNA mediated bystander effect produced from low dose irradiated endothelial cells", *Mutation Research*, vol. 712, pp. 1–10, 2011.

[FOU 17] FOURNIER L., Effets sanitaires d'une exposition chronique à de faibles doses de rayonnements ionisants : contribution à l'estimation des risques radio-induits de cancers dans une cohorte française de travailleurs du nucléaire, Thesis, Université Paris-Saclay, 2017.

[GAS 18] GASKIN J., COYLE D., WHYTE J. et al., "Global estimate of lung cancer mortality attributable to residential radon", *Environmental Health Perspectives*, vol. 126, no. 5, p. 057009, 2018.

[GER 05] GERMONNEAU P., TILLAUT H., GOMES D. et al., "Aspects méthodologiques de l'investigation des signalements d'agrégats spatio-temporels de maladies non infectieuses", *Environnement, Risques & Santé*, vol. 4, pp. 329–334, 2005.

[GHI 04] GHIASSI-NEJAD M., ZAKERI F., ASSAEI R.G. et al., "Long-term immune and cytogenetic effects of high level natural radiation on Ramsar inhabitants in Iran", *Journal of Environmental Radioactivity*, vol. 74, pp. 107–116, 2004.

[GIL 17] GILLIES M., RICHARDSON D.B., CARDIS E. et al., "Mortality from circulatory diseases and other non-cancer outcomes among nuclear workers in France, the United Kingdom and the United States (INWORKS)", *Radiation Research*, vol. 188, pp. 276–290, 2017.

[GLA 18] GLAIT H.M., SZENKIERMAN J., ECHEGOYEN L.A., "Controversia Hormesis o Modelo Lineal sin Umbral: Revisión o inacción", *Alasbimn Journal*, Feb 2018 issue, pp. 1–16, 2018.

[GRN 99] GRNC (GROUPE RADIOÉCOLOGIE NORD COTENTIN), Estimation des niveaux d'exposition aux rayonnements ionisants et des risques de leucémies associés de populations du Nord-Cotentin, Synthesis report, July 1999.

[GOL 08] GOLDBERG M., QUANTIN C., GUÉGUEN A. et al., "Bases de données médico-administratives et épidémiologie : intérêts et limites", *Courrier des statistiques*, vol. 124, pp. 59–70, 2008.

[GRA 17] GRANT E.J., BRENNER A., SUGIYAMA H. et al., "Solid cancer incidence among the life span study of atomic bomb survivors: 1958–2009", *Radiation Research*, vol. 187, pp. 513–537, 2017.

[GUÉ 19] GUÉGUEN Y., BONTEMPS A., EBRAHIMIAN T.G., "Adaptive responses to low doses of radiation or chemicals their cellular and molecular mechanisms", *Cellular and Molecular Life Sciences*, vol. 76, no. 7, pp. 1255–1273, 2019.

[GUI 97] GUIZARD A.V., SPIRA A., TROUSSARD X. et al., "Incidence des leucémies de 0 à 24 ans dans le Nord-Cotentin", *Revue d'Épidémiologie et de Santé Publique*, vol. 45, pp. 530–535, 1997.

[HAM 15] HAMRA G.B., RICHARDSON D.B., CARDIS E. et al., "Cohort profile: The international nuclear workers study (INWORKS)", *International Journal of Epidemiology*, Access published 6 July 2015.

[HAU 13] HAURI, D., SPYCHER B., HUSS A. et al., "Domestic radon exposure and risk of childhood cancer: A prospective census-based cohort study", *Environmental Health Perspectives*, vol. 121, pp. 1239–1244, 2013.

[HUA 14] HUANG W.Y., MUO C.H., LIN C.Y. et al., "Paediatric head CT scan and subsequent risk of malignancy and benign brain tumour: A nation-wide population-based cohort study", *British Journal of Cancer,* vol. 110, pp. 2354–2360, 2014.

[ICR 05] ICRP, "ICRP Publication 99, Low-dose extrapolation of radiation-related cancer risk", *Annals of the ICRP*, vol. 35, no. 4, pp. 1–141, 2005.

[INS 20] INSERM, *Essais nucléaires et santé. Conséquences en Polynésie française. Synthèse et recommandations*, EDP Sciences, Montrouge, 2020.

[JAR 17] JARGIN S., "Radiation safety standards and hormesis", *SDRP Journal of Earth Sciences & Environmental Studies*, vol. 2, no. 4, pp. 238–247, 2017.

[JAR 18] JARGIN S.V., "Hormesis and radiation safety norms: "Comments for an update"", *Human and Experimental Toxicology*, pp. 1–11, 2018.

[JOU 14] JOURNY N., REHEL J.L., DUCOU LE POINTE H. et al., "Are the studies on cancer risk from CT scans biased by indication? Elements of answer from a large-scale cohort study in France", *British Journal of Cancer*, vol. 112, pp. 185–193, 2015.

[KAB 20] KABILAN U., GRABER T.E., ALAIN T., KLOKOV D., "Ionizing radiation and translation control: A link to radiation hormesis?", *International Journal of Molecular Sciences*, vol. 21, p. 6650; 2020.

[KEN 13] KENDALL G.M., LITTLE M.P., WAKEFORD R.W. et al., "A record-based case control study of natural background radiation and the incidence of childhood leukemia", *Leukemia*, vol. 27, no. 1, pp. 3–9, 2013.

[KRE 13] KRESTININA L.Y., DAVIS G., SCHONFELD S. et al., "Leukaemia incidence in the Techa River Cohort: 1953–2007", *British Journal of Cancer*, vol. 109, no. 11, pp. 2886–2893, 2013.

[KRI 15] KRILLE L., DREGER S., SCHINDEL R. et al., "Risk of cancer incidence before the age of 15 years after exposure to ionising radiation from computed tomography: Results from a German cohort study", *Radiation and Environmental Biophysics*, vol. 54, pp. 1–12, 2015.

[KUD 15] KUDRYASHEVAA N.S., ROZHKOB T.V., "Effect of low-dose ionizing radiation on luminous marine bacteria: Radiation hormesis and toxicity. Review", *Journal of Environmental Radioactivity*, vol. 142, pp. 68–77, 2015.

[LAN 10] LANE R.S.D., FROST S.E., HOWEC G.R. et al., "Mortality (1950–1999) and cancer incidence (1969–1999) in the cohort of Eldorado uranium workers", *Radiation Research*, vol. 174, pp. 773–785, 2010.

[LAN 19] LANE R.S.D.,·TOMÁŠEK L., ZABLOTSKA L.B. et al., "Low radon exposures and lung cancer risk: Joint analysis of the Czech, French, and Beaverlodge cohorts of uranium miners", *International Archives of Occupational and Environmental Health*, doi: 10.1007/s00420-019-01411-w, 2019.

[LAU 00] LAURIER D., MARTIN J.M., HUBERT P., Etudes épidémiologiques dans des zones à haut niveau de radioactivité naturelle, IPSN, SEGRILEADS technical note, 2000/91.

[LEU 15] LEURAUD K., RICHARDSON D.B., CARDIS E., "Ionising radiation and risk of death from leukaemia and lymphoma in radiation-monitored workers (INWORKS): An international cohort study", *Lancet Haematology*, vol. 2, p. e281, 2015.

[LIT 10] LITTLE M.P., TAWN E.J., TZOULAKI I. et al., "Review and meta-analysis of epidemiological associations between low/moderate doses of ionizing radiation and circulatory disease risks, and their possible mechanisms", *Radiation and Environmental Biophysics*, vol. 49, pp. 139–153, 2010.

[LIT 12] LITTLE M.P., AZIZOVA T.V., BAZYKA D. et al., "Systematic review and meta-analysis of circulatory disease from exposure to low-level ionizing radiation and estimates of potential population mortality risks", *Environmental Health Perspectives*, vol. 120, pp. 1503–1511, 2012.

[LIT 16] LITTLE M.P., "Radiation and circulatory disease", *Mutation Research,* vol. 770, pp. 299–318, 2016.

[LIT 18] LITTLE M.P., KITAHARA C.M., CAHOON E.K. et al., "Occupational radiation exposure and risk of cataract incidence in a cohort of US radiologic technologists", *European Journal of Epidemiology*, vol. 33, pp. 1179–1191, 2018.

[LIT 20] LITTLE M.P., CAHOON E.K., KITAHARA C.M. et al., "Occupational radiation exposure and excess additive risk of cataract incidence in a cohort of US radiologic technologists", *Occupational and Environmental Medicine*, vol. 77, pp. 1–8, 2020.

[LIU 06] LIU G.W., GONG P.S., ZHAO H.G. et al., "Effect of low-level radiation on the death of male germ cells", *Radiation Research*, vol. 165, pp. 379–389, 2006.

[LUM 21] LUMNICZKY K., IMPENS N., ARMENGOL G. et al., "Low dose ionizing radiation effects on the immune system", *Environment International*, vol. 149, p. 106212, 2021.

[MAG 07] MAGUIRE P., MOTHERSILL C., MCCLEAN B. et al., "Modulation of radiation responses by pre-exposure to irradiated cell conditioned medium", *Radiation Research*, vol. 167, no. 4, pp. 485–492, 2007.

[MAJ 21] MAJEWSKI M., OSTHEIM P., GLUZMAN-POLTORAK Z. et al., "Gene expression changes in male and female rhesus macaque 60 days after irradiation", *PLOS One*, doi: 10.1371/journal.pone.0254344, 2021.

[MAS 00] MASSE R., "Rayonnements ionisants", *Comptes rendus de l'Academie des sciences. Serie III, Sciences de la vie*, vol. 323, pp. 633–640, 2000.

[MAS 02] MASSE R., "Pathogenic effects of low dose irradiation: Dose–effect relationships", *Comptes rendus physique*, vol. 3, pp. 1049–1058, 2002.

[MAT 13] MATHEWS J.D., FORSYTHE A.V., BRADY Z. et al., "Cancer risk in 680 000 people exposed to computed tomography scans in childhood or adolescence: Data linkage study of 11 million Australians", *BMJ*, vol. 346, p. f2360, 2013.

[MEL 07] MELNOV S., MAROZIK P., DROZD T., "Clastogenic factors bystander effects and genomic instability in vivo", in C. MOTHERSILL, I. MOSSE, C. SEYMOUR (eds), *Multiple Stressors: A Challenge for the Future*, Springer, Dordrecht, pp. 171–182, 2007.

[MIL 17] MILLER A.C., RIVAS R., TESORO L. et al., "Radiation exposure from depleted uranium: The radiation bystander effect", *Toxicology and Applied Pharmacology*, vol. 331, pp. 135–141, 2017.

[MOL 16] MØLLER A.P., MOUSSEAU T.A., "Are organisms adapting to ionizing radiation at Chernobyl? Review", *Trends in Ecology & Evolution*, vol. 31, no. 4, pp. 281–289, 2016.

[MOR 02] MORTAZAVI S.M.J., GHIASSI-NEJAD M., IKUSHIMA T., "Do the findings on the health effects of prolonged exposure to very high levels of natural radiation contradict current ultra-conservative radiation protection regulations?", *International Congress Series*, vol. 1236, pp. 19–21, 2002.

[MOR 07] MORGAN W.F., SOWA M.B., "Non-targeted bystander effects induced by ionizing radiation", *Mutation Research*, vol. 616, pp. 159–164, 2007.

[MOR 11] MORGAN W.F., "Radiation induced genomic instability", *Health Physics*, vol. 100, no. 3, pp. 280–281, 2011.

[MOR 15] MORGAN W.F., SOWA M.B., "Non-targeted effects induced by ionizing radiation: Mechanisms and potential impact on radiation induced health effects", *Cancer Letters*, vol. 356, no. 1, pp. 7–21, 2015.

[MOS 14] MOSEEVA M.B., AZIZOVA T.V., GRIGORYEVA E.S. et al., "Risks of circulatory diseases among Mayak PA workers with radiation doses estimated using the improved Mayak Worker Dosimetry System 2008", *Radiation and Environmental Biophysics*, vol. 53, no. 2, pp. 469–477, 2014.

[MOT 16] MOTHERSILL C., SEYMOUR C., "Genomic instability and the spectrum of response to low radiation doses", in KOVALCHUK I., KOVALCHUK O. (eds), *Genome Stability, From Virus to Human Application*, Elsevier, Amsterdam, pp. 606–614, 2016.

[MOT 18] MOTHERSILL C., SEYMOUR C., "Old data—new concepts: Integrating "indirect effects" into radiation protection", *Health Physics*, vol. 115, no. 1, pp. 170–178, 2018.

[NCR 01] NCRP, Evaluation of the linear-non threshold dose-response model for ionizing radiation, NCRP report no. 136, 2001.

[NCR 17] NCRP, Implications of recent epidemiologic studies for the linear nonthreshold model and radiation protection, NCRP SC 1-25 Draft Commentary, September 15, 2017.

[NRC 05] NRC, Health risk from exposure to low levels of ionizing radiation. Pre-publication version, National Research Council of the National Academies of USA, BEIR VII, 2005.

[NRC 06] NRC (NATIONAL RESEARCH COUNCIL), Health risks from exposure to low levels of ionizing radiation, BEIR VII Phase 2, The National Academies, Washington, DC, 2006.

[OZA 12] OZASA K., SHIMIZU Y., SUYAMA A. et al., "Studies of the mortality of atomic bomb survivors, Report 14, 1950–2003: An overview of cancer and non cancer diseases (LSS-14)", *Radiation Research*, vol. 177, pp. 229–243, 2012.

[PEA 12] PEARCE M.S., SALOTTI J.A., LITTLE M.P. et al., "Radiation exposure from CT scans in childhood and subsequent risk of leukaemia and brain tumours: A retrospective cohort study", *Lancet*, vol. 380, pp. 499–505, 2012.

[PET 02] PETCU I., MOISOI N., SAVU D. et al., "Effect of pre-exposure to b rays of tritium on some biochemical parameters measured in organs of rats subsequently irradiated with fast neutrons", *Canadian Journal of Physiology and Pharmacology*, vol. 80, pp. 742–749, 2002.

[POB 97] POBEL D., VIEL J.F., "Case control study of leukaemia among young people near La Hague nuclear reprocessing plant: The environmental hypothesis revisited", *BMJ*, vol. 314, pp. 101–106, 1997.

[PRI 03] PRISE K.M., FOLKARD M., MICHAEL B.D., "A review of the bystander effect and its implications for low-dose exposure", *Radiation Protection Dosimetry*, vol. 104, no. 4, pp. 347–355, 2003.

[RIC 09] RICHARDSON D., SUGIYAMA H., NISHI N. et al., "Ionizing radiation and leukemia mortality among Japanese Atomic Bomb Survivors, 1950–2000", *Radiation Research*, vol. 172, pp. 368–382, 2009.

[RIC 15] RICHARDSON D.B., CARDIS E., DANIELS R.D. et al., "Risk of cancer from occupational exposure to ionising radiation: retrospective cohort study of workers in France, the United Kingdom, and the United States (INWORKS)", *BMJ*, vol. 351, p. h5359, 2015.

[ROB 21] ROBERT J., L'instabilité génomique, moteur de l'oncogenèse, 2021. Available at: https://sfc.asso.fr/wp-content/uploads/2020/05/instabilit%C3%A9-g%C3%A9nomique-moteur-pp_jr.pdf.

[ROD 08] RODGERS B.E., HOLMES K.M., "Radio-adaptive response to environmental exposures at Chernobyl", *Dose-Response*, vol. 6, pp. 209–221, 2008.

[SAM 18] SAMSON E., LEURAUD K., RAGE E. et al., "Bilan de la surveillance épidémiologique des travailleurs du cycle électronucléaire en France", *Radioprotection*, vol. 53, no. 3, pp. 175–184, 2018.

[SAN 10] SANDERS C.L., *Radiation Hormesis and the Linear-No-Threshold Assumption*, Springer-Verlag, Berlin, 2010.

[SAN 17] SANDERS C.L., *Radiobiology and Radiation Hormesis. New Evidence and Its Implications for Medicine and Society*, Springer, 2017.

[SCH 13] SCHONFELD S.J., KRESTININA L.Y., EPIFANOVA S. et al., "Solid cancer mortality in the Techa River cohort (1950–2007)", *Radiation Research*, vol. 179, no. 2, pp. 183–189, 2013.

[SCH 18] SCHOFIELD P.N., KONDRATOWICZ M., "Evolving paradigms for the biological response to low dose ionising radiation; the role of epigenetics", *International Journal of Radiation Biology*, vol. 94, no. 8, pp. 769–781, 2018.

[SHA 19] SHARMA S., SINGLA N., CHADHA V.D. et al., "A concept of radiation hormesis. Stimulation of antioxidant machinery in rats by low dose ionizing radiation", *Hellenic Journal of Nuclear Medicine*, vol. 22, no. 1, pp. 43–48, 2019.

[SHI 18a] SHIBAMOTO Y., NAKAMURA H., "Overview of biological, epidemiological, and clinical evidence of radiation hormesis", *International Journal of Molecular Sciences*, vol. 19, p. 2387, 2018.

[SHI 18b] SHIMIZU M., MISUMI M., YAMADA M. et al., "Choice reaction time and grip strength as predictors of cardiovascular mortality in middle-aged and elderly Japanese: From the radiation effects research foundation adult health study", *Internal Medicine Journal*, vol. 48, pp. 1331–1336, 2018.

[SHI 10] SHIMIZU Y., KODAMA K., NISHI N. et al., "Radiation exposure and circulatory disease risk: Hiroshima and Nagasaki atomic bomb survivor data, 1950–2003", *BMJ*, vol. 340, no. 5349, p. 193, 2010.

[SHO 18] SHORE R.E., BECK H.L., BOICE J.D.Jr. et al., "Implications of recent epidemiologic studies for the linear non threshold model and radiation protection", *Journal of Radiological Protection*, vol. 38, no. 3, pp. 1217–1233, 2018.

[SMI 16] SMITH R.W., SEYMOUR C.B., MOCCIA R.D. et al., "Irradiation of rainbow trout at early life stages results in trans-generational effects including the induction of a bystander effect in non-irradiated fish", *Environmental Research*, vol. 145, pp. 26–38, 2016.

[SPY 15] SPYCHER B.D., LUPATSCH J.E., ZWAHLEN M. et al., "Background ionizing radiation and the risk of childhood cancer: A census-based nationwide cohort study", *Environmental Health Perspectives*, vol. 123, pp. 622–628, 2015.

[SUG 16] SUGIE C., SHIBAMOTO Y., HASHIMOTO S. et al., "Biological effects of a radiation hormesis sheet emitting very low-dose-rate γ rays", *International Journal of Radiation Oncology, Biology, Physics*, vol. 96, no. 2S, pp. E561–E562, Poster Viewing 3376, 2016.

[TAN 18] TANG F.R., LOGANOVSKY K., "Low dose or low dose rate ionizing radiation-induced health effect in the human", *Journal of Environmental Radioactivity*, vol. 192, pp. 32–47, 2018.

[THI 15] THIERRY-CHEF I., RICHARDSON D.B., DANIELS R.D. et al., "Dose estimation for a study of nuclear workers in France, the United Kingdom and the United States of America: Methods for the International Nuclear Workers Study (INWORKS)", *Radiation Research*, vol. 183, pp. 632–642, 2015.

[TIL 17] TILL J.E., BECK H.L., GROGAN H.A. et al., "A review of dosimetry used in epidemiological studies considered to evaluate the linear no threshold (LNT) dose-response model for radiation protection", *International Journal of Radiation Biology*, doi: 10.1080/09553002.2017.1337280, 2017.

[TUB 05] TUBIANA M., AURENGO A., AVERBECK D. et al., Dose–effect relationships and the estimation of the carcinogenic effects of low doses of ionizing radiation, Joint Report no. 2, Académie Nationale de Médecine, Institut de France–Académie des Sciences (March 2005), 2005. Available at: www.academiemedecine.fr/actualites/rapports.asp.

[TUB 06a] TUBIANA M., AURENGO A., AVERBECK D. et al., "The debate on the use of linear no threshold for assessing the effects of low doses", *Journal of Radiological Protection*, vol. 26, pp. 317–324, 2006.

[TUB 06b] TUBIANA M., AURENGO A., AVERBECK A. et al., "Recent reports on the effect of low doses of ionizing radiation and its dose–effect relationship", *Radiation and Environmental Biophysics*, vol. 44, no. 4, pp. 245–251, 2006.

[TUB 07] TUBIANA M., MASSE R., De VATHAIRE F. et al., "La controverse sur les effets des faibles doses de rayonnements ionisants et la relation linéaire sans seuil", *Radioprotection*, vol. 42, pp. 133–161, 2007.

[UNS 93] UNSCEAR, Sources, Effects and Risks of Ionizing Radiation, United Nations, New York, 1993.

[UNS 00a] UNSCEAR, Sources and effects of ionising radiation, UNSCEAR 2000 report to the general assembly, with scientific annexes, Volume II: Effects, United Nations, New York, 2000.

[UNS 00b] UNSCEAR, Sources, Effects and Risks of Ionizing Radiation, Annex I Epidemiological evaluation of radiation-induced cancer, United Nations, New York, 2000.

[UTA 21] UTADA M., BRENNER A.V., PRESTON D.L. et al., "Radiation risk of ovarian cancer in atomic bomb survivors: 1958–2009", *Radiation Research*, vol. 195, pp. 60–65, 2021.

[VAC 09] VACQUIER B, ROGEL A, LEURAUD K. et al., "Radon associated lung cancer risk among French uranium miners: Modifying factors of the exposure-risk relationship", *Radiation and Environmental Biophysics*, vol. 48, pp. 1–9, 2009.

[VAI 18] VAISERMAN A., KOLIADA A., ZABUGA O. et al., "Health impacts of low-dose ionizing radiation: Current scientific debates and regulatory issues", *Dose-Response: An International Journal*, vol. 16, no. 3, p. 1559325818796331, July–September 2018.

[VAN 11] VAN DILLEN T., DEKKERS F., BIJWAARD H. et al., "Lung cancer from radon: A two-stage model analysis of the WISMUT cohort, 1955–1998", *Radiation Research*, vol. 175, no. 1, pp. 119–130. 2011.

[VIE 90] VIEL J.F., RICHARDSON S.T., "Childhood leukaemia around the La Hague nuclear waste reprocessing plant", *BMJ*, vol. 300, pp. 580–581, 1990.

[VIE 95] VIEL J.F., POBEL D., CARRE A., "Incidence of leukaemia in young people around the La Hague nuclear waste reprocessing plant: a sensitivity analysis", *StatMed*, vol. 14, pp. 2459–2472, 1995.

[VUI 19] VUILLEZ J.P., "Faibles doses de rayonnements : quand l'hormésis et l'effet abscopal se rencontrent …", *Médecine Nucléaire*, vol. 43, nos. 5–6, pp. 354–359, 2019.

[WAK 04] WAKEFORD R., "The cancer epidemiology of radiation", *Oncogene*, vol. 23, pp. 6404–6428, 2004.

[WAN 15] WANG H., YU K.N., HOU J. et al., "Radiation-induced bystander effect: Early process and rapid assessment. Mini-review", *Cancer Letters*, vol. 356, pp. 137–144, 2015.

[WHO 09] WHO, *Handbook on Indoor Radon: A Public Health Perspective*, WHO, 2009.

[YAM 04] YAMADA M., WONG F.L., FUJIWARA S. et al., "Noncancer disease incidence in atomic bomb survivors, 1958–1998", *Radiation Research*, vol. 161, no. 6, pp. 622–632, 2004.

[YAM 18] YAMASHITA S., SUZUKI S., SUZUKI S. et al., "Lessons from Fukushima: Latest findings of thyroid cancer after the Fukushima Nuclear Power Plant accident", *Thyroid*, vol. 28, no. 1, pp. 11–22, 2018.

[YOU 05] YOUNG R.W., KERR G.D., Reassessment of the atomic bomb radiation dosimetry for Hiroshima and Nagasaki – Dosimetry System 2002 (DS02), RERF, Report of the Joint US–Japan Working Group, 2005.

[ZAB 16] ZABALLA I., EIDEMÜLLER M., "Mechanistic study on lung cancer mortality after radon exposure in the Wismut cohort supports important role of clonal expansion in lung carcinogenesis", *Radiation and Environmental Biophysics*, vol. 55, pp. 299–315, 2016.

[ZHA 06] ZHANG H., LIU B., ZHOU Q. et al., "Alleviation of pre-exposure of mouse brain with low-dose $^{12}C^{6+}$ ion or $^{60}Co$ gamma-ray on male reproductive endocrine damages induced by subsequent high-dose irradiation", *International Journal of Andrology*, vol. 29, pp. 592–596, 2006.

# 12

# International and French Regulatory Values

## 12.1. Introduction

Exposure to ionizing radiation can be harmful. Natural exposures are unavoidable, while artificial exposures can be eliminated or reduced, but are sometimes useful and necessary, as in medicine for example.

Radiation protection is based on three basic principles: justification, optimization and compliance with dose limits or levels that must not be exceeded. Therefore, activities involving exposure to ionizing radiation must be justified by the benefits they bring, after taking into account all the advantages and disadvantages. Exposures must not only be justified, but also be kept as low as reasonably achievable. This optimization principle is also known as the "ALARA" (As Low As Reasonably Achievable) principle.

## 12.2. International expertise and organization

An international network of scientific and regulatory organizations is dedicated to protecting people and the environment from ionizing radiation. There are four levels to the process of establishing recommendations for threshold values (Figure 12.1). Other international organizations are involved on a more ad hoc basis, but are not included in this figure. These include the ICRU (International Commission on Radiation Units and Measurements), the FAO (Food and Agriculture Organization of the United Nations), the ILO (International Labor Organization), the WHO (World Health Organization), the OECD (Organization for Economic Cooperation and Development) and the NEA (Nuclear Energy Agency).

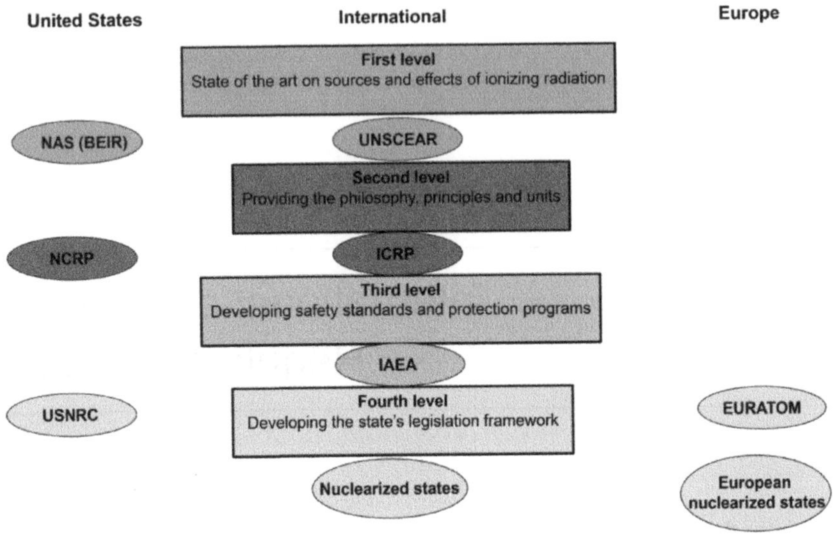


**Figure 12.1.** *Simplified, theoretical diagram of the international organization of radiation protection and the development of national legislation to protect the public from ionizing radiation. IAEA: International Atomic Energy Agency; BEIR: Biological Effects of Ionizing Radiation; ICRP: International Commission on Radiological Protection; EURATOM: European Atomic Energy Community; NAS: National Academy of Sciences; NCRP: National Council on Radiation Protection and Measurements; UNSCEAR: United Nations Scientific Committee on the Effects of Atomic Radiation; USNRC: US Nuclear Regulatory Commission*

The first level consists of establishing the state of knowledge on the sources and effects of ionizing radiation on humans. Its scientific foundations are provided by the United Nations Scientific Committee on the Effects of Atomic Radiation (UNSCEAR). Formed in 1955, this Committee is the official international authority on the sources and effects of ionizing radiation. It publishes assessment reports on the latest knowledge of radiation sources and the risks they may pose to human health and the environment. The latest UNSCEAR reports (2020/2021 reports) concern the assessment of human exposure to ionizing radiation due to medical uses [UNS 22a], the consequences of the Fukushima accident [UNS 22b] and the effects of low doses of radiation on the risk of developing cancer [UNS 22c]. The US National Academy of Sciences plays a similar role with its assessments of the biological effects of ionizing radiation (BEIR reports, most recently BEIR VII in 2006) [NRC 06].

At the second level, UNSCEAR's scientific findings are intended to feed into the activities of the International Commission on Radiological Protection (ICRP), whose objectives are the prevention of cancer and other diseases and effects arising from exposure to ionizing radiation, and the protection of the environment. The ICRP created, maintained and developed the International System of Radiation Protection (ISRP), which serves as the common basis worldwide for radiation protection standards, laws, guidelines, programs and practices applied to the industrial and medical uses of nuclear technology and substances. The latest major reform of radiation protection proposed by the ICRP is contained in its 2007 publication 103 [ICR 07]. Various recommendations concerning various tissues and organs were published in 2012 [ICR 12]. In the United States, the National Council on Radiation Protection and Measurements (NCRP) plays a similar role to the ICRP [NCR 20].

The third level is carried out by the International Atomic Energy Agency (IAEA). This United Nations agency develops safety standards and protection programs aimed at the safe, secure and peaceful use of nuclear technologies by its Member States. The IAEA is concerned with a wide range of civil nuclear activities, including the operation of nuclear facilities, the production, transport and use of radioactive materials, and the management of radioactive waste. IAEA safety standards are published in three series: Safety Fundamentals, Safety Requirements and Safety Guides. The first series establishes the fundamental safety objective and the principles of protection and safety. The second sets out the requirements that must be met to ensure the protection of people and the environment, now and in the future. Finally, the Safety Guides provide recommendations and advice on how to meet these requirements.

The fourth level is achieved at the level of each nuclearized state, which adopts its own legislation to protect its population and its nuclear workers. Radiation protection is organized on a state-by-state basis (Table I.1, national authorities in charge of nuclear safety and radiation protection). In the United States, for example, this is the USNRC (US Nuclear Regulatory Commission), and in Canada, the CNSC (Canadian Nuclear Safety Commission). The CNSC relies on the information provided by this international network to develop Canada's regulatory framework. In turn, the federal government supports the activities of these organizations, in addition to the CNSC regulatory staff and scientists who participate in their work.

In the case of the European Union, this fourth level is divided into two sub-stages. Firstly, the European Atomic Energy Community (EAEC or EURATOM) adopts Directives applicable to all Member States. Then, each State transcribes these Directives into its legislative and regulatory system (codes, laws, decrees, orders, etc.) within a time frame set by each Directive.

Most international regulatory bodies have adopted the radiation protection standards proposed by the ICRP. The ICRP dose limits are based on the linear non-threshold model, as recommended by UNSCEAR.

However, this approach is only theoretical, and in practice, a number of abuses have been observed, many of which are not specific to the nuclear field, but can be found in many physical and chemical risk assessments. The first consequence is that, in reality, the definition of standards is almost essentially the responsibility of the ICRP, which monopolizes the process, hardly ever referring to the work of UNSCEAR and the other international and national organizations that endorse ICRP decisions[1]. The second consequence stems from the conflicts of interest of certain members of international organizations (UNSCEAR, ICRP, IAEA) who work in the nuclear industry, although this trend is diminishing nowadays. The third consequence concerns the selection of scientific work taken into account by international organizations when defining radiation protection standards. Far too many scientific advances are overlooked. This was particularly glaring following the Chernobyl accident[2] where UNSCEAR completely ignored the Belarusian and Russian work published in the Annals of the American Academy of Sciences [YAB 09]. It is also regrettable that the WHA-12-40 agreement of May 28, 1959, between the WHO and the IAEA severely restricts the role of the WHO to that of the IAEA in the field of human health.

## 12.3. International and European recommendations and their evolution

The main IAEA recommendations adopted by the majority of nuclearized states concern public exposure and exposure of nuclear workers.

The threshold value is 1 millisievert per year for members of the public (this limit does not take into account justified medical practices or exposure to natural radioactivity, such as radon). It is 6 millisieverts per year for trainees and students aged 16 to 18 who, because of their studies, are required to use ionizing radiation sources.

This value is 20 millisieverts per 12 consecutive sliding months for occupationally exposed persons (e.g. workers in a nuclear power plant or employees in an interventional radiology department).

---

1 The ICRP states in the preface to publication 103: “The United Nations agencies, with the International Atomic Energy Agency as the lead agency, are using the 2007 ICRP Recommendations as a major input to their project of revising the International Basic Safety Standards, and likewise the European Commission uses the 2007 Recommendations as a major input to their revision of the European Basic Safety Standards”.

2 The 2012 interview with Dr Patrick Smeesters, Belgian radiation protection specialist and UNSCEAR member, “RADIOACTIVITÉ. Quand un scientifique de l’UNSCEAR met en cause... l’UNSCEAR”, *Médiapart,* Philips Michel, July 10, 2014, is eloquent.

These radiation protection principles are applied more strictly to groups of people who are more sensitive to the effects of ionizing radiation, such as fetuses, pregnant women and children. In the world of work, the unborn child is considered to be a member of the public involuntarily exposed to ionizing radiation. Their protection must therefore be equal to that of members of the public.

## 12.4. Radiation protection standards and regulatory limits

The IAEA, after examining the UNSCEAR syntheses and ICRP recommendations, issues its Fundamental Standards and their requirements. The latest are dated from 2016 [AIE 16].

For occupational exposure of workers over the age of 18, the dose limits are as follows. An effective dose of 20 mSv per year averaged over five consecutive years (100 mSv in five years), not exceeding 50 mSv in a single year. An equivalent dose to the crystalline lens of 20 mSv per year averaged over five consecutive years (100 mSv in five years), and a higher effective dose of up to 50 mSv may be authorized by the competent authority in any one year[3]. An equivalent dose to the extremities (hands and feet) or skin should be of 500 mSv per year.

For public exposure, the dose limits are as follows: an effective dose of 1 mSv per year; an equivalent dose to the crystalline lens of 15 mSv per year and an equivalent dose to the skin of 50 mSv per year.

Secondly, each state sets its own radiation protection standards. They are relatively similar.

### 12.4.1. *Some examples of international radiation protection standards*

In the United States, radiation protection standards are proposed by the NRC and published in the *Code of Federal Regulations* (Title 10, Chapter 1, part 20, section C for professionals and D for the public). The latest version dates from November 2013.

In Canada, the Canadian Radiation Protection Regulations limit the exposure of nuclear energy workers to a maximum annual effective dose of 50 mSv and to less than 100 mSv over a five-year period [CNS 17].

3 These doses seem excessive.

### 12.4.2. *European directives and transfer into French law*

In the case of the European Union, EURATOM [EUR 14] issues its Directives, which are then incorporated into the various national laws and regulations. This is done by a group of European experts set up under Article 31 of the EURATOM Treaty.

The effective dose for occupational exposure is limited to 20 mSv. For trainees and students (aged 16–18), this limit is lowered to 6 mSv. In addition to the effective dose limits, the following equivalent dose limits apply: 15 mSv per year for the crystalline lens, 150 mSv per year for the skin, averaged over a surface area of 1 $cm^2$ and 150 mSv per year for the extremities. These equivalent dose values also apply to the public. In addition, for the public, the effective dose limit for exposure is 1 mSv per year.

In France, the Ministry of Labor, after consultation with the French Nuclear Safety Authority (ASN), is responsible for drawing up radiation protection regulations and monitoring their application, via labor inspectors. The ASN issues an opinion on the Ministry of Labor's text, and, on behalf of the State, monitors nuclear safety and radiation protection to protect people and the environment through its radiation protection inspectors[4].

The Ministry of Labor used to have a Higher Council for Occupational Risks (*Conseil Supérieur des Risques Professionnels*, CSPRP), which became the Guidance Council on Occupational Conditions (*Conseil d'Orientation sur les Conditions de Travail*, COCT) by decree in 2008. These two structures organized work in specialized commissions. The Chemical, Physical and Biological Risks Commission (*Commission risques chimiques, physiques et biologiques*), now the Health Risks Commission (*Commission risques pour la santé*), was responsible for transposing European directives on various occupational risks, including ionizing radiation, into regulatory texts. Each commission has 24 members, including 10 representatives of the social partners.

---

4 The French nuclear safety authority (*Autorité de sûreté nucléaire* – ASN) appoints radiation protection inspectors from among its own staff to ensure compliance with radiation protection regulations at secret basic nuclear installations (such as nuclear submarines). These inspectors are part of a general corps of inspectors (medical public health inspectors, health and social affairs inspectors, health engineers, etc.). The Minister of Defense may also appoint radiation protection inspectors to monitor facilities and activities of interest to national defense. Radiation protection inspectors are appointed and sworn in under conditions defined by decree by the Conseil d'Etat. They are all bound by professional secrecy.

The procedure is democratic, insofar as the Ministry proposes its draft text in the form of a decree, which the commission examines article by article. Decree No. 86-1103 of October 2, 1986, on the protection of workers against the dangers of ionizing radiation, was the first regulatory text to refer, in the field of radiation protection, to a EURATOM Directive of the Council of the European Communities: Directive No. 80-836 of July 15, 1980[5], and Directive No. 84-467 amending the former with new dosimetric estimates of internal doses due to ingestion and inhalation of radioactive substances.

### 12.4.3. *Changes in radiation protection standards over time*

Exposure limits tend to be lowered as scientific knowledge of radioactive risk improves. In 1991, for example, in the United States, the annual effective dose limit for professionals was 50 mSv, with equivalent dose limits of 500 mSv for individual organs, except for the crystalline lens, where the limit was 150 mSv.

## 12.5. Radionuclide toxicity and radiotoxicity

The provisional environmental quality standard (*norme de qualité environnementale provisoire*, NQEp) proposed by the European Union for uranium in continental waters is 0.3 $\mu g.L^{-1}$ (circular of May 7, 2007, DCE/23) [RF 07].

Plutoniums are only regulated with regard to their radiotoxicity, unlike uranium, whose chemical toxicity must be taken into account as it is more significant than its radiological toxicity. The WHO drinking water guidelines for $^{238}Pu$ and $^{239}Pu$ are 1 $Bq.L^{-1}$ and for $^{241}Pu$ 10 $Bq.L^{-1}$ [WHO 04]. Health Canada has calculated its drinking water quality guidelines so that the daily consumption of two liters of water for one year should not exceed an effective dose of 0.1 mSv for each radionuclide considered. Since uranium's chemical toxicity is greater than its radiological toxicity, the CNSC [CNS 17] applied the guideline of 20 $\mu g.L^{-1}$ for total uranium (i.e. 0.5 $Bq.L^{-1}$ natural uranium) in this assessment.

The hazardousness of a radionuclide can be chemical and/or radioactive. The IRSN classifies radionuclides into toxicity groups. Limit values below which there is exemption from declaration are also proposed according to European Directive no. 96/29/EURATOM of May 13, 1996, replaced by Council Directive 2013/59/ EURATOM of December 5, 2013 [EUR 14] (Table 12.1).

5 This Directive took account of the fundamental rules of radiation protection laid down by the ICRP60.

| Radionuclide | $^{14}$C | $^{60}$Co | $^{90}$Sr | $^{137}$Cs | $^{129}$I |
|---|---|---|---|---|---|
| Bq | $10^7$ | $10^5$ | $10^4$ | $10^4$ | $10^5$ |
| | $^{192}$Ir | $^{237}$Np | $^{234}$Th | $^{241}$Am | $^{242}$Cm |
| Bq | $10^4$ | $10^3$ | $10^5$ | $10^4$ | $10^5$ |

**Table 12.1.** *Exemption value for various radionuclides (activity below which a practice is exempt from declaration according to European Directive 2013/59/EURATOM of December 5, 2013) (from [EUR 14])*

## 12.6. Food standards

Various regulatory values for the concentration of certain radionuclides in foodstuffs have been laid down in several countries and in the European Union, particularly following the Chernobyl accident. The same process was repeated in Japan after the Fukushima accident [AMI 18].

For radionuclides considered to be major in the event of a nuclear accident, the *Codex Alimentarius* Commission of the joint FAO/WHO program has adopted food contamination standards [WHO 11]. Guideline limits for radionuclide concentrations are set out in the *Codex Alimentarius* [COD 95], which has been revised several times, most recently in 2015 [COD 15]. At the seventh session of the joint FAO/WHO program held in Moscow from April 8–12, 2013, limit values for some twenty radionuclides present in baby foods and adult foods (Table 12.2) were adopted.

For products from countries affected by the Chernobyl accident, the European Commission set the acceptable contamination level for two radionuclides ($^{134}$Cs and $^{137}$Cs) at 370 Bq.kg$^{-1}$ for dairy products and baby food, and 600 Bq.kg$^{-1}$ for other products intended for human consumption. Since 1987 (Euratom, 1989), the European Union has set maximum permissible levels of radioactive contamination in foodstuffs for all radiological emergencies, for several radionuclides ($^{241}$Am, $^{90}$Sr and $^{137}$Cs). In addition, for $^{137}$Cs, marketing prohibition standards have been laid down for ready-to-eat animal feed (Euratom, 1990). These two regulations were revised in 2016 [CE 16] (Tables 12.3 and 12.4).

| Radionuclide | Food for general consumption ($Bq.kg^{-1}$) | Milk and baby food ($Bq.L^{-1}$ or $Bq.kg^{-1}$) |
|---|---|---|
| $^{238}Pu$, $^{239}Pu$, $^{240}Pu$, $^{241}Am$ | 10 | 1 |
| $^{90}Sr$, $^{106}Ru$, $^{129}I$, $^{238}U$ | 100 | 100 |
| $^{35}S$, $^{60}Co$, $^{89}Sr$, $^{103}Ru$, $^{134}Cs$, $^{144}Ce$, $^{192}Ir$ | 1,000 | 1,000 |
| $^{3}H$, $^{14}C$, $^{99}Tc$ | 10,000 | 1,000 |

**Table 12.2.** *Permissible limits for food activity in accidental situations (FAO, WHO) (after [COD 15])*

| | Plutonium isotopes ($^{239}Pu$, $^{241}Am$) | Cesium isotopes ($^{134}Cs$, $^{137}Cs$) | Strontium isotopes ($^{90}Sr$) | Iodine isotopes ($^{131}I$, $^{129}I$) |
|---|---|---|---|---|
| Baby food | 1 | 400 | 75 | 150 |
| Dairy products | 20 | 1,000 | 125 | 500 |
| Other foodstuffs | 80 | 1,250 | 750 | 2,000 |
| Liquids for consumption | 20 | 1,000 | 125 | 500 |

**Table 12.3.** *Maximum permissible levels of radioactive contamination for foodstuffs following a nuclear accident or other radiological emergency ($Bq.kg^{-1}$) (modified from [CE 16])*

| Food | $^{137}Cs$ |
|---|---|
| Pork | 1,250 |
| Poultry, lamb, veal | 2,500 |
| Other | 5,000 |

**Table 12.4.** *Maximum permissible levels of radioactive contamination for ready-to-eat animal feed following a nuclear accident or other radiological emergency ($Bq.kg^{-1}$) (modified from [CE 16])*

For water, the guide value for $^{134}Cs$ and $^{137}Cs$ is 10 $Bq.L^{-1}$ [SOU 09]. For iodine, the guide values in water are 1 and 10 $Bq.L^{-1}$ for $^{129}I$ and $^{131}I$, respectively.

## 12.7. References

[AIE 16] AIEA, Radioprotection et sûreté des sources de rayonnements : Normes fondamentales internationales de sûreté, Prescriptions générales de sûreté Partie 3, collection normes de sûreté de l'AIEA No. GSR Part 3, AIEA, Vienna, 2016.

[AMI 18] AMIARD J.C., "Radionucléides et aliments", in: CAMEL V., RIVIERE G., LE BIZEC B. (eds), *Risques chimiques liés aux aliments : principes et applications*, Lavoisier, Tec&Doc, Paris, pp. 387–410, 2018.

[CE 16] CE (CONSEIL DE L'UNION EUROPEENNE), "Règlement (Euratom) 2016/52 du Conseil du 15 janvier 2016 fixant les niveaux maximaux admissibles de contamination radioactive pour les denrées alimentaires et les aliments pour animaux après un accident nucléaire ou dans toute autre situation d'urgence radiologique", *Journal officiel de l'Union européenne*, L13/2–L13/11, 20.1.2016.

[CNS 17] CNSC (CANADIAN NUCLEAR SAFETY COMMISSION), Évaluation de la pertinence d'ajouter les radionucléides comme produits chimiques sources de préoccupations mutuelles à l'annexe 3 de l'Accord Canada – États-Unis relatif à la qualité de l'eau dans les Grands Lacs, CC172-187/2018F, 2017.

[COD 95] CODEX ALIMENTARIUS, Codex general standard for contaminants and toxins in food and feed. CODEX STAN 193-1995, 1995. Available at: http://www.fao.org/fileadmin/user_upload/livestockgov/documents/1_CXS_193e.pdf.

[COD 15] CODEX ALIMENTARIUS, General standard for contaminants and toxins in food and feed (CODEX STAN 193-1995), 2015. Available at: www.fao.org/input/download/standards/17/CXS_193e_2015.pdf.

[EUR 14] EURATOM, "Directive 2013/59/Euratom du Conseil du 5 décembre 2013 fixant les normes de base relatives à la protection sanitaire contre les dangers résultant de l'exposition aux rayonnements ionisants", *Journal officiel de l'Union européenne*, L13/1 à L13/73 du 17.1.2014.

[ICR 07] ICRP, "ICRP Publication 103, The 2007 Recommendations of the International Commission on Radiological Protection", *Annals of the ICRP*, vol. 37, nos. 2–4, 2007.

[ICR 12] ICRP, "ICRP Publication 118, ICRP statement on tissue reactions and early and late effects of radiation in normal tissues and organs – threshold doses for tissue reactions in a radiation protection context", *Annals of the ICRP*, vol. 41, nos. 1–2, pp. 3–322, 2012.

[NCR 20] NCRP, Approaches for integrating information from radiation biology and epidemiology to enhance low-dose health risk assessment, Report No. 186, 2020.

[NRC 06] NRC, Health risks from exposure to low levels of ionizing radiation, BEIR VII Phase 2, The National Academies Press, Washington, DC, 2006.

[RF 07] RF, Circulaire du 7 mai 2007 définissant les "normes de qualité environnementale provisoires (NQEp)" des 41 substances impliquées dans l'évaluation de l'état chimique des masses d'eau ainsi que des substances pertinentes du programme national de réduction des substances dangereuses dans l'eau, DCE/23, 2007.

[SOU 09] SOUIDI M., BERTHO J.M., LECOMPTE-PRADINES C. et al., in MENAGER M.T., GARNIER-LAPLACE J., GOYFFON M. (eds), *Toxicologie nucléaire environnementale et humaine*, Lavoisier, Paris, pp. 469–488, 2009.

[UNS 22a] UNSCEAR, Sources, effects and risks of ionizing radiation, Volume I, Annex A: Evaluation of medical exposure to ionizing radiation, United Nations, New York, 2022.

[UNS 22b] UNSCEAR, Sources, effects and risks of ionizing radiation, Volume II, Annex B: Levels and effects of radiation exposure due to the accident at the Fukushima Daiichi Nuclear Power Station: implications of information published since the UNSCEAR 2013 Report, United Nations, New York, 2022.

[UNS 22c] UNSCEAR, Sources, effects and risks of ionizing radiation, Volume III, Annex C. Biological mechanisms relevant for the inference of cancer risks from low-dose and low-dose-rate radiation, United Nations, New York, 2022.

[WHO 04] WHO, Uranium in drinking-water, Background document for development of WHO guidelines for drinking-water quality, 2004.

[WHO 11] WHO, Préoccupations liées au nucléaire au Japon – FAQ, 2011. Available at: http://www.who.int/hac/crises/jpn/faqs/fr/index7.html.

[YAB 09] YABLOKOV A.V., NESTRENKO V.B., NESTRENKO A.V., "Chernobyl: consequences of the catastrophe for people and the environment", *Annals of the New York Academy of Sciences*, vol. 1181, no. 1, pp. 221–286, 2009.

# 13

# General Conclusions and Perspectives

## 13.1. Introduction

At the end of this book, several questions deserve to be addressed and developed. Is our knowledge of the harmful effects of ionizing radiation on humans sufficient? Have the mechanisms of action of ionizing radiation on living matter been clarified? What is the value of the estimated radioactive risk to humans? Does this risk estimate have any limitations or weaknesses? How can the current estimate of radioactive risk to humans be improved?

Officially and internationally, the estimation of radioactive risk to humans is based mainly on two international bodies: UNSCEAR and the ICRP (see Chapter 12)

Two criticisms can be leveled at these bodies: the appointment or co-option of their members, some of whom have obvious conflicts of interest since they are employees of the nuclear industry. Furthermore, in its scientific summaries, UNSCEAR makes a selection of the publications available. This process, which is common to most national and international expert bodies, has serious drawbacks, the main one being that expert bodies frequently have opposing views on a scientific problem. There are numerous examples of this in many fields. This fact has been amply demonstrated in two European publications [AEE 01, AEE 13]. This is due to the fact that many official scientific bodies, such as the Academies of Science or expert organizations, are ultra-conservative and have difficulty in questioning scientific facts. This is exacerbated by the pressure exerted by manufacturers, who want to maintain their practices at all costs, even though they are harmful, in order to keep their products manufactured and sold. The most blatant examples of this are the failure to recognize the toxicity of tobacco, asbestos, chlordecone, pesticides, etc. The same phenomenon can be seen in the actual estimation of radioactive risk.

In the case of the health impact of Chernobyl, for example, UNSCEAR did not take into account publications by doctors in Ukraine and Belarus.

## 13.2. Knowledge of the effects of ionizing radiation on humans

Our knowledge of the effects of ionizing radiation on humans has progressed considerably. In recent years, certainties have been acquired. On the contrary, controversy persists in a number of areas. We therefore need to make further progress in our understanding of the molecular, cellular and individual defense mechanisms against ionizing radiation.

The French Academies of Science and Medicine have always been opposed to recognizing the effect of low doses of ionizing radiation. Pleas to this effect were made by both Academies in 2005 [AUR 04] and Tubiana et al. [TUB 07]. Tubiana et al. [TUB 07] asserted that "defense against ionizing radiation involves the cellular microenvironment and the immune system, and that defense mechanisms against low doses of ionizing radiation are different and more effective". Since then, knowledge has shown this hypothesis to be wrong.

### 13.2.1. *Assured knowledge of the effects of ionizing radiation on humans*

Since the harmfulness of ionizing radiation was first recognized at the beginning of the 20th century, knowledge has progressed considerably. Consensus has emerged on the absence of a harmful threshold, and more recently on the fact that low doses have harmful effects, and that these effects are not limited to cancerous diseases. There is almost complete consensus that the response to ionizing radiation is, depending on the case, linear or linear-quadratic.

#### 13.2.1.1. *UNSCEAR summaries*

At irregular intervals, UNSCEAR produces scientific summaries on various subjects. To date, approximately 30 reports have been published. Many reports deal with the consequences of the two major nuclear accidents at Chernobyl and Fukushima.

The 2012 UNSCEAR white paper [UNS 12] concerns the biological mechanisms of action of low-dose radiation. The major advances reported are that understanding of the mechanisms of so-called non-targeted and delayed effects is improving and that there is evidence of differential responses in gene and protein expression for high- and low-dose radiation exposures. Similarly, for immune and inflammatory responses, there is a clearer association with disease.

Appendix B of UNSCEAR'S 2013 report [UNS 13] concerns the effects of radiation on children. It is correct that children may be two to three times more sensitive to radiation than adults for certain health effects. Generally speaking, for approximately 25% of tumor types, children are significantly more radiosensitive regardless of the model. These include leukemia, thyroid, skin, breast and brain cancer. For lung cancer, on the contrary, children are more resistant than adults. For many deterministic effects, there are large variations in amplitude over the period from infancy to childhood and adolescence. CT exposure during childhood is correlated with the induction of neoplasms. Exposure in childhood is more risky than exposure in adulthood (e.g., for cognitive defects, cataracts and thyroid nodules).

Appendix B of the 2018 report [UNS 18b] evaluated epidemiological studies reporting cancer risk by unit dose on the basis of individual cumulative doses resulting from exposure to low-LET (linear energy transfer) radiation from environmental sources. Previous studies, including those on Japanese atomic bomb survivors, as well as medical and occupational exposures, show that some types of cancer are more strongly associated with radiation exposure than others. In the LSS (Life Span Study), for example, the excess relative risk (ERR) per unit dose after childhood exposure is significantly higher for leukemia than for solid cancers. Similarly, the ERR per unit dose for thyroid cancer in people exposed in childhood is significantly higher than for other solid cancers.

An assessment of thyroid cancer data in regions affected by the Chernobyl accident has been carried out by UNSCEAR [UNS 18a]. The total number of cases and the crude incidence rate (number of cases per 100,000 person-years) increased uniformly over the period 2006–2015. It approached 20,000. This number is almost three times higher than the number of thyroid cancer cases recorded in the same cohort during the period 1991–2005.

New results have been published from studies on the incidence of leukemia in people exposed to radiation during childhood and adolescence [UNS 20a]. The strength of the effect is greater if the dose is received in childhood than in adulthood. Among these studies, those on children undergoing diagnostic CT examinations are the most powerful.

The committee [UNS 20b] has previously considered the issue of the sources and effects of exposure to radon ($^{222}$Rn) and thoron ($^{220}$Rn) and their descendants with regard to workers and the public in several assessments. The committee confirmed that inhalation of radon and its descendants is a known lung carcinogen (section 11.3.1.5). However, this does not appear to be the case for Swiss children exposed to high natural doses.

There is very strong and reliable evidence that, following the induction of DNA damage, incomplete, defective or dysfunctional responses to this damage contribute to induced mutation and chromosomal damage. These responses are thus involved in the occurrence of cancers after exposure to all the doses and dose rates studied (section 8.2.1.4). They concern (a) direct DNA damage in the form of DNA double-strand breaks (DSBs) and complex lesions and (b) damage attributable to the generation of reactive oxygen and related species through effects on mitochondria [UNS 21].

Chromatin remodeling pathways are probably also relevant to the risk of cancer from ionizing radiation (section 8.2.2). Similarly, this risk may arise through action on epigenetic regulatory genes. Modeling studies supported by experimental work indicate that low-dose exposures can promote the growth of pre-existing precancerous cells and clones in tissues. This is a pathway in addition to the direct induction of mutations in cells by radiation, in which low-dose exposures are predicted to increase cancer risk [UNS 21].

Some studies of gene expression following radiation exposure suggest different responses at low versus moderate/high doses. In addition, there is long-standing evidence that the number of mutational steps required for leukemia is lower than for solid cancers [UNS 21].

According to UNSCEAR in 2021 [UNS 21], a global threshold for cancer induction is unlikely, and there is evidence that known mechanisms work at least up to 10 mGy. Moreover, the mutational mechanism would imply a non-threshold dose–risk relationship. At the lowest dose levels, where DNA double-strand breaks are induced in around one cell in 10, or even less (approximately 2 mGy of exposure at low LET), effects mediated by reactive oxygen species (ROS) are likely to predominate, and these include the potential promotional action of radiation.

#### 13.2.1.2. *The effects of low-dose ionizing radiation are harmful*

Estimating the risks, in particular carcinogenic risks, to which subjects exposed to very low doses of irradiation (below 20 mSv) could be subjected is important in medicine, since virtually all radiological examinations deliver doses that do not exceed this limit. Irradiations to which workers are exposed, or people living in regions where natural irradiation is high, are also of this order or lower [AUR 04].

The UNSCEAR synthesis [UNS 21] is clear on this point: low doses of radiation have harmful effects (see section 12.2.1.1). Examples abound of adverse effects occurring at doses below 1 mSv.

#### 13.2.1.3. *Dose response is linear or linear-quadratic*

The dose–response relationships for mutation (LOH, Loss of Heterozygosity) and micronucleus formation are linear in the low-dose range up to at least 50 and 10 mGy of low-LET radiation, respectively. Similarly, the dose response for activation of the DNA damage response (DDR) is linear up to 10 mGy of low-LET radiation [UNS 21].

In the low-exposure regime, it is legitimate to assume that doses are independent. The effect of an additional dose therefore does not depend on the previous dose. The notion of cumulative dose is of little importance, since a new exposure is unlikely to affect the same cell, and the repair mechanisms at work should have time to play their role if this should nevertheless occur. The linear-quadratic model is one of the key tools of radiobiology and physics. It provides a simple relationship between cell survival and delivered dose and has been widely used to analyze and predict responses to ionizing radiation both in vitro and in vivo. Despite its ubiquity, questions remain about its interpretation and wider applicability. The main question is whether it is a practical empirical fit or representative of a deeper mechanistic behavior. A recent review by McMahon [MCM 19a, b] sheds some light [LAU 23].

#### 13.2.1.4. *The relationship between human cancer-initiating effects and epidemiological studies*

There are two ways of linking information on biological processes and mechanisms deemed relevant to human cancer with existing epidemiological data on disease incidence in exposed populations. The first is to apply biomarkers of radiation-related disease in epidemiological risk investigations. The second approach is to integrate qualitative and quantitative biological data into mechanistic modeling of cancer risk. As it happens, quantitative mechanistic modeling of radiological risk is more advanced than that currently employed in the field of chemicals [UNS 21].

### 13.2.2. *Imperfect knowledge of the effects of ionizing radiation on humans*

Despite considerable progress in our understanding of the harmful effects of ionizing radiation, there are still a number of gaps in our knowledge. These include, in particular, molecular and cellular defense mechanisms.

#### 13.2.2.1. *Defense mechanisms against ionizing radiation*

These mechanisms are numerous and relatively effective. However, there is ample evidence that they fail.

The cell does not passively undergo the accumulation of radiation damage. It reacts by at least three mechanisms (section 8.4.4). The first is the fight against the active forms of oxygen created by oxidative stress. The elimination of damaged, potentially mutagenic cells relies on two mechanisms: i) apoptosis, which can be triggered by doses of the order of a few mSv and which can be used to kill cells whose genome has been altered or is malfunctioning, and ii) mitotic death of cells whose damage has not been repaired. The third mechanism involves DNA repair systems that are stimulated or activated by doses of the order of ten mSv [AUR 04].

#### 13.2.2.2. *The possible existence of hormesis*

A few experiments demonstrate a hormesis effect, but many others contradict this phenomenon. All the major organizations concerned with radiation protection (UNSCEAR, ICRP, BEIR, etc.) have consistently rejected hormesis. Indeed, even though hormesis does exist in certain cases, the precautionary principle takes precedence.

This controversy is still with us today. It can be summarized by the debate between Doss and Little [DOS 14]. Doss asserts that when free radical production is increased, for example from exposure to low-dose radiation (or increased physical/mental activity), our body responds with enhanced defenses consisting of, among other things, increased antioxidants, DNA repair enzymes, immune system response, confirming adaptive protection. He adds that there is considerable evidence from animal studies supporting the hypothesis that low-dose radiation reduces the likelihood of cancer as well as non-malignant diseases. This translates into better tumor control and reduced metastasis during radiotherapy of non-Hodgkin's lymphoma patients.

For his part, Little (in [DOS 14]) notes that there is evidence of an excess incidence of cancer of most types associated with radiation exposures of the order of 10–20 mGy resulting from diagnostic X-ray exposure in a British study (*Oxford Survey of Childhood Cancers*) and in various other groups exposed in utero (see also section 9.2.2.5 [PEA 12, MAT 13]). This implies that in utero fetal doses of the order of 10 mSv significantly increase the risk of childhood cancer (see section 9.2.1.3). At slightly higher doses, increased risks of leukemia and brain cancer have been observed in patients who were exposed in childhood to multiple CT scans resulting in doses of approximately 60 mGy to the respective tissues (red bone marrow, brain). The excessive risks in all these studies are consistent with those in the Japanese data on atomic bomb survivors. In addition, a recent systematic review and meta-analysis have suggested an excess risk associated with radiation at occupational and environmental dose levels (cumulative whole-body dose <0.5 Sv). There is also growing evidence, from Japanese atomic bomb survivors and various other groups exposed to moderate and low doses, of the existence of an excess risk of cataracts [DOS 14]. These examples demonstrate the non-existence of hormesis.

### 13.2.2.3. *Low- and high-LET effects*

Are the mechanisms that operate and may be associated with cancer tumor development similar after low- and high-LET exposures?

Given the higher frequency of complex/cluster damage induction after high-LET exposure, the DDR to these lesions will play a more important role after low-LET exposures. One study demonstrates that low doses (50 mGy) of low-LET radiation act to promote the growth of the Trp53 mutation in esophageal epithelial cells (section 8.4.5.2). Modeling that indicates an inhomogeneous distribution of radon (and its descendants) in the lungs leads to prolonged exposure at a high dose rate of a small population of cells, and this prolonged irradiation is likely to have an impact on lung carcinogenesis and radon-related risks [UNS 21].

The yield of strand breaks per unit dose is almost constant over a wide LET range (see Chapter 8). For low-LET radiation, approximately 30% of double-strand breaks are complex, due to additional breaks. The proportion of such clustered damage increases with LET, reaching approximately 70% for high-LET radiation [NIK 01].

The frequency of DNA mismatch following X-ray-induced double-strand breaks is nonlinearly related to dose, with less probability of mismatch at low doses than at high doses. The dose dependence for high-LET particles, on the contrary, is closer to linear, with higher frequencies of misjoining than for X-rays, particularly at lower doses. However, the main determinant of the probability of misjoining is the distance between two double breaks present simultaneously [RYD 05].

### 13.2.2.4. *Effects of low-dose ionizing radiation on non-cancerous diseases*

For decades, the only harmful effects of ionizing radiation considered were cancerous diseases. Since then, however, it has become clear that various non-cancerous diseases are caused by radiation. These include cardiovascular disease and cataracts, all at very low radiation doses. This radically changes our view of the radioactive risk to humans.

The potential contributions of phenomena such as transmissible genomic instability, bystander phenomena and adaptive response are still unclear [UNS 21]. Substantial new data on low-dose and low-dose rate risk have been obtained from epidemiological investigations, in particular from occupational and medical cohorts. These studies have added to the epidemiological evidence underlying the estimation of cancer risk at low dose and low-dose rate. Furthermore, UNSCEAR [UNS 21] concluded that there is good justification for the use of a no-threshold model for risk inference given the sound knowledge of the role of mutations and chromosomal aberrations in carcinogenesis.

Bystander effects and adaptive responses are still unclear according to UNSCEAR [UNS 21]. Some studies have suggested thresholds for the induction of transmissible instability and bystander effects at approximately 100 mGy of low-LET radiation. If this is true, it would indicate that the phenomena are irrelevant for the inference of low-dose cancer risk. Furthermore, adaptive response studies remain without a confirmed mechanistic basis and have mixed results. Similarly, studies from people living in areas with high levels of natural background radiation, which are interpreted by some as providing evidence of an adaptive response, are not sufficiently consistent to be adopted for risk assessment purposes according to UNSCEAR [UNS 21].

#### 13.2.2.5. *Transmission of the effects of ionizing radiation to descendants*

The implications of studies on the transmissible induction of genomic instability are not clear-cut [UNS 21]. A contradiction appears between the results at the genetic level, where mutations in the DNA sequence and epigenetic modifications of this molecule are transmitted to descendants (see Chapter 8) without apparently having any consequences on morbidity (see Chapter 9). Research is needed to resolve this discrepancy.

#### 13.2.2.6. *Radiosensitivity*

The notion of individual radiosensitivity to radiation refers to hypersensitivity to ionizing radiation. This variability in the effect of the same dose depending on the individual receiving it is still an area of research [ASN 13]. It can vary by a factor of 10 between individuals (see Chapter 6). The phenomenon of individual radiosensitivity probably affects between 5% and 15% of the population [BOU 12].

Recent immunofluorescence techniques have lowered the detection threshold for DNA DSBs to 1 mGy (a factor of 100). Exposures of cultured human mammary epithelium cells have shown that the phenomenon of individual hyper-radiosensitivity also exists at low mammography doses. The mechanisms of individual radiosensitivity are linked to abnormalities in the detection and repair of DNA damage [BOU 12].

However, according to the Nuclear Energy Agency [NEA 16], differences in individual sensitivity to ionizing radiation due to genetic, epigenetic, age and gender variations do not justify modifying the radiation protection system. This assertion seems highly debatable. In fact, the effects of radiation on women are more severe than on men. Furthermore, as intestinal transit time increases with age, and even doubles between the ages of 20 and 50–60, the internal dose is higher for the latter.

### 13.2.3. *Towards a better understanding of the effects of ionizing radiation on humans*

The various questions raised in the previous section clearly demonstrate the need to deepen our knowledge of the effects of ionizing radiation on humans. In particular, we need to understand the molecular mechanisms of the response to radiation [CLA 99, NEA 16]. Several avenues need to be explored, such as the use of biomarkers.

Following irradiation, the induction of biological effects in cells involves several complex stages (see Chapter 8). The first stage concerns interactions between radiation and the cellular environment. These consist of physical and chemical reactions that produce ions, excited molecules and radical species. Excitations and ionizations are complete in approximately 10–15 s, and are followed by chemical thermal equilibrium of the species produced in 10–12 s [POU 01].

#### 13.2.3.1. *Biomarkers*

The development of possible biomarkers for radiation-induced diseases would greatly assist epidemiological studies (section 8.5). It is necessary to integrate molecular, epidemiological and systemic biological studies, for example using the Organization for Economic Cooperation and Development (OECD) chemical risk approach [NEA 16].

Radiation biodosimetry aims to determine characteristic early biomarkers of radiation exposure. This is necessary for effective triage after a radiological incident. Radiation metabolomics can meet this objective by assessing metabolic perturbations following exposure [PAN 19].

In addition to the biomarkers specific to thyroid cancers presented in Chapter 11, other biomarkers are likely to provide early detection of the onset of other cancers. This is the case for squamous cell carcinoma of the head and neck, where molecular and genetic determinants of tumor cell response to radiation have been studied, and several potential biomarkers are emerging that could further improve the application and efficacy of radiotherapy for this tumor [OW 15]. Similarly, research is underway to find biomarkers marking the radiosensitivity of certain individuals [SUB 21].

The hypothesis was that biomarkers of genotoxicity should be higher in healthcare workers exposed to ionizing radiation than in those not exposed. The frequencies of chromosomal aberrations and micronuclei were significantly different between ionizing radiation-exposed and unexposed workers (3.19, 95% CI 1.46–4.93 and 1.41, 95% CI 0.97–1.86, respectively, for total number of aberrant cells and micronuclei frequencies), which was not the case for ring chromosomes and nucleoplasmic bridges [BAU 21].

However, the majority of biomarker research is carried out for lung cancers. With an increasing number of predictive biomarkers emerging in non-small cell lung carcinoma (NSCLC), immunohistochemistry (IHC) is being used as a rapid and cost-effective tool for screening and detection of many of these markers [JAI 19]. Various studies have demonstrated the significant values of cytokines such as TGF-β1, IL-6, KL-6, surfactant proteins and IL-1ra in predicting radiation-induced induction of lung cancer [KON 08]. There are now several subgroups of lung cancer patients defined by biomarkers [PEN 19]. Programmed expression of death-ligand 1 protein by IHC is now widely used for the selection of programmed cell death inhibitor 1 therapy in NSCLC patients [SHO 20]. Indoleamine 2, 3-dioxygenase (IDO) activity is inversely related to the risk of distant metastasis and survival in stage III NSCLC. The study by Wang et al. [WAN 20] validated the fact that elevated IDO activity correlated with poorer survival outcomes in patients with early-stage NSCLC treated definitively with three-dimensional radiotherapy.

#### 13.2.3.2. *DNA break repair mechanisms*

Knowledge of early radiobiological events has evolved considerably in recent years. A better understanding of DNA repair mechanisms will open up the range of therapeutic treatments for cancer by inducing mutation of the molecules essential for repair in cancer cells [PIL 19].

Alterations in DDR pathways are characteristic of cancer. Incorrect repair of DNA damage often leads to genomic instability. Ataxia telangiectasia, mutated (ATM), a central component of the DNA repair system, is activated to enhance the homologous recombination repair pathway during DNA double-strand breaks. Other molecules involved in DDR include PARP (Poly (Adp-Ribose) Polymerase) and ATR (Ataxia Telangiectasia and Rad3 Related) [JIN 19].

The main biomarkers of DNA double-strand break repair are the signaling molecules most specific to these breaks, namely ATM, DNA-PKcs, 53BP1 and γ-H2AX [VAL 17], and also BRCA1, BRCA2, Ku70 and Ku80 [GRA 06].

Grabsch et al. [GRA 06] carried out the first study to show a relationship between expression of the DNA repair proteins ATM and BRCA1 after a double break and survival in colorectal cancer patients.

A novel MRE11-dependent DNA double-strand break repair pathway was observed. Dysfunction of the BRCA1 protein in response to radiation enabled Bencokova [BEN 07] to propose a preliminary model involving the BRCA1 and NF1 proteins.

Several molecules are involved in triggering DNA break repair. Among these is the ATM kinase. The ATM protein is a major player in the response to ionizing radiation. It diffuses from the cytoplasm to the nucleus after irradiation. The demonstration of this cytonuclear transit of the ATM protein has provided a coherent biological interpretation of the linear-quadratic model in its most mathematical aspects, validated by around a hundred cases [BOD 16].

The role of the ATM, BRCA1 and TP53 DNA DSB checkpoint/repair genes in sporadic breast cancer needs to be clarified, as ATM and BRCA1 mutations are rare in sporadic tumors [DIN 04].

#### 13.2.3.3. *Immunology*

The immune system of vertebrates, including humans, has evolved considerably since that of other organisms. It comprises two main pathways: innate and adaptive responses, each employing a large number of cellular components (Figure 13.1).

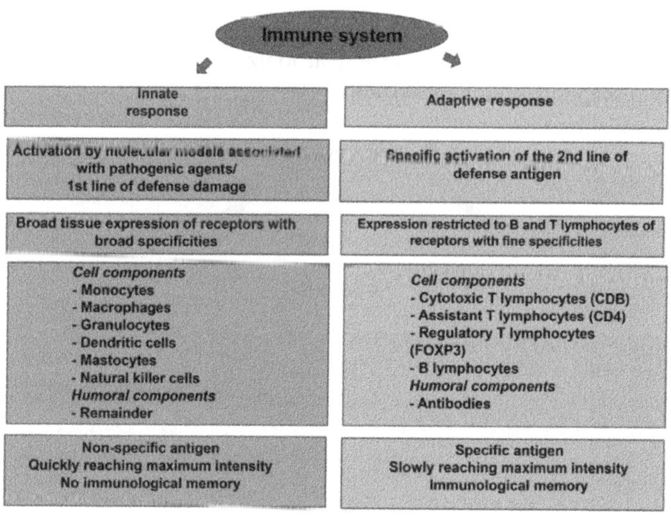


**Figure 13.1.** *Schematic representation of the human immune system and its main features (according to [LUM 21])*

Ionizing radiation interacts with the immune system in many ways, reflecting the complexity of the immune system itself. Immune system responses to ionizing radiation can be good biomarkers (Table 13.1).

| Effects of ionizing radiation | Possible immunological biomarker |
|---|---|
| Imbalance of peripheral blood mononuclear cells | Changes in the number of B, T and NKT lymphocytes<br>Changes in T lymphocyte subpopulations |
| Accelerated immuno-aging | Reduction of native T lymphocytes<br>Expansion of memory T cells<br>Thymus involutions<br>Reduced telomere length in leukocytes |
| Humoral immune response | Changes in immunoglobulin levels (IgA, IgM, IgO, IgE) |
| Inflammation | Pro-inflammatory cytokine response |

**Table 13.1.** *Summary of immune effects described by radiation exposure (according to [LUM 21])*

Depending on the dose of ionizing radiation, immune system responses will vary (Figure 13.2). The central conclusion is that ionizing radiation fundamentally and permanently reshapes the immune system.

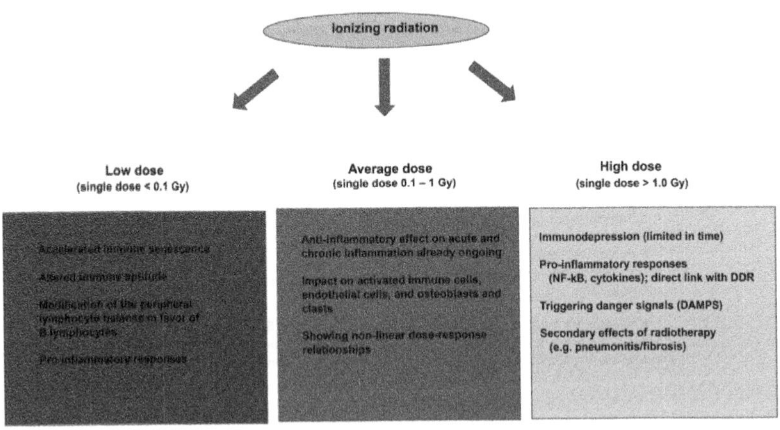


**Figure 13.2.** *Schematic representation of the most important immune and inflammatory procedures developing after low-, medium- and high-dose irradiation based on available epidemiological, clinical and experimental data (according to [LUM 21]). DDR: DNA damage response; DAMPs: damage-associated molecular patterns*

#### 13.2.3.4. *Mechanisms for combating oxidative stress*

After exposure to ionizing radiation, cellular functions and antioxidant defense are altered. There is an increase in the production of ROS, an increase in malondialdehyde (MDA) and a decrease in superoxide dismutase (SOD), glutathione peroxidase and catalase (CAT), i.e. a reduction in defenses against oxidative stress [AKB 19].

#### 13.2.3.5. *Mechanistic models*

In radiobiology, most models commonly used in the clinic are abstract and empirical and do not offer significant scope for mechanistic interpretation or prediction in new cell lines or therapies. Relevant mechanistic models need to take account of the various physical, chemical and biological steps involved (Figure 13.3).

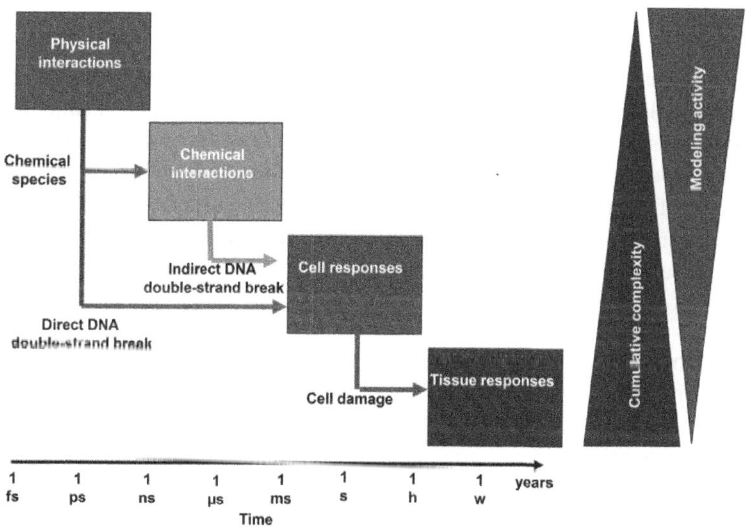


**Figure 13.3.** *Illustration of the stages of a complete mechanistic response model and approximate associated timescale (according to [MCM 19b])*

Initial responses are dominated by physical interactions, whose models generate distributions of reactive chemical species, with direct DSBs typically generated in less than a nanosecond. These chemical species can also be included in a chemical interaction model, which generates indirect DSB predictions, on a scale of 1 ms or less. Responses at the cellular level then unfold over hours or days, as cells repair this damage and potentially respond by shutting down, dying or seeing other changes in function. Finally, tissue-level responses integrate these responses on a scale of hours to years, giving rise to clinically observable effects. Mechanistic modeling activity in radiotherapy has focused mainly at the beginning of this timescale, on early physico-chemical and cellular responses, with relatively little modeling of whole-tissue responses. This is partly due to the great cumulative complexity of describing all aspects of these responses in a single system [MCM 19a].

## 13.3. Strengths and weaknesses of the radioactive risk assessment method

In this book, radioactive risk assessment has been carried out in eight stages. Each of these stages has its own shortcomings, which are listed below.

### 13.3.1. *Shortcomings in identifying radioactive hazards*

The radioactive danger posed to humans is not unique, but multiple. Indeed, many radionuclides are likely to have harmful effects on human health. Ionizing radiation comes in a variety of forms. For some, interactions with living matter are well understood, while for others, our knowledge is much more incomplete. Among these types of radiation, the least studied are those caused by Auger electrons [IRS 10]. There are many ways in which humans are exposed to ionizing radiation, some of natural origin and others man-made.

### 13.3.2. *Shortcomings in quantifying radioactive contamination of the anthroposphere*

Radionuclide metrology is currently highly efficient, sensitive and accurate. The only remaining uncertainties concern alpha-emitting radionuclides, where quantification is more difficult.

The main gap in this field concerns the inaccuracy of quantifying the microdistribution of radionuclides in human cells. This microdistribution is essential for alpha particles and low-energy beta emitters (soft beta). Depending on whether this type of radionuclide is located inside or outside the cell nucleus, the consequences are considerably different. The main radionuclides concerned by this issue are therefore all alpha emitters such as uranium and all transuranics [ICR 19], and especially beta emitters such as tritium and carbon-14 [HOF 20]. The average path of tritium in living matter is 0.6 μm, with a maximum of 6 μm – far less than the size of a human cell.

### 13.3.3. *Shortcomings in the identification of exposure pathways*

Humans are exposed to ionizing radiation via a wide variety of pathways. The natural origin of radionuclides derives from two sub-origins, telluric and cosmic, while the man-made origins are military, industrial and medical. There are no gaps in the inventory at this level.

### 13.3.4. *Shortcomings in estimating exposure doses*

External exposure doses are fairly well understood and quantified. On the contrary, internal exposure doses are poorly estimated, particularly for low-energy alpha and beta emitting radionuclides (known as soft beta) due to the heterogeneity of their intracellular microdistribution. This depends on whether the radionuclide is located on the cell membrane or in the nucleus. Furthermore, an organ is frequently an assembly of several types of cells which, with regard to ionizing radiation, will be more or less radiosensitive, have different physiological functions, more or less important for the health of the individual and a bioaccumulation that may be specifically different. In addition, assessment by organ is not valid for this type of ionizing radiation, which travels very small distances in living matter.

### 13.3.5. *Gaps in the assessment of adverse effects at various biological levels*

The harmful effects of ionizing radiation occur at all levels of biological organization, from the molecule to the individual, and affect a wide range of functions. Yet officially, the only effects used to estimate the radioactive risk are the appearance of various cancers. This excludes all other health effects, whether physical, such as circulatory system diseases, or psychological, such as anxiety attacks.

### 13.3.6. *Shortcomings of the dose versus adverse effect relationship*

The linear relationship between dose and adverse effect is well-documented and frequently verified for high doses of radiation and to a lesser extent for medium doses. On the contrary, for low doses of radiation, numerous controversies have arisen concerning the dose–response relationship.

The first controversy concerns the form of this relationship. Is it linear, infra-linear or supra-linear? The second controversy concerns the presence or absence of a threshold[1] for any harmful effect of ionizing radiation. The third controversy concerns the presence or absence of a restorative effect of low doses (hormesis effect). The fourth controversy concerns the importance of bystander effects on cells not targeted by ionizing radiation, but which are nevertheless damaged and/or stressed. The latest work in this field supports a linear response, with no threshold and no restorative effect, and suggests a reassessment of the bystander effect.

### 13.3.7. *Shortcomings in the choice of no-adverse-effect values*

Unlike chemical risk assessment for metals or metalloids naturally present in the environment, such as mercury, lead or cadmium, radioactive risk assessment does not take into account the smallest threshold effect of adverse effects occurring at lower biological levels, such as molecules or cells. The most common threshold is the no observed adverse effect level (NOAEL) [AMI 17]. If this approach were applied to nuclear power, it is clear that the nuclear industry would be severely limited in its development, as is the case for the mercury, lead and cadmium industries.

### 13.3.8. *Shortcomings in the organization of official standards setting*

Chapter 12 described the four main phases involved in setting radiation protection standards. This approach is based on international organizations (UNSCEAR, ICRP, IAEA). However, as pointed out in the introduction to this book, not all these institutions are free from conflicts of interest. For example, the IAEA is closely linked to nuclear technology, and various members of UNSCEAR and ICRP are also linked to the nuclear industries. An effort must be made to ensure that there is no conflict of interest between those who assess risks and set international radiation protection standards, and those who use ionizing radiation in their professional activities.

### 13.3.9. *Shortcomings in radioactive risk assessment*

Assessing a physical or chemical risk seems straightforward, since all that is required is to relate exposure dose estimates to the doses of adverse effects observed. In the case of radioactive risk assessment, this presupposes that both the

1 In toxicology, the threshold dose is the dose at which a measurable reaction of an adverse effect occurs. Below this threshold value, no perceptible reaction is measurable, even though an adverse reaction does exist. Liminal effects are therefore the very first observable signs of a harmful phenomenon. The most classic threshold is the no observable effect concentration (NOEC).

radiation dose received and the adverse effects are correctly estimated. The comments above clearly demonstrate that this is not the case. The dose estimate is underestimated, and this necessarily results in an underestimation of the radioactive risk. Furthermore, the ICRP (ICRP 2012) proposes overestimated estimates of the threshold adverse effect for the various physiological systems and organs. The extent of this underestimation will depend on the underestimation of the dose and the overestimation of the threshold effect. Given the current state of knowledge, it is impossible to answer this question. It is therefore necessary to pursue approaches in this field, striving to fill the gaps identified above. These gaps are summarized in Figure 13.4.

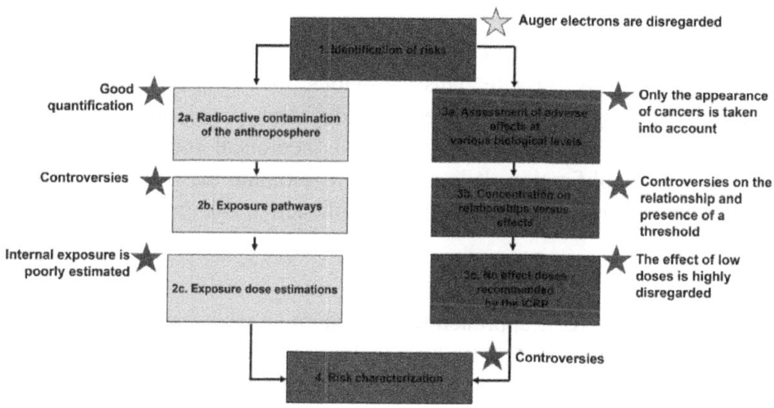


**Figure 13.4.** *Gaps in human radioactive risk estimation. The color of the stars corresponds to the level of the gap: low green, medium yellow and high red*

## 13.4. Towards a new approach to estimating radioactive risk

The ICRP's current approach has its limitations, even though there has been a marked improvement in recent years.

### 13.4.1. *Limitations of the ICRP approach*

The main limitations of the ICRP's approach concern its conception of a standard average person and questionable parameters.

#### 13.4.1.1. *The average standard person*

The ICRP's approach is based on estimating the risk for the average male. This approach is too limited, as even for an adult there is considerable individual variability. Furthermore, half of all adults are women, and more than half of the world's human population are not adults. Furthermore, the standard male varies from one population to another, being very young in Africa and much older in Western Europe. It would be better to think in terms of the most radiosensitive individual – what is known in all physical and chemical risk assessments as a "worst case".

However, as in all animal species and for the vast majority of physical and chemical hazards, the youngest individuals are the most sensitive, and the younger they are, the more sensitive they are. This is true for radiosensitivity, although it seems that the infant is more radiosensitive than the fetus or embryo.

At every age, the parameters likely to have an impact on radiosensitivity are characteristic and highly variable. This is obvious, for example, in the case of diet, which varies with age, and also between different populations, and within a population between individuals. The standard theoretical individual approach is therefore an oversimplification. It is essential to estimate the extreme variations in parameters that will enable us to estimate the radioactive risk envelope.

#### 13.4.1.2. *Questionable priorities*

For too long, the ICRP's approach was based on an adult male. This priority was questionable, and we should have started by estimating the radioactive risk to embryos, fetuses and young children. Effective radiation protection must protect the entire human population. It must therefore identify which groups within the general population are the most radio-sensitive. For the harmful effects of ionizing radiation, as for most physical, chemical and biological hazards, the groups most at risk are children, from infancy to adolescence, as well as embryos and fetuses. Radiation protection standards must therefore be set to protect these groups. However, the ICRP has only substantially taken into account the specificity of young children since 1989 [ICR 89a, ICR 89b] and that of the embryo and fetus only in 2001 and 2004 [ICR 01, ICR 04].

The cancer risks associated with prenatal and postnatal exposure to ionizing radiation for medical diagnosis in children were assessed by means of a meta-analysis, including 24 studies. For prenatal exposure (X-rays or tomography), no significant increase in risk was reported for all cancers, leukemias and brain tumors. For postnatal exposure, an increased risk was observed only for tomography, mainly for leukemia and brain tumors [ABA 21]. However, this must be qualified for other types of exposure, such as the mortality and morbidity observed in the children of Hiroshima and Nagasaki (section 9.2.3.2). Similarly, other studies (section 9.2.2.3) indicate significant risks for cancers of the digestive system and for cancers specific to women. Prenatal radiation exposure is a risk factor that strongly increases solid cancer mortality in women, but not in men. In contrast, mortality from non-cancerous diseases is increased for both sexes (section 9.2.2.4).

Following the Chernobyl accident, the prevalence of Down's syndrome (or chromosome 21 trisomy) rose sharply in 1987/1988. Based on the fact that maternal meiosis is an error-prone process, the hypothesis of a causal relationship between low-dose irradiation and non-disjunction is the most likely explanation for the observed increase in this syndrome [SPE 12]. This increase in chromosome 21 trisomy was already observed, as early as 1972, in Chinese people living in a naturally high background radiation area (HBRA) [LUX 90].

Similarly, populations subject to the impact of natural ionizing radiation in areas enriched with uranium or thorium have not been studied rigorously enough (see Chapter 11 [GHI 04]). However, only comparisons between geographical regions with high and low natural irradiation and where living conditions are similar could provide information for doses below 20 mSv and low-dose rates, so the results of current surveys in Kerala (India) and China [AUR 04] should be followed closely. Since then, numerous studies have been carried out, with the results showing no effect from natural exposure. This must be strongly qualified, as the numbers involved are always very small, and generally, the number of exposed individuals is greater than the number of control individuals, which is curious. Most authors speak of radio-adaptation, but without providing any arguments in support of this hypothesis.

### 13.4.2. *Questionable parameters*

The example of digestive transit time is illustrative. In 1975, ICRP 23 ([ICR 75], page 352) only gave 24-hour values, without clearly indicating the transit period. However, it has long been known that digestive transit is influenced by many factors, such as age, sex and food.

The duration of digestive transit varies greatly with age. In infants, it is very short compared with adults (10–12 h), due to the liquid nature of the food (milk) [FAB 64]. Likewise, older individuals have a slower colonic transit than younger individuals ($P = 0.0008$). Indeed, normal aging seems to reduce the propulsive capacity of the colon [MAD 04]. Transit in the small intestine varies for an adult from 5 to 6 h. Transit in the large intestine is 2–5 h in the right colon, 2–8 h in the left colon and 14–24 h in the sigmoid colon. Total transit of the large intestine varies from 14 to 48 h [FAB 64]. According to other sources, colonic transit is slow (30–40 h) [DRA 08]. In adults, normal colonic transit lasts 36 h and becomes abnormal if longer than 72 h [CAM 98]. Moreover, gastric evacuation time varies according to the food. For example, 90% of fats are eliminated in 4h 30 min, whereas gastric emptying time for a normal meal varies from 5 to 7 h [FAB 64].

It was not until 2006 that the ICRP carried out a review of variations in digestive transit time [ICR 06]. Gender influences stomach transit, which is found to be longer in women (95 min) than in men (70 min). The same applies to the total transit time for all three parts of the colon, with men taking 36 h and women 48 h [ICR 06].

### 13.4.3. *The evolution of the ICRP's concept of radiation protection*

The ICRP's concept of radiation protection has evolved over time, which is to be expected as scientific knowledge in the field advances. Originally (Figure 13.5), radiation effects were classified into stochastic and non-stochastic or deterministic effects. Stochastic effects include fatal and non-fatal cancers, while non-stochastic or deterministic effects include radiation effects with a threshold-type dose–response model. For radiation protection purposes, mortality and incidence were assessment criteria used to manage the risk of stochastic and non-stochastic or deterministic effects, respectively, while an "emergency dose level" for the thyroid was assigned based on its cancer incidence [HAM 14].

Today, the concept has evolved but has not been fundamentally altered. This evolution is due in part to the Chernobyl accident, which clearly demonstrated its inadequacy in view of the fact that the health of the site's nuclear workers and the population impacted by this accident was not only developing various radiation-induced cancers, and also other physiological illnesses, as well as psychological illnesses with, in particular, symptoms of anguish and anxiety. This concept was also evolving under the pressure of new discoveries such as the "bystander" effects discovered by university teams who had entered the nuclear research field in the wake of this major nuclear accident. Deterministic effects, now called tissue reactions, now include life-threatening circulatory diseases, although no dose limit has been set (Figure 13.5).

The current ICRP system has several drawbacks. The distinction between stochastic lethal effects and tissue reactions is far from obvious for many effects, such as thyroid and skin cancers, cataracts and circulatory diseases. In many cases, the presence of a threshold of effect is increasingly called into question.

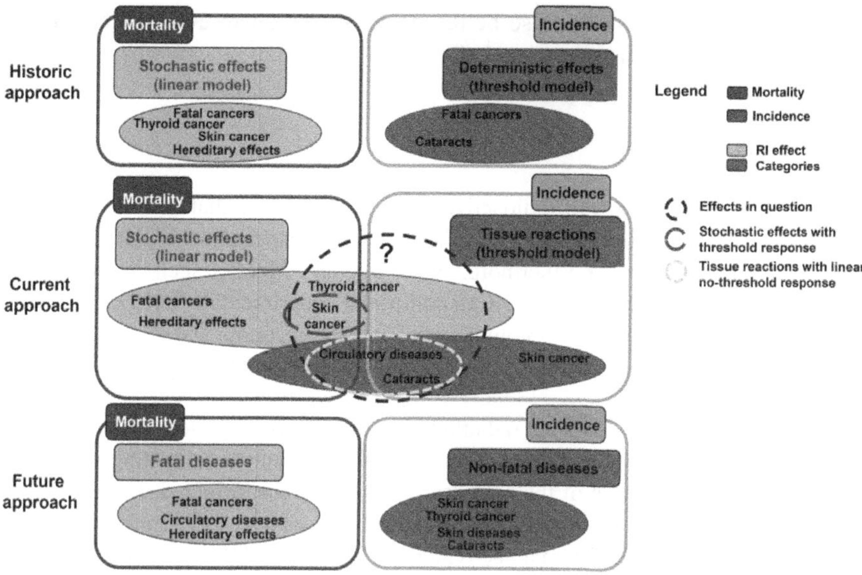


**Figure 13.5.** *Classification of ionizing radiation effects and radiation protection parameters according to three periods: historical, current and proposed future (modified from [HAM 14]). For a color version of this figure, see www.iste.co.uk/amiard/risk.zip*

### 13.4.4. *A new approach to radiation protection*

Serious changes are also needed to the concepts currently used by the ICRP to define both adverse effects and radiation protection parameters.

The diagram in Figure 13.5 shows the original definition, the current situation and Hamada and Fujimichi's [HAM 14] proposed classification of radiation effects and associated radiation protection parameters. The risk of radiation effects is managed either in terms of mortality (red line) or incidence (green lines). The solid orange and blue areas represent the categories of radiation effects. The yellow dotted line shows radiation effects classified as tissue reactions but with a linear non-threshold (LNT) dose response, while the red dotted line shows radiation effects classified as stochastic effects but with a dose similar to a threshold response. A purple dotted line further accentuates radiation effects that blur the boundary between stochastic effects and tissue responses.

The current radiation protection situation can be improved if radiation effects are classified into fatal and non-fatal diseases, for which the risk management assessment criteria are mortality and incidence, respectively. Overall, the need to limit hereditary effects should also be reconsidered, as human evidence has so far shown no detectable transgenerational effects [HAM 14].

## 13.5. Conclusions

The main shortcomings of radioactive risk assessment are that it is based on the occurrence of cancers and neglects all other risks, such as cardiovascular disease or the onset of cataracts. Furthermore, non-targeted effects, such as chemical communication between cells, have been confirmed experimentally at low doses, but while their significance for radiological risk remains uncertain [NEA 16], they are ignored when the basis of risk estimation should be the precautionary principle.

Human exposure to ionizing radiation due to human activities peaked around 1963–1964, due to heavy atmospheric fallout following intensive testing of atomic bombs, with the population of the northern hemisphere most affected. Another peak, mainly affecting European populations, occurred following the Chernobyl accident. Since then, exposure due to human activities has fallen significantly. However, there are three exceptions: tritium, carbon-14 and krypton-85 releases from the nuclear industry, which, instead of falling in line with the ALARA principle and the Sintra Accords signed on July 23, 1998, are rising sharply. While exposures from the nuclear industry are declining in industrialized countries, and in France in particular, exposures from medical uses are increasing considerably, hence the numerous recommendations made by the ICRP and the IAEA on this subject. Each use should be subject to a benefit–risk study.

The assessment of all physical and chemical risks follows an identical destiny: as scientific knowledge on the subject improves, the risk increases and the limits imposed on this risk have to be revised downwards. This is true for radioactive risk, and new, more restrictive thresholds have been applied to all populations, public and professional alike. New population categories have been defined, such as children and pregnant women, with much stricter limits. New risks for non-cancerous diseases have also been introduced.

## 13.6. References

[ABA 21] ABALO K.D., RAGE E., LEURAUD K. et al., "Early life ionizing radiation exposure and cancer risks: systematic review and meta-analysis", *Pediatric Radiology*, vol. 51, no. 1, pp. 45–56, 2021.

[AEE 01] AEE, *Late Lessons from Early Warnings: The Precautionary Principle 1896–2000*, AEE, Copenhagen, 2001.

[AEE 13] AEE, *Late Lessons from Early Warnings: Science, Precaution, Innovation*, 2013. Available at: http://www.eea.europa.eu/publications/latelessons-2.

[AKB 19] AKBARI A., JELODAR G., NAZIFI S. et al., "Oxidative stress as the underlying biomechanism of detrimental outcomes of ionizing and non-ionizing radiation on human health: Antioxidant protective strategies", *Zahedan Journal of Research in Medical Sciences*, vol. 21, no. 4, p. e85655, 2019.

[ALI 15] ALIYU A.S., RAMLI A.T., "The world's high background natural radiation areas (HBNRAs) revisited: A broad overview of the dosimetric, epidemiological and radiobiological issues", *Radiation Measurements*, vol. 73, pp. 51–59, 2015.

[AMI 17] AMIARD J.C., *Les risques chimiques environnementaux. Méthodes d'évaluation et impacts sur les organismes*, 2nd edition, Lavoisier, Tec&Doc, Paris, 2017.

[ASN 13] ASN, L'état de la sûreté nucléaire et de la radioprotection en France, report, 2013.

[ASN 23] ASN, Acteurs et bases de la réglementation en radioprotection, 2023.

[AUR 04] AURENGO A., AVERBECK D., BONNIN A. et al., La relation dose-effet et l'estimation des effets cancérogènes des faibles doses de rayonnements ionisants, Académie des Sciences, Académie nationale de Médecine, 20 October 2004.

[BAU 21] BAUDIN C., BERNIER M.O., KLOKOV D. et al., "Biomarkers of genotoxicity in medical workers exposed to low-dose ionizing radiation: Systematic review and meta-analyses", *International Journal of Molecular Sciences*, vol. 22, p. 7504, 2021.

[BEN 07] BENCOKOVA Z., Signalisation et réparation des cassures double-brin de l'ADN dans les gliomes : Modulation de la réponse aux traitements chimio-radiothérapeutiques, Thesis, Université Joseph-Fourier, Grenoble I, 2007.

[BOD 16] BODGI L., CANET A., GRANZOTTO A. et al., "L'énigme de l'interprétation biologique du modèle linéaire-quadratique enfin résolue ? Une synthèse pour les non-mathématiciens", *Cancer/Radiothérapie*, vol. 20, no. 4, pp. 314–321, 2016.

[BOU 12] BOURGUIGNON M., FORAY N., COLIN C. et al., "Radiosensibilité individuelle et risque aux faibles doses médicales", *Médecine Nucléaire*, vol. 36, pp. 424–428, 2012.

[CAM 98] CAMILIERI M., FORD M.J., "Review article colonic sensorimotor physiology in health, and its alteration in constipation and diarrhoeal disorders", *Alimentary Pharmacology & Therapeutics*, vol. 12, no. 4, pp. 287–302, 1998.

[CLA 99] CLARKE R., "Control of low-level radiation exposure: Time for a change?", *Journal of Radiological Protection*, vol. 19, no. 2, pp. 107–115, 1999.

[DIN 04] DING S.L., SHEU L.F., YU J.C. et al., "Abnormality of the DNA double-strand-break checkpoint/repair genes, ATM, BRCA1 and TP53, in breast cancer is related to tumour grade", *British Journal of Cancer*, vol. 90, 1995–2001, 2004.

[DOS 14] DOSS M., LITTLE M.P., ORTON C.G., "Point/Counterpoint: Low-dose radiation is beneficial, not harmful", *Medical Physics*, vol. 41, no. 7, p. 070601, 2014.

[DRA 08] Dray X., Marteau P., "Physiologie gastro-intestinale de l'homme", in Roberfroid M.B., Coxam V., Delzenne N.M. (eds), *Aliments fonctionnels*, Lavoisier Tec&Doc, Paris, pp. 409–427, 2008.

[FAB 64] Fabry C., Schéma anatomo-physiologique du tractus gastro-intestinal à prendre en considération pour le calcul des niveaux de contamination radioactive, CEA-R-2413 report, 1964.

[GHI 04] Ghiassi-Nejad M., Zakeri F., Assaei R.G. et al., "Long-term immune and cytogenetic effects of high level natural radiation on Ramsar inhabitants in Iran", *Journal of Environmental Radioactivity*, vol. 74, pp. 107–116, 2004.

[GRA 06] Grabsch H., Dattani M., Barker L. et al., "Expression of DNA double-strand break repair proteins ATM and BRCA1 predicts survival in colorectal cancer", *Clinical Cancer Research*, vol. 12, no. 5, pp. 1494–1500, 2006.

[HAM 14] Hamada N., Fujimichi Y., "Classification of radiation effects for dose limitation purposes: History, current situation and future prospects", *Journal of Radiation Research*, vol. 55, pp. 629–640, 2014.

[HOF 20] Hofmann W., Li W.B., Friedland W. et al., "Internal microdosimetry of alpha-emitting radionuclides", *Radiation and Environmental Biophysics*, vol. 59, no. 1, pp. 29–62, 2020.

[ICR 75] ICRP, ICRP Publication 23, Report of the Task Group on Reference Man, Pergamon, Oxford, 1975.

[ICR 89a] ICRP, "ICRP Publication 56, Age-dependent doses to members of the public from intake of radionuclides, Part. 1", *Annals of the ICRP*, vol. 20, no. 2, 1989.

[ICR 89b] ICRP, "ICRP Publication 67, Age-dependent doses to members of the public from intake of radionuclides, Part 2", *Annals of the ICRP*, vol. 23, nos. 3–4, 1989.

[ICR 01] ICRP, "ICRP Publication 88, Doses to the embryo and fetus from intakes of radionuclides by the mother (corrected version May 2002)", *Annals of the ICRP*, vol. 31, nos. 1–3, 2001.

[ICR 04] ICRP, "ICRP Publication 95, Doses to infants from ingestion of radionuclides in mothers' milk", *Annals of the ICRP*, vol. 34, nos. 3–4, 2004.

[ICR 06] ICRP, "ICRP Publication 100, Human alimentary tract model for radiological protection", *Annals of the ICPR*, vol. 36, nos. 1–2, pp. 1–336, 2006.

[ICR 19] ICRP, "ICRP Publication 141, Occupational intakes of radionuclides: Part 4", *Annals of the ICRP*, vol. 48, nos. 2–3, pp. 1–514, 2019.

[IRS 10] IRSN, L'évaluation et la gestion des risques associés aux expositions aux radionucléides émetteurs Auger et bêta, avis et propositions de pistes de recherche, Report DS-2010-03, 2010.

[JAI 19] Jain D., Nambirajan A., Borczuk A. et al., "Immunocytochemistry for predictive biomarker testing in lung cancer cytology", *Cancer Cytopathology*, vol. 127, pp. 325–339, 2019.

[JIN 19] JIN M.J., OHA D.Y., "ATM in DNA repair in cancer", *Pharmacology & Therapeutics*, vol. 203, p. 107391, 2019.

[KON 08] KONG F.M., AO X., WANG L. et al., "The use of blood biomarkers to predict radiation lung toxicity: A potential strategy to individualize thoracic radiation therapy", *Cancer Control*, vol. 15, no. 2, pp. 140–150, 2008.

[LAU 23] LAURIER D., BILLARAND Y., KLOKOV D. et al., "Fondements scientifiques de l'utilisation du modèle linéaire sans seuil (LNT) aux faibles doses et débits de dose en radioprotection", *Radioprotection*, vol. 58, no. 4, pp. 243–260, 2023.

[LUM 21] LUMNICZKY K., IMPENS N., ARMENGOL G. et al., "Low dose ionizing radiation effects on the immune system", *Environment International*, vol. 149, p. 106212, 2021.

[LUX 90] LUXIN W., TONGRU Z., ZUFAN T. et al., "Epidemiological investigation in high background radiation areas in Yangjiang, China", in SOHRABI M., AHMED J.U., DURRANI S.A. (eds), *High Levels of Natural Radiation, Proceedings of an International Conference on High Levels of Natural Radiation*, Ramsar (Iran), pp. 523–547, 1990.

[MAD 04] MADSEN JL, GRAFF J., "Effects of ageing on gastrointestinal motor function", *Age and Ageing*, vol. 33, pp. 154–159, 2004.

[MAT 13] MATHEWS J.D., FORSYTHE A.V., BRADY Z. et al., "Cancer risk in 680 000 people exposed to computed tomography scans in childhood or adolescence: data linkage study of 11 million Australians", *BMJ*, vol. 346, 2013.

[MCM 19a] MCMAHON S.J., "The linear quadratic model: usage, interpretation and challenges", *Physics in Medicine & Biol*ogy, vol. 64, p. 01TR01, 2019.

[MCM 19b] MCMAHON S.J., PRISE K.M., "Mechanistic modelling of radiation responses", *Cancers*, vol. 11, p. 205, doi:10.3390/cancers11020205, 2019.

[NEA 16] NEA, Report radiological protection science and application, OECD, NEA No.7265, 2016.

[NIK 01] NIKJOO, H., O'NEILL, P., WILSON, W.E. et al., "Computational approach for determining the spectrum of DNA damage induced by ionizing radiation", *Radiation Research*, vol. 156, pp. 577–583, 2001.

[OW 15] OW T.J., PITTS C.E., KABARRITI R. et al., "Effective biomarkers and radiation treatment in head and neck cancer", *Archives of Pathology & Laboratory Medicine*, vol. 139, pp. 1379–1388, 2015.

[PAN 19] PANNKUK E.L., LAIAKIS E.C., GIRGIS M. et al., "Temporal effects on radiation responses in nonhuman primates: Identification of biofluid small molecule signatures by gas chromatography–mass spectrometry metabolomics", *Metabolites*, vol. 9, no. 5, p. 98, doi: 10.3390/metabo9050098, 2019.

[PEA 12] PEARCE M.S., SALOTTI J.A., LITTLE M.P. et al., "Radiation exposure from CT scans in childhood and subsequent risk of leukaemia and brain tumours: A retrospective cohort study", *Lancet*, vol. 380, no. 9840, pp. 499–505, 4 August 2012.

[PEN 19] PENNELL N.A., ARCILA M.E., GANDARA D.R. et al., "Biomarker testing for patients with advanced non–small cell lung cancer: Real-world issues and tough choices", *ASCO Educational Book*, vol. 39, pp. 531–542, 2019.

[PIL 19] PILIE P.G., TANG C., MILLS G.B., et al., "State-of-the-art strategies for targeting the DNA damage response in cancer", *Nature Reviews. Clinical Oncology*, vol. 16, pp. 81–104, 2019.

[POU 01] POUGET J.P., MATHER S.J., "General aspects of the cellular response to low- and high-LET radiation", *European Journal of Nuclear Medicine and Molecular Imaging*, vol. 28, pp. 541–561, 2001.

[RYD 05] RYDBERG B., COOPER B., COOPER P.K. et al., "Dose-dependent misrejoining of radiation-induced DNA double-strand breaks in human fibroblasts: Experimental and theoretical study for high and low LET radiation", *Radiation Research*, vol. 163, no. 5, pp. 526–534, 2005.

[SHO 20] SHOLL L.M., HIRSCH F.R., HWANG D. et al., "The promises and challenges of tumor mutation burden as an immunotherapy biomarker: A perspective from the International Association for the Study of Lung Cancer Pathology Committee", *Journal of Thoracic Oncology*, vol. 15, no. 9, pp. 1409–1424, 2020.

[SPE 12] SPERLING K., NEITZEL H., SCHERB H., "Evidence for an Increase in Trisomy 21 (Down Syndrome) in Europe after the Chernobyl Reactor Accident", *Genetic Epidemiology*, vol. 36, pp. 48–55, 2012.

[SUB 21] SUBEDI P., GOMOLKA M., MOERTL S. et al., "Ionizing radiation protein biomarkers in normal tissue and their correlation to radiosensitivity: A systematic review", *Journal of Personalized Medicine*, vol. 11, p. 140, 2021.

[TUB 07] TUBIANA M., MASSE R., DE VATHAIRE F. et al., "La controverse sur les effets des faibles doses de rayonnements ionisants et la relation linéaire sans seuil", *Radioprotection*, vol. 42, no. 2, pp. 133–161, 2007.

[UNS 12] UNSCEAR, Biological mechanisms of radiation actions at low doses, 2012 White Paper, United Nations, New York, 2012.

[UNS 13] UNSCEAR, Sources, effects and risks of ionizing radiation, Volume II, Annex B - Effects of radiation exposure of children, United Nations, New York, 2013.

[UNS 18a] UNSCEAR, Evaluation of data on thyroid cancer in regions affected by the Chernobyl accident, White Paper, United Nations, New York, 2018.

[UNS 18b] UNSCEAR, Sources, effects and risks of ionizing radiation, Annex B - Epidemiological studies of cancer risk due to low-dose-rate radiation from environmental sources, United Nations, New York, 2018.

[UNS 20a] UNSCEAR, Sources, effects and risks of ionizing radiation, Annex A – Evaluation of selected health effects and inference of risk due to radiation exposure, United Nations, New York, 2020.

[UNS 20b] UNSCEAR, Sources, effects and risks of ionizing radiation, Annex B – Lung cancer from exposure to radon, United Nations, New York, 2020.

[UNS 21] UNSCEAR, Sources, effects and risks of ionizing radiation, Volume III, Annex C, Biological mechanisms relevant for the inference of cancer risks from low-dose and low-dose-rate radiation, United Nations, New York, 2021.

[VAL 17] VALLARD A., RANCOULE C., GUY J.B. et al., "Radiothérapie et biomarqueurs de la réparation de l'ADN", *Bulletin du cancer*, vol. 104, no. 11, pp. 981–987, 2017.

[WAN 20] WANG W., HUANG L., JIN J.Y. et al., "A validation study on IDO immune biomarkers for survival prediction in non-small cell lung cancer: Radiation dose fractionation effect in early stage disease", *Clinical Cancer Research*, vol. 26, no. 1, pp. 282–289, 2020.

# Glossary

**Allele:** a variant of a gene resulting from mutation and inherited, which performs the same function as the original gene, but in its own way. Any gene can have several alleles, which often determine the appearance of different hereditary traits.

**Amino acid:** a unitary molecule that, when combined with similar molecules in a precise order, forms a protein. Each amino acid gives the protein specific chemical properties, and the order in which it is assembled gives it a precise function.

**Chloroplast:** a complex organelle of green plants containing chlorophyll.

**Diploid:** a cell with two pairs of homologous chromosomes (2n), as opposed to a haploid cell with only one (n).

**Eukaryote:** an organism whose cells have a structured nucleus (as opposed to the prokaryote).

**Exosome:** a 30–90 nm lipid vesicle discharged by a cell into its environment.

**Gamete:** a male or female reproductive cell containing a single chromosome.

**Haploid:** a cell in which each pair of chromosomes is duplicated after splitting.

**Heterozygous:** a cell or individual with two different genes (recessive and dominant) on each chromosome of the same pair (opposite of homozygous).

**Meiosis:** a fundamental mechanism of sexual reproduction. It occurs during the process of gametogenesis, i.e. the formation of male (sperm) and female (egg) reproductive cells or gametes.

**Nucleoside:** a building block of nucleic acids, consisting of a nitrogenous base associated with a sugar (ribose for RNA and deoxyribose for DNA).

**Nucleotide:** a building block of a nucleic acid molecule (DNA, RNA). A nucleotide is formed by the combination of three substances: a purine base (adenine or guanine) or a pyrimidine base (cytosine, thymine or uracil), a sugar (ribose or deoxyribose) and a phosphoric acid.

**Organelle:** a differentiated cellular element with a specific function. Organelles are the cell's analogue to the organs of the human body. Like organs, they have complex structures and specialized functions.

**Oxidative stress:** the damage caused to cells by molecules derived from oxygen. The most common are free radicals. Oxidative stress is caused by excessive oxidation of the body's cells. For example, if we cut an apple and leave the cut pieces in the air, a brown pigment is formed. It is the action of certain molecules, coupled with oxygen, that leads to the production of this pigment through oxidation. If this oxidation, which is a reaction between oxygen and another chemical substance, occurs in an organism, it will release free radicals that can attack healthy cells. Molecules known as antioxidants slow down or prevent oxidation and thus the production of free radicals.

**Purine:** a heterocyclic nitrogen molecule consisting of a pyrimidine ring fused to an imidazole ring.

**Zygote:** a fertilized egg, before the first segmentation of the egg or the diploid cell in plants, resulting from the fusion of two gametes.

# Index

## E, F



## G, H, I



## L, M

Other titles from

iSTE

in

Ecological Science

## 2024

AKBAR Saima, KUMAR SINGH Suraj, KANGA Shruti
*Disaster Management: Preparedness and Mitigation*

FLEURENCE Joël
*Aquatic Biotechnologies: From Genetic Engineering to Enzymatic or Fermentation Engineering*

MATAGNE Patrick
*Birth of Scientific Ecology: Eugenius Warming (1841–1924)*

## 2023

GAID Kader
*Drinking Water Treatment 1: Water Quality and Clarification*
*Drinking Water Treatment 2: Chemical and Physical Elimination of Organic Substances and Particles*
*Drinking Water Treatment 3: Organic and Mineral Micropolluants*
*Drinking Water Treatment 4: Membranes Applied to Drinking Water and Desalination*
*Drinking Water Treatment 5: Calco-carbonic Equilibrium and Disinfection*

## 2022

AMIARD Jean-Claude
*Marine Radioecology*
*(Radioactive Risk Set – Volume 6)*

GODET Laurent, DUFOUR Simon, ROLLET Anne-Julia
*The Baseline Concept in Biodiversity Conservation: Being Nostalgic or Not in the Anthropocene Era*

GRISON Claude, CASES Lucie, LE MOIGNE Mailys, HOSSAERT-MCKEY Martine
*Photovoltaism, Agriculture and Ecology: From Agrivoltaism to Ecovoltaism*

LÉVÊQUE Christian
*Biodiversity Erosion: Issues and Questions*

## 2021

AMIARD Jean-Claude
*Disarmament and Decommissioning in the Nuclear Domain*
*(Radioactive Risk Set – Volume 4)*
*Management of Radioactive Waste*
*(Radioactive Risk Set – Volume 5)*

FLEURENCE Joël
*Microalgae: From Future Food to Cellular Factory*

JOLY Fernand, BOURRIÉ Guilhem
*Mankind and Deserts 1: Deserts, Aridity, Exploration and Conquests*
*Mankind and Deserts 2: Water and Salts*
*Mankind and Deserts 3: Wind in Deserts and Civilizations*

PIOCH Sylvain, SOUCHE Jean-Claude
*Eco-design of Marine Infrastructures: Towards Ecologically-informed Coastal and Ocean Development*

## 2020

BRUSLÉ Jacques, QUIGNARD Jean-Pierre
*Fish Behavior 1: Eco-ethology*
*Fish Behavior 2: Ethophysiology*

ROSSIGNOL Jean-Yves
*Climatic Impact of Activities: Methodological Guide for Analysis and Action*

## 2019

AMIARD Jean-Claude
*Industrial and Medical Nuclear Accidents: Environmental, Ecological, Health and Socio-economic Consequences*
*(Radioactive Risk Set – Volume 2)*
*Nuclear Accidents: Prevention and Management of an Accidental Crisis*
*(Radioactive Risk Set – Volume 3)*

BOULEAU Gabrielle
*Politicization of Ecological Issues: From Environmental Forms to Environmental Motives*

GIRAULT Yves
*UNESCO Global Geoparks: Tension Between Territorial Development and Heritage Enhancement*

KARA Mohamed Hichem, QUIGNARD Jean-Pierre
*Fishes in Lagoons and Estuaries in the Mediterranean 2: Sedentary Fish*
*Fishes in Lagoons and Estuaries in the Mediterranean 3A: Migratory Fish*
*Fishes in Lagoons and Estuaries in the Mediterranean 3B: Migratory Fish*

OUVRARD Benjamin, STENGER Anne
*Incentives and Environmental Policies: From Theory to Empirical Novelties*

## 2018

AMIARD Jean-Claude
*Military Nuclear Accidents: Environmental, Ecological, Health and Socio-economic Consequences*
*(Radioactive Risk Set – Volume 1)*

FLIPO Fabrice
*The Coming Authoritarian Ecology*

GUILLOUX Bleuenn
*Marine Genetic Resources, R&D and the Law 1: Complex Objects of Use*

KARA Mohamed Hichem, QUIGNARD Jean-Pierre
*Fishes in Lagoons and Estuaries in the Mediterranean 1: Diversity, Bioecology and Exploitation*

## 2016

BAGNÈRES Anne-Geneviève, HOSSAERT-MCKEY Martine
*Chemical Ecology*

## 2014

DE LARMINAT Philippe
*Climate Change: Identification and Projections*